Handbook for Rhizobia

P. Somasegaran H.J. Hoben

Handbook for Rhizobia

Methods in Legume-*Rhizobium* Technology

With 85 Illustrations

Springer-Verlag
New York Berlin Heidelberg London Paris
Tokyo Hong Kong Barcelona Budapest

Padma Somasegaran
Heinz J. Hoben
University of Hawaii
NifTAL Project
1000 Holomua Road
Paia, HI 96779-9744 USA

Library of Congress Cataloging-in-Publication Data
Somasegaran, P. (Padmanabhan)
Handbook for rhizobia : methods in Legume-rhizobium technology /
P. Somasegaran and H.J. Hoben.
p. cm.
Includes bibliographical references and index.

ISBN-13: 978-1-4613-8377-2 e-ISBN-13: 978-1-4613-8375-8
DOI: 10.1007/978-1-4613-8375-8

1. Rhizobium—Laboratory manuals. 2. Nitrogen-fixing microorganisms—Laboratory manuals. 3. Plant-microbe relationships—Laboratory manuals. I. Hoben, H.J. (Heinz J.) II. Title.
QR82.R45S655 1994
589.9'5—dc20 93-21236

Printed on acid-free paper.

Softcover reprint of the hardcover 1st edition 1994

An earlier version of this book entitled *Methods in Legume-Rhizobium Technology* was prepared by the authors under a US Agency for International Development contract.

Acquiring editor: Robert C. Garber.
Production managed by Impressions, Inc., and supervised by Ellen Seham; manufacturing supervised by Vincent Scelta.
Typeset from the authors' WordPerfect files by Impressions, a division of Edwards Brothers, Inc., Madison, WI.

9 8 7 6 5 4 3 2 1

Foreword

A good case can be made to support the claim that the N_2-fixing partnership between rhizobia and legumes is the most significant contribution that a soil bacterium can make to agricultural and sylvicultural practices. This symbiosis has the potential to free the host legumes from dependence on nitrogenous fertilizer, as well as opening up the opportunity for increasing soil fertility. The full realization of this potential depends on maximizing the contribution of each partner, attending to specificity in the association, and providing conditions for plant growth and nodule function. The N gain from the atmosphere will be no better than that permitted by the intrinsic capacity of the host and the effectiveness of the root nodules. To achieve the best result from this symbiosis, there needs to be a working link between those concerned primarily with the plant host (agronomist, plant breeder, botanists, etc.) and the rhizobiologist, responsible for the bacterial side. It is particularly to facilitate the contribution of the rhizobiologist that this volume has been produced as an update of the excellent previous volume.

It was my privilege to be associated with the early days of NifTAL and the courses it initiated some 18 years ago. Since then, I have admired NifTAL's continuing work in providing material and training resources for many countries that are developing their interest in biological N_2 fixation. The courses and manuals are intended to help the beginner get started, but are also designed to cover techniques on which research will be based. Undoubtedly, the manual's influence will extend beyond the immediate demands of a training course. Hopefully, individual workers will be able to use it to enter and extend their knowledge and practical application of this invaluable plant-bacterium association.

James M. Vincent
Emeritus Professor
Department of Microbiology
University of Sydney
Australia

Foreword to NifTAL Edition

This is the Foreword to the earlier version of this book, entitled Methods in Legume-*Rhizobium* Technology.

There is no doubt that in the near future the emerging biotechnology, based on genetic engineering and somatic cell fusion, will contribute significantly to solving agricultural problems. Presently, however, so much of the available technology, i.e., inoculum technology, is not being fully utilized in agriculture. It would be prudent to devote major efforts to its adoption. A serious obstacle to the adoption of modern technologies, especially in developing countries, is the shortage of trained personnel. Therefore, it is essential for all development support projects to include a training component.

This book is the culmination of several years of experience in training scientists and technicians from developing countries. The 6-week training course, for which this book is intended, was developed at NifTAL and, in the early years, was taught there. Subsequently, the course was taken to the field and offered at host institutions in Africa, Asia, and Latin America. P. Somasegaran and H.J. Hoben have done a commendable job of drawing from their experience with these courses. They have compiled an "All You Ever Wanted to Know about . . ." style book that is not only valuable to developing country scientists, but is also useful for technicians and graduate students starting work with the legume/*Rhizobium* symbiosis.

1985

B. Ben Bohlool
Professor, University of Hawaii
Director, NifTAL Project

Preface

In the last 10 years, basic research in biological nitrogen fixation (BNF) by the legume-rhizobia symbiosis has contributed much to our understanding of this agriculturally and environmentally significant plant-microbe interaction. Rapid advances in molecular cell biology and biochemistry have greatly enriched our knowledge of the symbiotic interaction at the gene level, and there are already indications that molecular approaches will influence future taxonomy and classification of rhizobia. Further, the detection and identification of rhizobia have been affected by the development of sensitive and more rapid immunodetection procedures such as the ELISA (enzyme-linked immunosorbent assay), the immunoblot, and others. These developments enhance the new approaches to teaching and doing research with rhizobia.

In this book, we present all the techniques and methods that we have taught in NifTAL's training courses in rhizobia technology. The contents will be useful to BNF technologists working with rhizobia, and will also provide material that easily may be adapted for teaching undergraduate courses in applied BNF technology.

The development of the book was natural because we had broadened the scope of the original 6-week course to include the newer methods that were gaining impetus in research and application with rhizobia. The contents are organized into modules for the following courses: Legume Inoculant Production and Quality Control, Rhizobia Microbiology and Genetic Technology, Monitoring Microorganisms in the Environment, and Rhizobia Technology.

ORGANIZATION

The text is divided into five sections and each section is divided into several chapters. Each section is preceded by an introduction. Key references and a materials list are given at the end of each chapter. References and recommended reading have been compiled for each section.

Section I covers isolation of rhizobia, microbiology, characterization, and enumeration of rhizobia in pure culture and in soil by direct and indirect methods. *Section II* is devoted to traditional serological methods and immunoassays used for strain identification, and the use of antibiotic markers and rhizobiophages. *Section III* is concerned with the evaluation of the N_2-fixing potential of rhizobia with the host legume under controlled conditions in the greenhouse and in the field. Some concepts on the ecology of introduced and indigenous rhizobia are addressed. *Section IV* consists of chapters that focus on small- and medium-scale fermentor-based mass culture techniques for rhizobia. Production of carrier-based inoculants and seed inoculation is covered. *Section V* has nine chapters that introduce basic analytical molecular biology methods for research with rhizobia.

Padma Somasegaran
Heinz J. Hoben

Acknowledgments

The authors gratefully acknowledge colleagues and staff at the University of Hawaii NifTAL Center and other institutions who made the preparation of this edition possible.

We would like to remember the support and encouragement of the late Dr. B. Ben Bohlool, who initiated the task of producing this edition.

The authors are especially indebted to colleagues who contributed towards Section V. Our sincere appreciation and thanks are extended to Dr. Dulal Borthakur (Biotechnology Program, University of Hawaii) for working together with us closely in writing and reviewing some of the chapters; Dr. Doug Rice and Kathy MacGlashan (NifTAL Center) for providing modified protocols, suggestions, and "lumigraphs"; and Dr. David Berryhill (North Dakota State University) for sharing protocols.

We appreciate the support of Dr. Paul Singleton (Director, NifTAL Project) in this endeavor; Professor James M. Vincent (Emeritus Professor) for his encouragement, technical review, and constructive suggestions; Dr. Harold Keyser for reviewing and improving the text in some sections; and Dr. Brian Holl (University of British Columbia) for his invaluable suggestions and comments for this edition.

The production of this edition was the work of a very special group of dedicated professionals in the NifTAL Communication Section, and we are especially grateful and indebted to them for their efforts. Patty Nakao provided helpful suggestions and coordination; Debra Hughes Lordan meticulously worked on the graphics and numerous illustrations, and updated illustrations done by Richard Gabrielson and Keith Avery; Princess Ferguson and Susan Hiraoka helped in the coordination and typing of this edition in its earlier stages; Ann Coopersmith proofread the text; and Sally Ekdahl completed the task by patiently editing and typing the entire text. We are also grateful to Surya Tewari and Bruce Martin for demonstrating procedures in several photographs. Once again, we extend our thanks to all the NifTAL Training Course participants and other readers who had conveyed suggestions and corrections to be incorporated into this edition. The authors gratefully acknowledge the financial support from the UNESCO/MIRCEN at NifTAL for the external technical review and editing of this edition. This work was made possible by the University of Hawaii NifTAL Center through support provided by the Office of Agriculture, Bureau for Research and Development, United States Agency for International Development under grant no. DAN-1311-G-00-1049-00.

Contents

SECTION I

General Microbiology of Rhizobia

CHARACTERISTICS OF THE RHIZOBIA

Rhizobia (the fast-growing *Rhizobium* spp. and the slow-growing *Bradyrhizobium* spp.) or root nodule bacteria are medium-sized, rod-shaped cells, 0.5–0.9 μm in width and 1.2–3.0 μm in length. They do not form endospores, are Gram-negative, and are mobile by a single polar flagellum or two to six peritrichous flagella. Uneven Gram staining is frequently encountered with rhizobia, depending on the age of the culture. Cells from a young culture and nodule bacteroids usually show even Gram staining while older and longer cells give a banded appearance with unstained areas. These unstained areas have been identified to be large granules of polymeric beta-hydroxybutyric acid (PHBA). The PHBA is refractile under phase-contrast microscopy. Rhizobia are predominantly aerobic chemoorganotrophs and are relatively easy to culture. They grow well in the presence of O_2 and utilize relatively simple carbohydrates and amino compounds. With the exception of a few strains, they have not been found to fix N in the free-living form except under special conditions. Some strains of rhizobia require vitamins for growth. Bradyrhizobia isolated from soybean (*Glycine max*) and cowpea (*Vigna unguiculata*) nodules were found to remain viable and able to rapidly nodulate their respective host legumes after being stored in purified water at ambient temperatures for periods of at least 1 year. However, *Rhizobium* spp. are likely to lose viability rapidly in water. Optimal growth of most strains occurs at a temperature range of 25–30°C and a pH of 6.0–7.0. Despite their usual aerobic metabolism, many strains are able to grow well under microaerophilic conditions at O_2 tensions of less than 0.01 atm. Generally, most rhizobia produce white colonies, but those that nodulate *Lotononis bainesii* produce a characteristic red nonheme carotenoid pigment when cultured in yeast-mannitol (YM) medium. Most rhizobia only weakly absorb congo red (diphenyldiazo-bis-α-naphthylaminesulfonate) dye, which is included in culture media for isolating rhizobia. However, if the culture medium is not buffered, acid-producing rhizobia cause the dye to turn purple. Other interesting and useful characteristics of rhizobia are their growth reactions in the standard YM medium containing bromthymol blue as the pH indicator. Fast-growing rhizobia produce an acid reaction in the YM medium containing bromthymol blue (pH 6.8) while slow growers produce an alkaline reaction.

FREE-LIVING RHIZOBIA IN THE SOIL

Rhizobia are facultative microsymbionts that live as normal components of the soil microbial population when not living symbiotically in the root nodules of the host legume. Outside the root nodule, rhizobia are mostly found on the root surface (rhizoplane), soil around and close to the root surface (rhizosphere), and, to a lesser extent, nonrhizosphere soil. The increase in numbers of rhizobia in the rhizosphere is a response to the excretion of nutrients by plant roots, especially the host legume. Besides the host legume, nonlegumes, moisture and temperature, soil acidity and alkalinity, and salt content of the soil affect rhizobial populations in the soil. Numbers of rhizobia in the soil can range from undetectable to 1,000,000 rhizobia g^{-1} soil. An indirect counting procedure based on plant infection and application of a most-probable-number (MPN) estimate determines the population of rhizobia in the soil. The soil is the major reservoir of free-living rhizobia and under dense or pure stands of legumes, multiplication in the rhizosphere and release of rhizobia from the senescing nodules recolonizes the soil. The various species of rhizobia are not found universally in all soils, but where they are absent, these bacteria may be introduced by seed or soil inoculation. Diverse gene pools of indigenous rhizobia are most likely to be found in the centers of diversity of the host legume. Soil rhizobia are susceptible to attack and lysis by specific bacterial viruses (bacteriophages). Rhizobial phages are as widespread as their host rhizobia and occur widely in soils. The Gram-negative *Bdellovibrio* can parasitize free-living rhizobia. These small, short-curved vibrios are obligate parasites that attach themselves to the larger rhizobial cell and live at the expense of the host.

Rhizobia are somewhat unique among soil microorganisms in their ability to form N_2-fixing symbioses with legumes and, exceptionally, a nonlegume (*Parasponia*). To enjoy the benefits of this partnership, any introduced rhizobia must not only exhibit saprophytic competence among other soil microorganisms, but they must out-compete other rhizobia for infection sites on legume roots. Therefore, potential for physiological versatility is an important trait contributing to their adaptation to the competitive and complex soil environment.

RHIZOBIA AS SYMBIONTS

The free-living rhizobia in the soil can enter the root hairs of the susceptible host legume by a complex series of interactions known collectively as the infection process. This begins with the adhesion of the specific rhizobia to the surface of the root hair. Adhesion is followed by deformation, and curling of the root hair, which results in the characteristic shepherd's crook appearance. The hypha-like infection thread develops gradually in the root hair as a tubular structure that is actually an invagination of the root hair wall. The infection thread contains large numbers of rhizobial cells, and the thread branches through the root cortex passing close to the host cell nuclei. The rhizobia are released from the tip of the infection thread into the cytoplasm of the host cells, where multiplication takes place. Before the release of the rhizobia, rapid host cell division takes place. The dividing host cells are tetraploid. The final structure is a central core containing the rhizobia and a cortical area that becomes occupied by the vascular system, which connects to the young root. The host cell membrane, which had enclosed the infection thread, buds off vesicles containing the rhizobia. The rhizobia divide and differentiate into the form known as bacteroids. The host cell membrane, now referred to as the peribacteroid membrane and the bacteroids, together form the peribacteroid unit. The peribacteroid membranes effectively separate the bacteroids from the plant cytoplasm and provide the plant with a means of regulating nutrient exchange with the bacteroids.

An infected cell from a nodule of a mature soybean plant may contain up to 10,000 peribacteroid units. The forms of bacteroids encountered in the nodules of legumes vary considerably. Branched rods (X- and Y-shaped) and large pear-shaped rounded forms are found in the nodules of *Medicago* spp., *Pisum* spp., and several other species. Perfectly spherical bacteroids are common in nodules of peanuts (*Arachis hypogaea*). The plant largely determines the size and shape of bacteroid, and the numbers in each peribacteroid unit. The synthesis of a protein called *leghemoglobin* in the nodule tissue characterizes effective symbiosis. The presence of leghemoglobin gives a pink/red color to the nodule interior. However, leghemoglobin is absent or present in small quantities in ineffective nodules that appear white when sliced open. The synthesis of leghemoglobin requires genetic information from the legume and the rhizobia.

The enzyme nitrogenase is a complex of two enzymes, an Fe-containing protein and an Fe-Mo protein. It is responsible for the conversion (reduction) of atmospheric N into NH_4^+, and is synthesized in the cytosol of the bacteroids. The legume utilizes NH_4^+ to convert certain precursor metabolites (e.g., α-ketoglutarate, phosphoenopyruvate) into amino acids, which, in turn, are synthesized into proteins. The complex biochemical reactions whereby the inert atmospheric nitrogen is enzymatically reduced into a utilizable form for the plant by the nitrogenase enzyme complex of the bacteroids is called biological nitrogen fixation (BNF).

CLASSIFICATION OF THE RHIZOBIA

Rhizobia are a genetically diverse and physiologically heterogeneous group of bacteria that are nevertheless classified together by virtue of their ability to nodulate members of the *Leguminosae*. The *Leguminosae* is divided into three subfamilies: *Caesalpinoideae*, *Mimosoideae*, and *Papilionoideae*.

The *Caesalpinoideae* consist mostly of woody plants that show nodulation in a very small number of species. The genus *Chamaecrista* of the *Caesalpinoideae* is well nodulated while nodulation has not been observed in the genera *Delonix*, *Tamarindus*, *Peltophorum*, and many others. The genus *Cassia* is also interesting. *Cassia* spp. grow well in poor soils, but nodulation has not been observed in most of them. *C. leschenaultiana*, found in Hawaiian soils, nodulates with some bradyrhizobia.

The *Mimosoideae* consist mostly of woody species and nodulation occurs at a higher frequency than in the *Caesalpinoideae*. Important genera in this subfamily include *Leucaena* spp., *Acacia* spp., and *Prosopis* spp. Species in certain genera are nodulated by *Rhizobium* and *Bradyrhizobium*. For example, *A. senegal*, *A. farnesiana*, and *A. pennatula* are nodulated by *Rhizobium* while *A. mearnsii*, *A. auriculiformis*, *A. mangium*, and *A. albida* are nodulated by *Bradyrhizobium*. Similarly, *Pithocellobium dulce* is nodulated by *Rhizobium* while *P. jiringa* is nodulated by *Bradyrhizobium*.

The subfamily *Papilionoideae* is well studied for nodulation. Most of the genera in this subfamily are nodulated. The present-day classification of rhizobia is based on earlier studies of the symbiosis with members of the *Papilionoideae*.

The ability of certain rhizobia to infect and nodulate particular group(s) of legume species is important in the classification of rhizobia. Rhizobia are generally classified according to a host-based system. In this host-based system, legume(s) have been assembled into cross-inoculation groups, which are useful in organizing the diverse legumes and their rhizobial partners. Essentially, a cross-inoculation group consists of a collection of legume species that will develop effective nodules when inoculated with the rhizobia obtained from the nodules from any member of that legume group. Classification by this system is by no means perfect, due to cross-inoculation(s) with rhizobia from outside the assigned group and failure to cross-inoculate within a group. The system is not a taxonomic one, but has some practical application. Certain legume-rhizobial associations are highly specific while others are promiscuous. For example, the *Cicer arietinum–Rhizobium* sp. symbiosis is highly specific. *C. arietinum* will nodulate effectively with inoculation with rhizobia isolated only from the nodules of *C. arietinum*. Another instance of specificity is between the forage species *Lotononis bainesii* and the red-pigment-producing *Bradyrhizobium* sp. Here, inoculation of *L. bainesii* sp. with the red strain is necessary for effective nodulation of *L. bainesii*.

At the other extreme are legumes where inoculation with a specific rhizobial strain may not be needed for effective nodulation. Examples of unspecialized or promiscuous groups of legumes are widespread among tropical legumes. Notable examples are *Psophocarpus tetragonolobus*, *Vigna* spp., *Crotalaria* spp., *Macroptilium* spp., *Lablab purpureus*, and *Cajanus cajan*.

Rhizobia belong in the family *Rhizobiaceae*, which consist of the following genera: Genus I—*Rhizobium*; Genus II—*Bradyrhizobium*; Genus III—*Agrobacterium*; and Genus IV—*Phyllobacterium*. Only Genera I and II fix N symbiotically in the root nodules of legumes. The species of rhizobia in Genera I and II, and the cross-inoculation groups of legumes nodulated by these rhizobia are summarized in Table I.1.

In Genus I are the fast-growing acid producers that develop pronounced turbidity in liquid media within 2–3 days and have a mean doubling time of 2–4 h. The cells are motile by two to six peritrichous flagella. They can grow on a wide range of carbohydrates, but usually grow best on glucose, mannitol, or sucrose. Rhizobia of this group are generally infective on temperate legumes.

TABLE I.1 Species of Rhizobia in Genera I and II, and the Cross-Inoculation Groups of Legumes Nodulated by These Rhizobia

Rhizobia	Cross-Inoculation Group	Legumes in Cross-Inoculation Group
Genus I: *Rhizobium*		
Rhizobium leguminosarum bv. viceae	Pea	Peas (*Pisum* spp.), vetches; (*Vicia* and *Lathyrus* spp.); lentils (*Lens esculenta*)
R. leguminosarum bv. trifolii	Clover	Clovers (*Trifolium* spp.)
R. leguminosarum bv. phaseoli	Bean	Common beans (*Phaseolus vulgaris*); scarlet runner bean (*Phaseolus coccineus*)
R. meliloti	Alfalfa	Alfalfa/medics (*Medicago* spp.); sweet clovers (*Melilotus* spp.); fenugreek (*Trigonella foenumgraecum*)
R. loti	Lotus	Trefoils (*Lotus corniculatus* and *L. tenuis*); lupine (*Lupinus densiflorus*); serradella (*Ornithopus sativus*); kidney vetch (*Anthyllis vulneraria*)
R. galegae		Goat's rue (*Galega orientalis*)
R. fredii	Soybean	Soybean (*Glycine max*)
Rhizobium spp.		Leucaena (*Leucaena* spp); *Gliricidia sepium, Sesbania grandiflora, Calliandra callothyrsus, Pithocellobium dulce, Prosopis pallida, P. juliflora, Acacia senegal, A. farnesiana, Robinia pseudoacacia*
Rhizobium sp.	Chickpea	Chickpea (*Cicer arietinum*)
Genus II: *Bradyrhizobium*		
Bradyrhizobium japonicum	Soybean	Soybean (*Glycine max*)
Bradyrhizobium spp.	Cowpea	Pigeon pea (*Cajanus cajan*); peanut/groundnut (*Arachis hypogaea*); *Acacia mearnsii, A. mangium, A. auriculiformis*; limabean (*Phaseolus lunatus*); winged bean (*Phosphocarpus tetragonoloba*); siratro (*Macroptilium atropupureum*); guar bean (*Cyamopsis tetragonolobus*); cowpea, mungbean, black/green gram, rice bean (*Vigna* spp.), *Desmodium* spp., *Stylosanthes* spp.; hyacinth bean (*Lablab purpureus*)

In Genus II are the slow-growing, alkali-producing rhizobia, collectively known as bradyrhizobia. They require 3–5 days to produce moderate turbidity in liquid media and have a mean doubling time of 6–8 h. Most strains in this group grow best with pentoses as their C source. The cells are motile by a single polar or subpolar flagellum. A large genera of tropical legume species are nodulated by bradyrhizobia.

Classification of rhizobia is becoming increasingly complex and is revised periodically because of new findings that propose new genera and new species. For example, soybeans are now known to be nodulated by a distinct group of fast-growing, acid-producing rhizobia. These rhizobia were classified in Genus I as *Rhizobium fredii*, but have now been renamed *Sinorhizobium fredii*. More recently, *R. huakuii* has been proposed for a new species of *Rhizobium* that nodulates *Astragalus sinicus*, a legume that is used as green manure in South China. In the bean group, *R. tropicii* has been proposed as a novel species nodulating *P. vulgaris* and *Leucaena* sp. A highly specific symbiosis is established between a specialized strain of fast-growing rhizobia and the tree legume, *Sesbania rostrata*. The nodules are formed on the stems of *S. rostrata* and the nodulating rhizobia have been named *Azorhizobium caulinodans*. A new genus named *Photorhizobium* has been proposed to a group of photosynthetically active rhizobia isolated from the stem nodules of *Aeschenomene*. Also, in the revised classification, *R. trifolii*, *R. phaseoli*, and *R. leguminosarum* are recognized as one species and designated *R. leguminosarum* consisting of three biovars, namely trifolii, phaseoli, and viceae. *R. meliloti* remains as before and *R. loti* has been assigned to the fast-growing *Lotus* rhizobia. Genetically related to *R. loti* are rhizobia from *Lotus corniculatus*, *Lotus tenuis*, *Cicer arietinum*, *Leucaena leucocephala*, and *Sophora microphylla*. Rhizobia nodulating *Vigna*, *Arachis*, *Desmodium*, *Macroptilium*, *Stylosanthes*, and many other tropical legumes are still unclassified at the species level, but are grouped as *Bradyrhizobium* spp. The non-legume *Parasponia* (previously called *Trema*) is also nodulated by a *Bradyrhizobium* sp. Besides *Leucaena*, there are other legumes (e.g., *Sesbania*, *Neptunia*, *Calliandra*, and *Acacia*) that are nodulated by fast-growing, acid-producing rhizobia. The taxonomic status of these organisms needs to be resolved in the future.

1

Collecting Nodules and Isolating Rhizobia

The purpose of this chapter is to become familiar with legumes in the field, examine their nodules, isolate rhizobia from nodules, and preserve the isolates. The subfamilies in the *Leguminosae* will be discussed and identifications will be made with the help of a botanical key. Nodules will be sectioned and examined. Simple stains of nodule smears will be examined under the microscope. Rhizobia will be isolated from nodules and grown on presumptive test media. The isolates will be authenticated on their original host plants and then preserved on ceramic beads.

KEY STEPS/OBJECTIVES

1. Identify legumes in the field, collect nodulated specimens, and preserve nodules.
2. Examine nodules and bacteroids under the microscope.
3. Surface sterilize nodules and isolate rhizobia on differential media.
4. Perform Gram stain and reisolate on differential media.
5. Store isolates on agar slants.
6. Surface sterilize and pregerminate seeds for authentication.
7. Plant and inoculate seedlings for authentication.
8. Examine plants periodically for nodulation.
9. Terminate experiment, examine nodules, and reisolate.
10. Prepare broth culture of authenticated isolate for desiccation on beads.
11. Prepare bead storage vials.
12. Impregnate sterilized beads with broth culture of rhizobia.
13. Regrow rhizobia stored on beads.

a. Recognizing Legumes and Identifying Them in the Field (Key Step 1)

Become familiar with the general taxonomic characters of the *Leguminosae*. Study the different flower types of the three subfamilies: *Caesalpinoideae*, *Mimosoideae*, and *Papilionoideae* (Appendix 1).

Note the main similarities among all legumes in their compound leaves and the seed placentation in pods as shown in Appendix 1, Figure A1.4 and A1.5. However, in many *Acacia* spp. (e.g., *Acacia auriculaeformis*, *Acacia mangium*, and *Acacia koa*), the compound leaves are only formed and seen in seedlings. The compound leaves are replaced by phyllodes as the plants mature (Figure A1.5). Compound leaves are also characteristic of numerous nonleguminous families such as: *Bignoniaceae* (e.g., *Jacaranda*, *Spathodea*); *Caprifoliaceae* (e.g., *Sambucas mexicana* var. *bipinnata*); *Solanaceae* (e.g., *Lycopersicon*, *Solanum tuberosum*), *Passifloraceae* (e.g., *Passiflora* spp.). Familiarize yourself with the basic characteristics of each subfamily as outlined in Appendix 1. Learn to identify legumes in the field and become familiar with the appearance of the most common agricultural legumes in your area. Use a suitable botanical key.

It is not essential to identify the less common legumes. Many aspects of classification within the *Leguminosae* are in dispute, even amongst plant taxonomists. The course that many collectors follow is to recover a good plant specimen (including flowers and fruits), dry and press it, and forward it to a reliable herbarium (Royal Botanical Garden, Kew, Richmond, Surrey TW 3 AE, England) for precise identification.

b. Recovering Nodules in the Field (Key Step 1)

Identify plants of several legume species in the field and select one representative of each for sampling. With a spade, describe a circle with a radius of approximately 15 cm around the plant and cut out this section to a depth of at least 20 cm. Still using the spade, slowly lift out the clump. Carefully remove the soil from the root material with your hands. Avoid detaching secondary roots from the plant, as nodules may be found on the lateral roots as well as the taproot. Carefully place the whole plant into a plastic bag. If the legume has seeds, collect the seeds and store them in the refrigerator for the authentication test. In the laboratory, place a sieve of an appropriate size and mesh under each root sample to catch nodules that may become detached from the root. Carefully wash the roots under a gentle stream of water from a tap or a hose.

The distribution of the nodules on the root system is dependent on the legume species and rhizobial strain as well as soil structure and composition. Examples of nodule types and distribution on some species are illustrated in Appendix 1. Record plant host, area and date of collection, and soil type, and keep it as a permanent record for the nodule isolate (Appendix 24, Table A24.2).

c. Preserving Nodules (Key Step 1)

Fresh nodules may be stored in the refrigerator overnight. Do not freeze nodules because ice crystals may rupture and kill the bacteroids. Frozen nodules may, however, be used for serological typing. For long-term storage, desiccation in glass vials is recommended. A preservation vial is shown in Appendix 2, Figure A2.1.

d. Examining Nodules and Bacteroids (Key Step 2)

Note the shape and size of the nodules recovered from the collected plants. Nodule size and shape vary with the rhizobia and host plant species. Large, round nodules may be found on cowpea (*Vigna unguiculata*) and soybean (*Glycine max*) plants. *Leucaena* and *Acacia* are among legumes that do not have round nodules. See Appendix 1, Figure A1.6 for nodule shape description.

Cut thin sections of nodules with a razor blade and float them on a drop of water on a microscope slide; use a cover glass and examine under low power (10×) and high power (40×) objectives. An active N-fixing nodule contains a protein called *leghemoglobin*. Its presence in the nodule can be noted by the characteristic pink, red, or brown coloration. Active nodules may also be black. Black nodules are not very common. They have been reported on *Lablab purpureus*, *Dolichos biflorus*, and *Vigna unguiculata* when inoculated with some strains of rhizobia. Senescent nodules are usually grayish green. When nodules on the soil surface are exposed to sunlight, they may develop a green exterior. This green color is due to chlorophyll development on the cortical region of the nodule. Most ineffective rhizobia cause nodules with white interiors that lack leghemoglobin.

Gently rub the cut surface of a nodule on a clean microscope slide to make a smear. Allow the smear to air dry and then pass the slide through a flame. Cool the slide and stain the smear with dilute carbolfuchsin (Appendix 4) for 10–20 s. Wash in water, blot off excess moisture, and air dry. Examine under the oil immersion objective. Note the difference in morphology between the bacteroids in this smear and bacteria of the same rhizobial species grown in pure culture. Note the size and shape of the bacteroids compared to the rod forms found in pure culture (Chapter 3, Figure 3.1).

e. Isolating Rhizobia from a Nodule (Key Step 3)

Wash roots thoroughly to remove soil. Collect about 10 nodules from each plant. Sever the nodule from the root by cutting the root about 0.5 cm on each side of the nodule. When moving the nodule, use forceps on the root appendages to reduce the risk of damaging the nodule.

Immerse intact, undamaged nodules for 5–10 s in 95% ethanol or isopropanol (to break the surface tension and to remove air bubbles from the tissue); transfer to a 2.5–3% (v/v) solution of sodium hypochlorite, and soak for 2–4 min. Rinse in five changes of sterile water using sterile forceps for transferring. Forceps may be sterilized quickly by dipping in alcohol and flaming. Utilize sterile glass or plastic Petri dishes as containers for the alcohol, sodium hypochlorite, and water. Alternatively, nodules may be placed into an Erlenmeyer flask (125 ml). The sterilizing and rinsing fluids may be changed as required, leaving the nodule in the flask each time.

An acidified mercuric chloride solution (0.1% w/v) or a solution of hydrogen peroxide (3% v/v) may be used for sterilizing nodules. However, mercuric chloride is highly toxic and hydrogen peroxide is expensive, making sodium hypochlorite (available as commercial bleach) the preferred choice. When hydrogen peroxide is used, the five to six rinses with sterile water may be omitted.

Desiccated nodules must be rehydrated before sterilizing. Place nodules into a small beaker with clean, cool water and leave in the refrigerator to imbibe overnight. A 1-h soaking at room temperature is sufficient for nodules that have been desiccated for only a short time. Crush the surface-sterilized nodule with a pair of blunt-tipped forceps in a large drop of sterile water in a petri dish. Alternatively, the nodule may be crushed in a sterile test tube with a sterile glass rod. Streak one loopful of the nodule suspension on a yeast-mannitol agar (YMA) plate containing congo red (CR). Similarly treat one loopful of the nodule suspension on a YMA plate containing bromthymol blue (BTB) (Appendix 3). The primary isolate may be streaked in one continuous motion, as shown in method 1 of Figure 1.1.

Well-isolated colonies may be obtained with method 2 (Figure 1.1), which is most commonly used with isolations from primary plates. It is performed as follows.

Deposit culture on agar with inoculation loop, then streak out to area 1. Resterilize loop, and cool by touching the agar surface near the side of the Petri dish, then streak from area 1 to area 2. Repeat the procedure until area 4 is reached.

The isolation procedure lends itself well to improvisation and many variations exist.

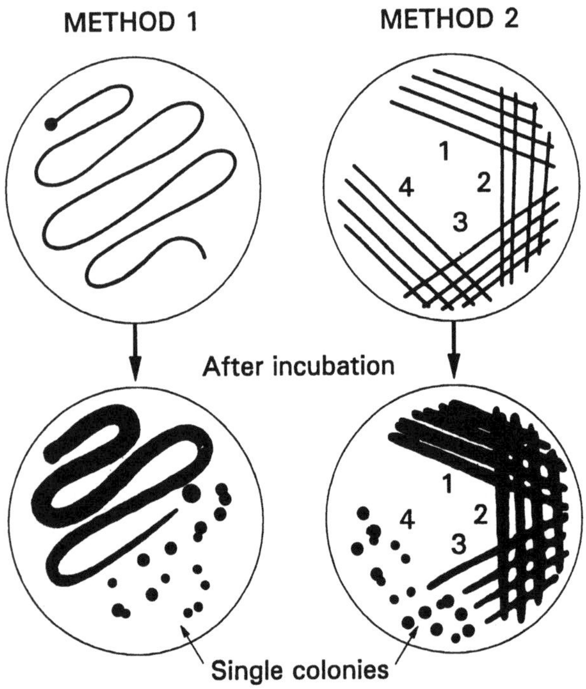

Figure 1.1 Streaking the plate.

Here are some variations—try them and compare your success at isolation by at least two methods. The needle method of isolation is especially useful with freshly harvested nodules 2 mm or larger in diameter. Wash the nodule first in water, then alcohol, then hold it with forceps and briefly pass it through a flame. Place this surface-sterilized nodule on a small piece of sterile filter paper (2 × 2 cm) in a sterile Petri dish. A new piece of filter paper should be used for each nodule. The same Petri dish can be used for several nodules. Dip the blunt-tipped forceps into 95% alcohol and flame momentarily. While holding the nodule with the forceps and resting the nodule on sterile filter paper, quickly slice off a small section with a flamed, hot scalpel. Still holding the nodule with the forceps on the filter paper, insert the tip of a sterile inoculation needle (with a 1-mm loop) into the cut surface. Load the loop with inoculum (Figure 1.2a). Streak directly onto a YMA plate containing CR and a YMA plate containing BTB. When using the needle method, the nodule can also be held in the fingers of one hand while inserting the needle with the other hand. Brace the heels of the hands together to steady them (Figure 1.2b).

Another method consists of serially diluting the nodule bacterial suspension and then pourplating it. This is done as follows. Lay out four sterile plastic Petri dishes marked A, B, C, and D. With a sterile Pasteur pipette, place two separated drops of water into each dish. Crush the sterilized nodule in a sterile Petri dish or test tube. Flame the transfer loop and cool it in drop 1 of dish A, then transfer the bacteroid suspension from the crushed nodule to drop 2 of dish A and mix. Next, flame the loop, cool it in drop 1 of dish B, and transfer one loopful from drop 2 of dish A to drop 2 of dish B and mix. Continue until drop 2 of each dish has been inoculated and mixed with the diluted nodule suspension of the previous one. Pour 15–20 ml liquid YMA (48°C) into the inoculum in each dish. Ensure mixing by gently moving the covered dish first clockwise and then counterclockwise on the table top. Allow three full circles for each movement. Continue mixing by moving the dish from the left to the right and from the right to the left three times. Then, without pausing, move the Petri dish forward and backward and backward and forward, also three times. Allow the agar to set before incubating. Invert the plates during incubation.

Additional procedures are illustrated in Figure 1.3.

f. Performing the Presumptive Test (Key Steps 4 and 5)

The plates prepared from the three methods described previously are referred to as primary isolation plates. Incubate these at 25–30°C in the dark. (Some slow-growing tropical rhizobia absorb CR when incubated in the light.)

After 4–10 days, look for well-isolated colonies. Pick off a single colony typical of rhizobia (Chapter 3) and perform a Gram stain (Chapter 3), then reisolate by streaking on:

1. YMA containing BTB
2. YMA containing CR
3. Peptone glucose agar

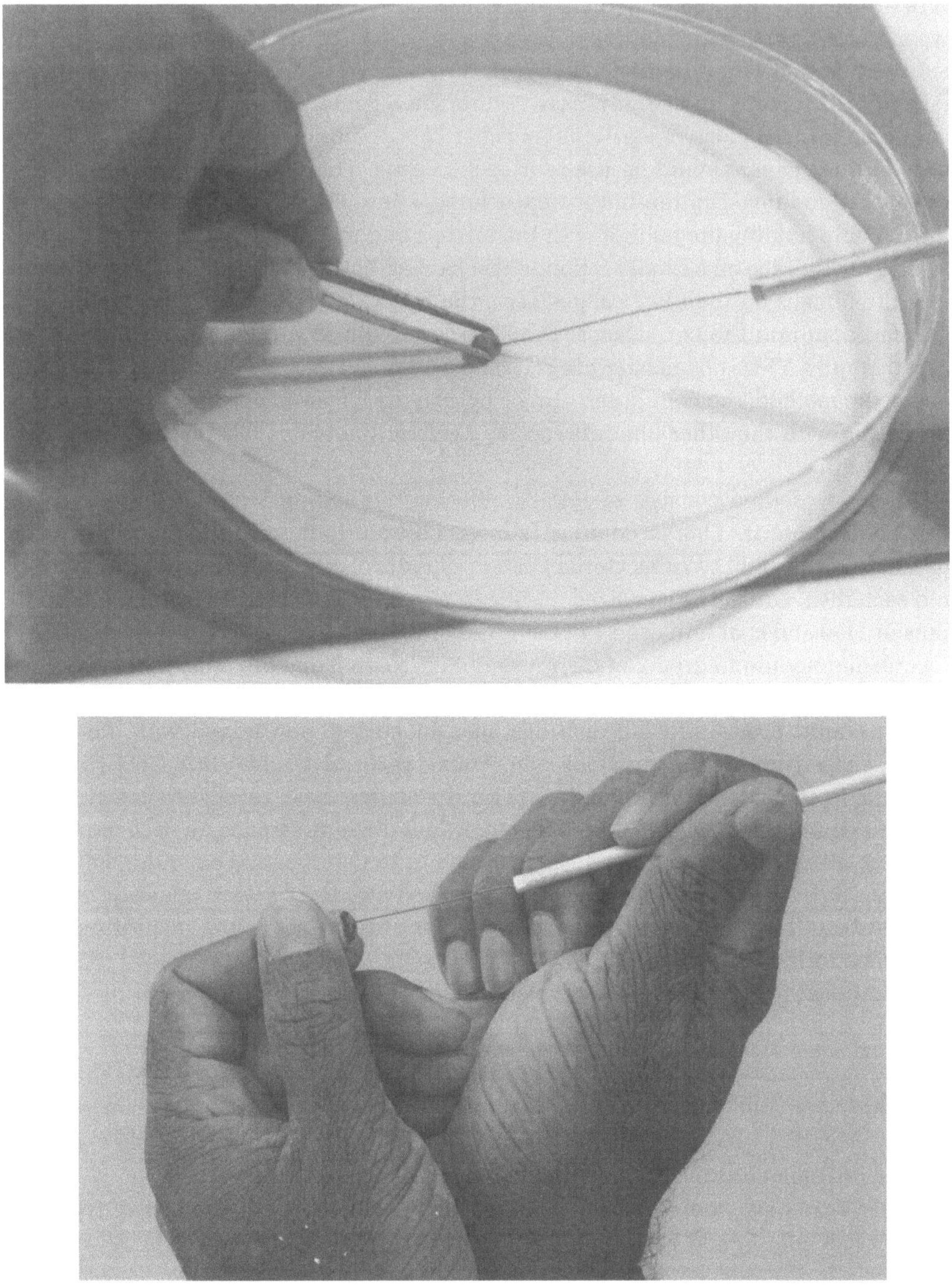

FIGURE 1.2 Taking a sample from inside a nodule using the needle method. Holding the nodule (a) with forceps in a Petri dish and (b) between thumb and forefinger.

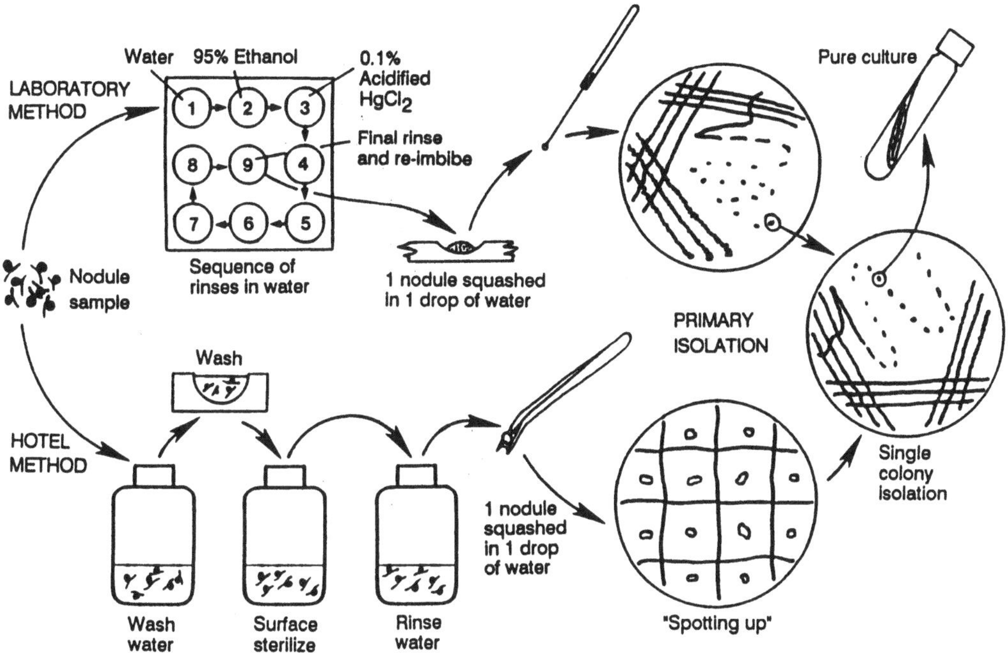

Figure 1.3 Isolation procedure used by Date and Halliday (1979b).

Select isolated typical colonies. It is possible that more than one colony type (e.g., small and large colonies, mucoid and dry, etc.) may appear on a plate streaked from a single nodule. Each of these should be streaked on the three media listed previously and should be considered an individual culture. More than one type of colony in a pure culture of rhizobia may indicate variants of the same strain or the occupancy of two different strains in the same nodule. If no isolated colonies develop, restreak a little of the confluent growth again onto each one of the three media.

Incubate and make daily observations for the appearance of colonies typical of rhizobia. Colonies should show little or no CR absorption when incubated in the dark. There are, however, exceptions (e.g., some strains of *R. meliloti* absorb CR strongly). A blue color, indicating an alkaline reaction on BTB, should be obtained with slow-growing *Bradyrhizobium* spp. A yellow color (acid) reaction is usually produced by the fast-growing *Rhizobium* spp. No growth or poor growth should be obtained on peptone glucose agar. Plates should be read for reactions after 3–5 days (fast growers) and 5–7 days (slow growers). (Unless one is definitely working with fast growers, an incubation of 7–10 days should be routine.) See Chapter 3 for details. Check secondary isolates for colony morphology typical of rhizobia, then perform a Gram stain (Chapter 3) to check for purity of culture. Transfer three separate colonies to culture tubes to be added to stock cultures. Stock cultures obtained at this time are considered presumptive rhizobia. The authenticity of these isolates as pure cultures of rhizobia is confirmed as shown in the next paragraph by the nodulation test (authentication) under bacteriologically controlled conditions. Select two representative colonies of the presumptive rhizobia from the isolation. Prepare 20–50 ml broth cultures in duplicates from each of the two colonies. Incubate on a shaker for use in the authentication tests.

g. Authenticating the Isolates as Rhizobia (Key Steps 6–9)

The importance of determining that the isolate is a pure culture which can form nodules on legume roots cannot be over stressed. It proves the authenticity of a pure culture of rhizobia. For large-seeded legumes like beans (*Phaseolus vulgaris*) and soybean, Leonard jars (Appendix 11) and growth pouches (Chapter 6) are recommended as growth units for authentication. Smaller-seeded legumes, like clovers (*Trifolium* spp.) and siratro (*Macroptilium atropurpureum*), may be grown in growth tubes (Appendix 7). Recommended hosts and growth systems to authenticate isolates are given in Appendix 9. Ideally, a rhizobial strain is tested for its ability to produce nodules on the legume species from which it was originally isolated. However, it may be more convenient to substitute another legume from the same cross-inoculation group, particularly when a small-seeded legume can be substituted for a large-seeded one. Chickpea (*Cicer arietinum*), although a large-seeded legume, can be successfully grown in tubes by excising the cotyledons. This process produces dwarfed chickpea plants. Siratro is used in authenticating most bradyrhizobia from tropical legumes because it nodulates with more than 90% of all bradyrhizobia. Rhizobia from specific hosts (e.g., soybean, *Lotononis*, chickpea, etc.) are not authenticated on siratro.

Set up two suitable growth units for each of the isolates plus at least two extra units that will serve as uninoculated controls. Consult Appendix 11 for the preparation of Leonard jars. Growth pouches are described in Chapter 6. Surface sterilize and pregerminate seeds as detailed in Appendix 10.

Inoculate 1 ml of broth culture for each isolate onto each of the pregerminated seeds in two growth units. The extra growth units are not inoculated and will serve as controls. Plant and inoculate in a clean area. Take precautions against wind drafts and insects, which may cause cross-contamination between treatments.

Examine plants for differences in vigor and color between the inoculated and uninoculated at 15–30 days of growth. Remove the plants from the rooting medium and note the presence or absence of nodules. The presence of nodules in the noninoculated treatment invalidates the test. Sparse nodulation or nodulation restricted to distal parts of the roots of control plants indicates external contamination and points to a need to improve general hygiene. The authentication test must be repeated with adequate bacteriological control.

If the presumptive tests are satisfactory, the isolates are regarded as fully authenticated cultures. The cultures of presumptive isolates are now confirmed as rhizobia and may be given collection numbers. When added to a culture collection, other relevant information should be added for each strain, e.g., parent host, site of collection, soil pH, etc., as shown on strain information form (Figures 1.4 and 1.5).

h. Preserving Rhizobial Cultures (Key Steps 10–13)

There are several satisfactory methods for preserving rhizobial cultures, including YMA slant in screw-cap tubes, desiccated on porcelain beads, lyophilized (freeze dried), and as frozen liquid suspension under liquid nitrogen. The choice of method will depend on facilities, experience, and finances (Table 1.1). The porcelain bead method is recommended for laboratories with limited resources.

To prepare for storage on beads, inoculate a loopful of culture from a YMA slant into 3 ml of sterile yeast–mannitol broth (YMB) and incubate to maximum turbidity on a rotary shaker. Place 20–30 ceramic beads (washed and oven dried) in a screw-cap test tube, cover the mouth of the tube with foil, and sterilize in the oven for 1–2 h at 160–170°C. Prepare storage tubes as depicted in Figure 1.6, using 6–7 g silica gel and sufficient cotton or glass wool to keep the silica gel in place. The rubber-lined caps for the tubes must be autoclaved separately in a rubber beaker, then dried in an oven at 80–90°C.

The glass wool may be oven sterilized in the storage tube with the silica gel. When cotton is used, it should be autoclaved in small balls in a foil-covered beaker. These cotton balls should be of a suitable size to facilitate easy aseptic transfer to the storage tube with forceps. Residual moisture is removed in the oven at 70–80°C before transferring it aseptically to the sterile storage tubes. The autoclaved caps are then added to the tubes.

Transfer the sterilized beads aseptically to the broth culture in the tubes and replug. Soak the beads for 1–2 h, then invert the tube and allow the excess broth culture to soak into the cotton plug. Transfer the beads impregnated with rhizobia into the storage tube aseptically, replace and tighten the screw caps securely. Examine the tubes after a day or so to ensure that the silica gel is still blue. If it turns pink or colorless, then too much moisture was transferred with the beads or an improper seal is permitting entry of moisture. To regenerate a culture, inoculate YMB with one or two beads. These are easily speared from the storage tube using a sterile needle with a small hook at its end. A week or more may be needed to obtain visual signs of growth. Once the broth becomes turbid, loopfuls should be streaked on presumptive test media to check for purity. Subculture from the broth onto YMA slants as desired.

RHIZOBIAL CULTURE RECORD SHORT FORM

A. CULTURE HISTORY: Nodule Isolate ( ) Donated ( )

1. Donor/Source: ________________ Culture no.: ________________
2. Parent legume: ________________ *Leguminosae* subfam.: ________________
3. Nodule collection site: ________________ Soil type: ________
4. Preserved nodule: yes ( ) no ( ) Surface sterilant: ________________
5. Isolation method: ________________ Date: ________
6. Isolation medium: YMA ( ) YMA + BTB ( ) YMA + CR ( ) Other: ________
7. BTB/CR reactions: ________________
8. Colony morphology: ________________
9. Authentication: ______ Legume: ________________ Growth unit: ________
10. Culture designation/preservation: ________________ Date: ________
11. Rhizobial species: ________________

B. EFFECTIVENESS DATA:

1. Test legume: ________________
2. Test unit/media: ________________ Date: ________
3. Field test: ________________ Date: ________
4. Test report: ________________

____

____

C. SUPPLEMENTARY INFORMATION:

1. Culture received as: ________________
2. Received from: ________________ Date: ________
3. Other culture designation(s): ________________
4. Comments: ________________

____

____

FIGURE 1.4 Rhizobial culture record sheet.

RHIZOBIAL CULTURE RECORD SHORT FORM

A. CULTURE HISTORY: Nodule Isolate (X) Donated ( )

1. Donor/Source: Dr. T. J. Wacek Culture no.: ________
2. Parent legume: *Phaseolus vulgaris* *Leguminosae* subfam.: *Papilionoideae*
3. Nodule collection site: Kula, Maui, Hawaii Soil type: mollisol
4. Preserved nodule: yes ( ) no (X) Surface sterilant: $HgCl_2$
5. Isolation method: Crush-method Date: 1976
6. Isolation medium: YMA ( ) YMA + BTB (X) YMA + CR (X) Other: ________
7. BTB/CR reactions: acid reaction on BTB/CR is not absorbed
8. Colony morphology: flat, white-opaque; 1.5–2.0 mm dia.
9. Authentication: + Legume: *Phaseolus vulgaris* Growth unit: Leonard jar
10. Culture designation/preservation: TAL 182/lyophilized Date: 1976
11. Rhizobial species: *Rhizobium leguminosarum* bv. *phaseoli*

B. EFFECTIVENESS DATA:

1. Test legume: *Phaseolus vulgaris* cv. Bountiful
2. Test unit/media: Leonard jar/Broughton & Dilworth, 1970 Date: 1976/77
3. Field test: Kuiaha acid soil site, Haiku, Maui Date: 1982
4. Test report: Highly effective on bean cultivars Bountiful, Pinto, and Kidney.

C. SUPPLEMENTARY INFORMATION:

1. Culture received as: ________
2. Received from: ________ Date: ________
3. Other culture designation(s): none
4. Comments: Highly competative against strains Kim5, and CIAT 899 in soil experiments. Recommended for bean inoculant production.

FIGURE 1.5 Completed rhizobial culture record sheet.

TABLE 1.1 Methods for Preservation of Strains of Rhizobia

Method	Expertise and Facilities Required	Length of Useful Storage Period	Advantages	Disadvantages	Remarks
Agar slopes in screw-cap tubes or agar covered with paraffin oil (Vincent 1970, p. 10)	Basic microbiological knowledge and facilities for pure culture (autoclave, clean transfer area, tubes, media, etc.)	1–2 years without transfer at 25–30°C but can be longer, if held at 5°C	Simplicity, low-cost, minimum facilities and expertise	Short storage time, increased chance of contamination and variants because of more frequent subculturing	Least desirable for long-term storage
Porcelain beads	As above, plus availability of beads, suitable airtight containers and dry sterilizing facilities for silica gel desiccant	3–15 years, with some rhizobia significantly shorter with others	Low-cost and longer storage time and therefore more time before rebeading. Facility for number of subcultures (i.e., one bead) from original	Not as long term as lyophilization and risk of contamination and variants when rebeading. Time required for rebeading	Good for 6–12 month storage[1]

Lyophilized or freeze dried	Basic microbiological facilities lyophilizing equipment (vacuum pump, freezing facility under vacuum) ampoules, glass blowing burner, etc.	Minimum 15–20 years experience suggests much longer	Once ampouled, minimal risk of variants or contamination. Virtually permanent storage. Can be at room temperature	Expensive for equipment and materials	Preferred
Liquid N storage	Expertise as above, plus cryostat and liquid N source	Years, but not much information available	Rapid operation	Very expensive; special precautions during freezing and thawing	None

[1]Poor survival with some fast-growing rhizobia (e.g., *R. leguminosarum* bv. phaseoli, and *Leucaena* and *Sesbania* rhizobia) fused $CaCl_2$ can be used as a substitute for silica gel.

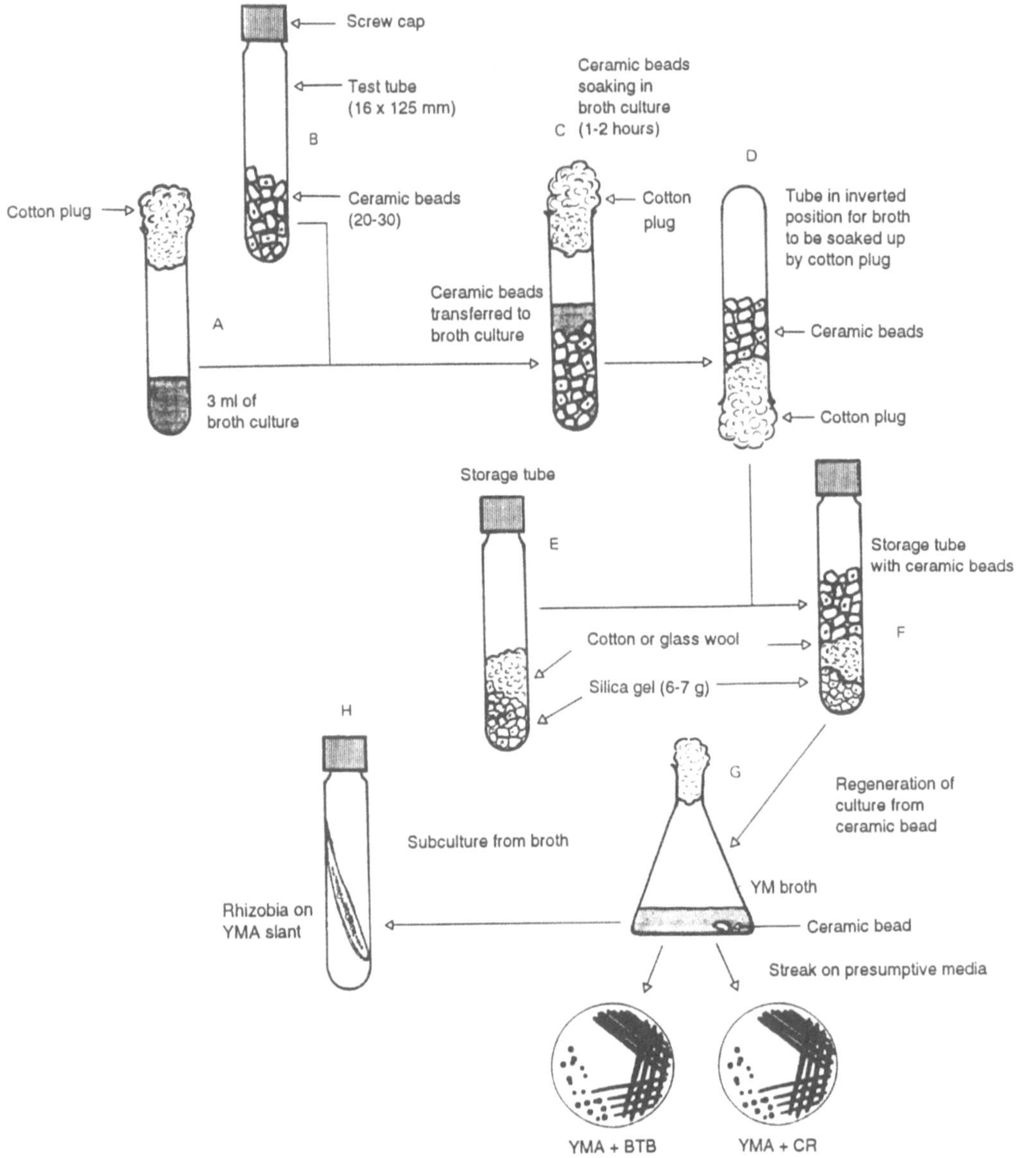

FIGURE 1.6 Ceramic bead method for storing rhizobia.

REQUIREMENTS

a. Recognizing Legumes and Identifying Them in the Field

A suitable botanical key describing the identification of legumes.

b. Recovering Nodules in the Field

Refrigerator
Spade
Sieve
Running water
Plastic bags for plants, smaller bags for seeds

c. Preserving Nodules

Refrigerator
Collection vial (Appendix 2)
Nodules from (b)

d. Examining Nodules and Bacteroids

Microscope
Bunsen burner
Microscope slides, coverslips, mounting fluid
Razor blade, inoculation loop
Distilled water
Carbolfuchsin stain (Appendix 3)
Rhizobial cultures
Nodulated plants from (b)

e. Isolating Rhizobia from a Nodule

Refrigerator
Scissors, forceps, inoculation loop, isolation needle, scalpel
Sterile Petri dishes
Sterile test tubes, glass rods
Erlenmeyer flask, 125 ml (optional); small beaker
Bunsen burner
Sterile Pasteur pipettes
Sterile filter paper (cut into small pieces)
Running water, sterile water, ethanol or isopropanol
Sodium hypochlorite solution, 3% (may be made from commercial bleach)
Plates of plain YMA, plates of YMA + CR and YMA + BTB
Liquid YMA (50°C)
Nodulated plants from (b)
Desiccated nodules

f. Performing the Presumptive Test

Transfer chamber
Incubator
Bunsen burner
Inoculation loop
YMA slants
Flasks, 125 ml with 50 ml of YMB
Plates of YMA + BTB, YMA + CR, and YMA and peptone glucose agar
Gram stain solutions

g. Authenticating the Isolates as Rhizobia

Transfer chamber
Greenhouse, growth room or shelf
Drying oven
Kjeldahl N determination equipment (optional)
Racks for growth pouches and growth tubes
Sterile pipettes, 10 ml
Alcohol burners
Scissors, paper bags for plant tops
Growth pouches, growth tubes (Chapter 5, Appendix 8)
Leonard jars (Appendix 11)
Materials and glassware for seed sterilization (Appendix 10)
Broth cultures from (f)

h. Preserving Rhizobial Cultures

Transfer chamber
Rotary shaker
Sterilizing oven, drying oven
Forceps, inoculation loop, flame, hooked needle
Test tube racks, screw-capped test tubes
Ceramic beads (washed and dried)
Silica gel (with indicator), absorbent cotton, aluminum foil
Beaker, 400 ml
Capped tubes with 3 ml of YMB, culture tubes with YMA slants
Cultures of rhizobia

KEY REFERENCES

Date, R.A. 1982. Collection, isolation, characterisation and conservation of *Rhizobium*. pp. 95–109. *In* J.M. Vincent (ed.) Nitrogen Fixation in Legumes. Academic Press, Sydney.

Fred, E.B., I.L. Baldwin, and E. McCoy. 1932. Root Nodule Bacteria and Leguminous Plants, University of Wisconsin, Madison.

Jordan, D.C. 1984. *Rhizobiaceae*. pp. 234–256. *In* N.R. Krieg and J.G. Holt (eds.) M. Bergey's Manual of Systematic Bacteriology, Vol. I. The Williams & Wilkins Co., Baltimore.

Vincent, J.M. 1970. A Manual for the Practical Study of Root Nodule Bacteria. IBP Handbook no. 15. Blackwell Scientific Publications, Oxford.

2

Observing the Infection Process

Clover (*Trifolium* spp.) rhizobia enter their host's roots through the root hairs. Infection is preceded by a deformation of root hairs and the forming of an infection thread that can be observed directly under the microscope. Root hair deformations may also be caused by nonnodulating strains of *Rhizobium*. Nonnodulating strains used in this chapter cause no infection threads to form. The purpose of this chapter is to observe the infection of legume roots by rhizobia. Clover seedlings are inoculated with rhizobia and stages of the infection process are then observed under the microscope.

KEY STEPS/OBJECTIVES

1. Culture strains of rhizobia in yeast-mannitol broth (YMB).
2. Sterilize and germinate clover seeds.
3. Mount seedling on microscope slide.
4. Incubate the seedlings in inoculated mineral medium.
5. Observe root hair deformation and infection threads.
6. Compare root hair deformations caused by different kinds of rhizobia strains.

a. Culturing Strains of Rhizobia in YMB (Key Step 1)

Inoculate 50-ml flasks or test tubes containing 20 ml of YMB in duplicate with the strains listed as follows:

1. *Rhizobium leguminosarum* bv. trifolii (TAL 382) isolated from nodules of *Trifolium semipilosum*.
2. *R. leguminosarum* bv. trifolii noninfective isolated from nodules of *Trifolium* sp.
3. *R. leguminosarum* bv. trifolii (TAL 1185) isolated from nodules of *Trifolium repens*.
4. *R. leguminosarum* bv. phaseoli (TAL 182) isolated from nodules of *Phaseolus vulgaris*.

5. *R. meliloti* (TAL 380) isolated from nodules of *Medicago sativa*.
6. *Bradyrhizobium* sp. (TAL 764) isolated from nodules of *Lupinus angustifolius*.

Other strains of the same species of rhizobia may be substituted. Incubate at 25–30°C for 5–7 days on a rotary shaker.

b. Germinating Seeds (Key Step 2)

Choose a small-seeded legume. Clover, especially *Trifolium repens* or *T. glomeratum*, is most suitable for this exercise. Surface sterilize seeds according to the procedure outlined in Appendix 10. Some clover species may need scarification with sulfuric acid. Others, like Nolan's white clover and strawberry clover (*Trifolium fragiferum*) germinate easily without scarification. Wash seeds with at least eight changes of sterile distilled water. Aseptically place the seeds onto water-agar plates for germination. Incubate the plates inverted for 48 h or more until roots are 6–8 mm long.

c. Preparing a Fahraeus Slide (Key Step 3)

Prepare 10 ml of Fahraeus C and N-free medium (Appendix 3) containing 0.6% agar in a 15-ml tube. Cool the liquid agar medium to 48°C in a water bath. For each strain of *Rhizobium* used, prepare two sterile 50-ml boiling tubes containing 25 ml of Fahraeus C and N-free medium without agar. Set up two additional tubes for uninoculated controls. Cover with 50-ml beakers. Materials needed for the slide culture of small-seeded legumes are shown in Figure 2.1.

Transfer approximately 0.2 ml of agar medium to a sterile microscope slide using a Pasteur pipette fitted with a rubber bulb. Leave one-half of the slide empty. This is best done by lining up the slide and a long coverslip side by side in a sterile Petri dish (Figure 2.1). Place the agar in five or six drops onto the bottom half of the slide. Immediately, transfer a well-formed seedling to the slide with a sterile inoculation loop. Place the seedling onto the slide in such a way that the root tip is immersed in the agar and the cotyledons are in the empty half of the slide. With sterile forceps, carefully place the long cover glass over the agar and the root tips. If the seed coat adheres to the cotyledons on the seedling, carefully remove it with sterile fine-tipped forceps. Transfer the slide-mounted seedlings to the tubes containing the Fahraeus mineral medium.

d. Inoculating Seedlings (Key Step 4)

Using the broth cultures which have been set up for this experiment in (a), inoculate two seedlings with each of the six strains of *Rhizobium* by adding five drops of the cell suspensions to individual tubes containing the mineral medium and the Fahraeus slides.

Alternatively, the seedlings may be inoculated by incorporating a cell suspension into the Fahraeus agar medium before the seedling is placed onto the slide. This speeds

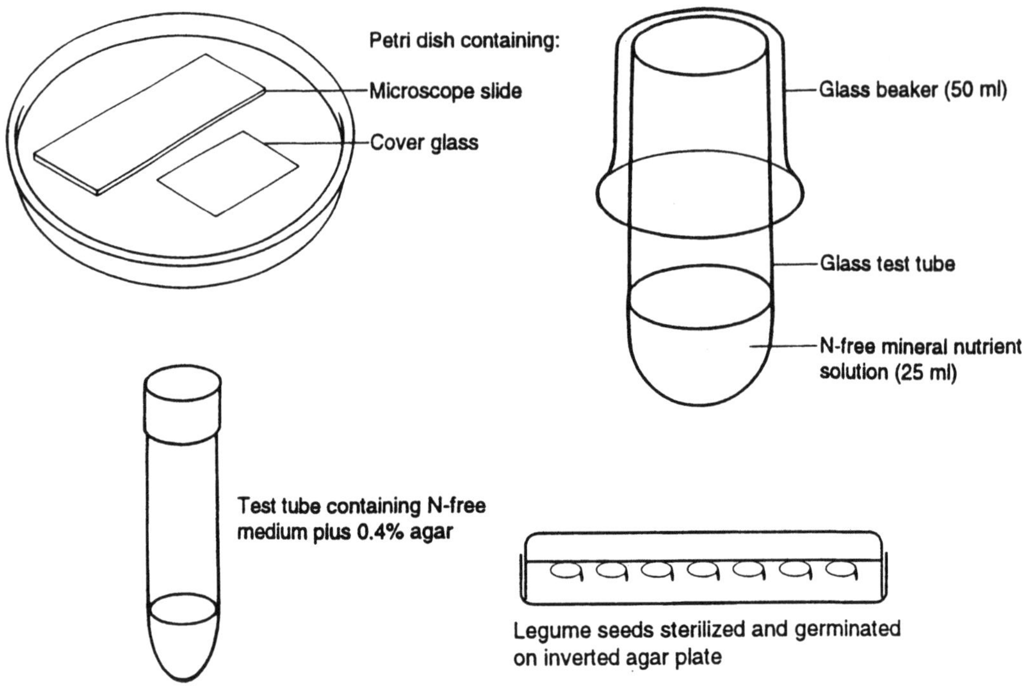

FIGURE 2.1 Materials for slide culture of small-seeded legumes (Fahraeus 1957). (Provided by D. Hubbell)

up the infection process. Add five drops of sterile broth medium to the controls. The full assembly is shown in Figure 2.2. Incubate at 25–30°C in a well-lighted environment.

e. Observing Root Hairs under the Microscope (Key Step 5)

After 24 h, remove the Fahraeus slide from the nutrient tube and examine it under the microscope. Remove the excess solution with absorbent filter paper. Observe with a phase-contrast or dark field microscope under low- and high-power magnifications. Search for root hair deformations and/or curling and infection threads.

Mark the position of your slide on the microscope stage so that the same spot may be found in later observations of the same root hair infections. Make observations in intervals of 12–24 h. Periodic observation may be made at shorter intervals if inoculation was done by including the cell suspension into the agar medium. Look for proliferation and colonization of rhizobia on root hairs as shown in Figure 2.3. Observe the progress of infection threads as illustrated in Figure 2.4. Return the slide to its tube between observations.

Take precautions against undue contamination when returning the slide to the mineral medium. Aseptic conditions cannot be maintained beyond the first observation.

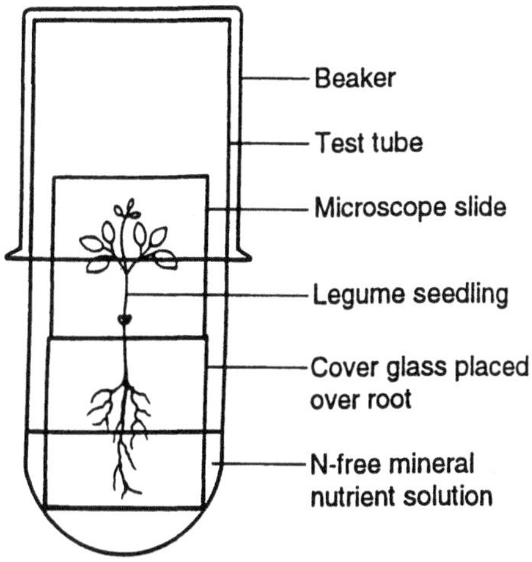

Figure 2.2 Full Fahraeus slide assembly with seedling.

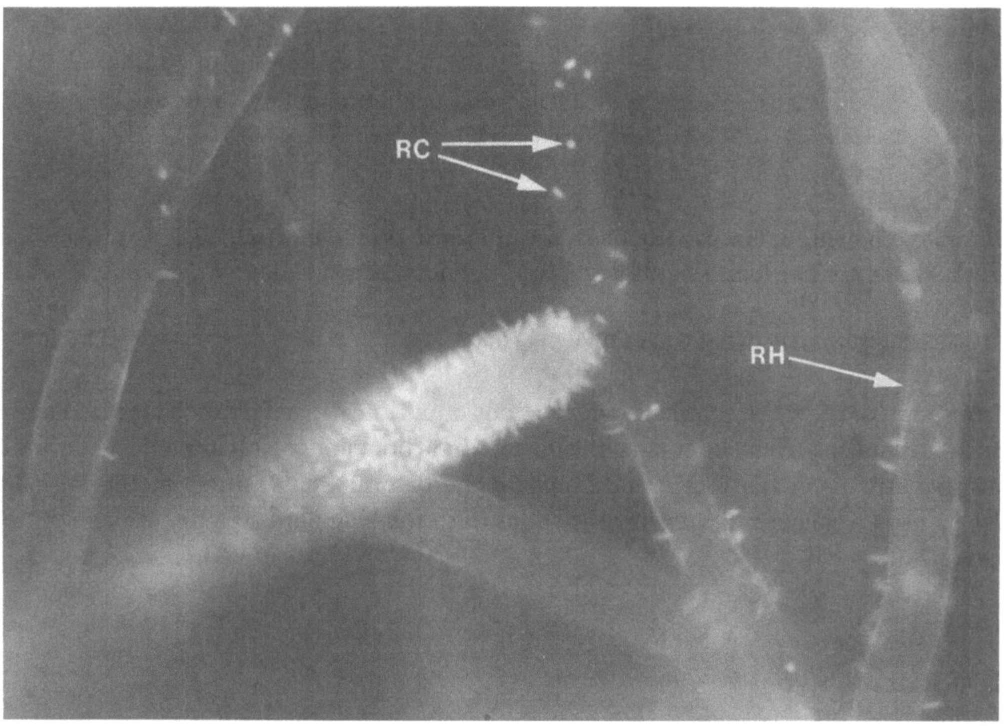

Figure 2.3 (Upper right) Selective proliferation and colonization of *R. leguminosarum* bv. trifolii (RC, rhizobial cells) on a root hair (RH) of its host legume (clover) in a Fahraeus slide system. (Photo contributed by B.B. Bohlool)

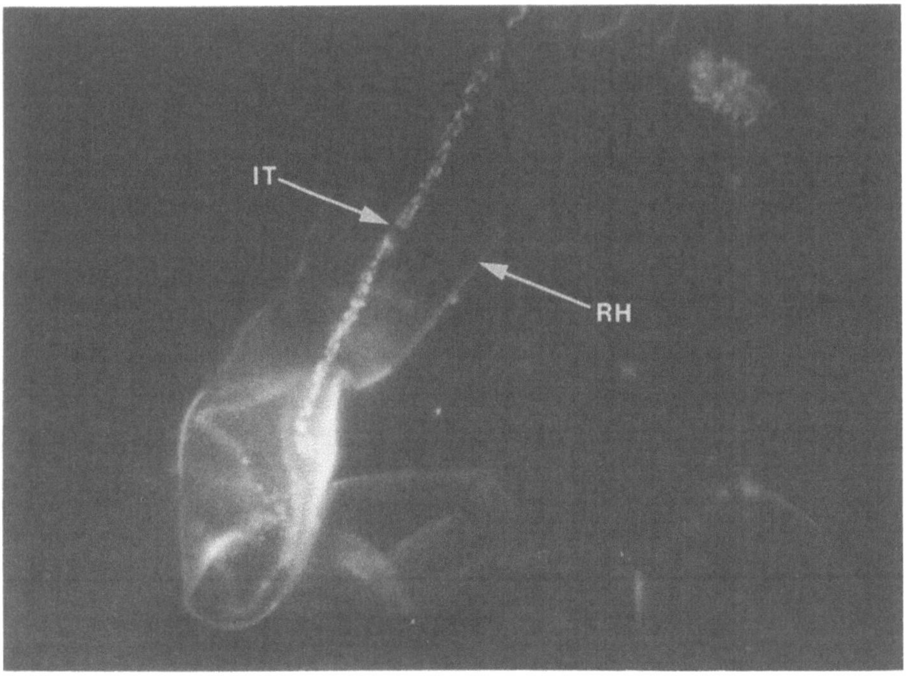

FIGURE 2.4 (Bottom) *R. leguminosarum* bv. trifolii inside infection thread (IT) in the root hair (RH) of its host legume (clover) (*Trifolium repens*). (Photo contributed by B.B. Bohlool)

However, contamination usually does not interfere, provided the root hairs chosen for observations are not located at the edges of the microscope slide.

f. Comparing Root Hair Deformations (Key Step 6)

Photograph or draw the root hair deformations or curling caused by each strain. Distinguish full curling from slight curling and root hair branching. Note the effects of non-invasive strains on the root hairs. Compare the deformations caused by the various strains used. Typical root hair deformations, like the shepherd's crook, are shown in Figure 2.5.

REQUIREMENTS

a. Culturing Strains of Rhizobia in YMB

Rotary shaker
Flasks, 50-ml, (or tubes) (12) containing 20 ml of culture broth each

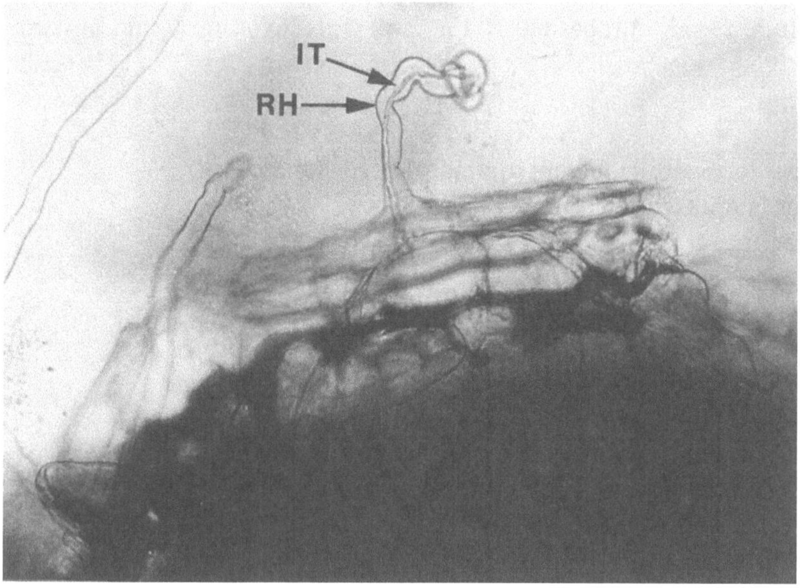

FIGURE 2.5 (Upper left) Deformed white clover root hair (RH) infected with *R. leguminosarum* bv. trifolii 0403. Note the shepherd's crook and the infection thread (IT). (Photo contributed by F. Dazzo)

Inoculation loop, flame
Slant cultures of clover rhizobia strains TAL 382, TAL 1185, TAL 182, TAL 380, TAL 386, noninfective strain of clover rhizobia

b. Germinating Seeds

Incubator
Materials and tools for sterilizing seeds (Appendix 10)
Plates of water agar (7.5 g agar per liter of distilled water)
Seeds of clover (*Trifolium repens*, *T. glomoratum*, or other)

c. Preparing a Fahraeus Slide

Water bath
Sterile microscope slides (1 × 24 × 40 mm)
Coverslips (kept in sterile Petri dishes)
Pasteur pipettes (sterile), rubber bulbs
Inoculation loop, forceps, flame
Fahraeus C and N-free medium (Appendix 3)
Fahraeus medium plus 0.6% agar in a 15-ml tube

Clover seedlings
Fahraeus medium, 25 ml, in tubes (39 × 150 mm) with covering 50-mm beakers

d. Inoculating Seedlings

Growth chamber (or well-lighted environment) at 25–30°C
Pasteur pipettes (sterile), rubber nipples
Tubes with seedlings from (c)

e. Observing Root Hairs under the Microscope

Microscope with phase or dark field condenser
Forceps
Filter paper (sterile and absorbent)
Seedlings in inoculated Fahraeus solution from (d)

f. Comparing Root Hair Deformations

Microscope as in (e) with camera attachment

KEY REFERENCES

Dart, P.J. 1974. The infection process. pp. 381–429. *In* A. Quispel (ed.) The Biology of Nitrogen Fixation. North Holland Publishing Company, Amsterdam, Holland.

Fåhraeus, A. 1957. The infection of clover root hairs by Nodule bacteria studied by a simple glass slide technique. J. Gen. Microbiol 16:374–381.

Hubbell, D.H. 1981. Legume infection by *Rhizobium*: A conceptual approach. BioScience 31:832–837.

Yao, P.Y., and J.M. Vincent. 1969. Host specificity in the root hair "curling factor" of *Rhizobium* spp. Aust. J. Biol. Sci. 22:413–423.

3

Cultural Properties, Cell Morphology, and Nutritional Requirements of Rhizobia

The aims of this chapter are to learn to distinguish rhizobia from other microorganisms by cell morphology, staining reactions, and growth responses on various media, and to show how media for rhizobia can be modified. Root nodule bacteroids, cultures of rhizobia and bradyrhizobia, and other microorganisms are examined under the microscope using staining techniques. They are also cultured on indicator media. The growth reactions of five strains of rhizobia and bradyrhizobia are observed on media containing combinations of C and N from different sources.

KEY STEPS/OBJECTIVES

1. Subculture rhizobia and other bacteria.
2. Observe cell morphology of rhizobia and other bacteria under phase-contrast microscopy.
3. Examine rhizobia and other bacteria for cell morphology using a simple stain (carbolfuchsin) and the Gram stain.
4. Culture rhizobia and other bacteria on indicator media.
5. Observe colony morphology and growth reactions on the indicator media.
6. Prepare agar media with different C and N sources.
7. Inoculate media with rhizobia.
8. Observe growth reactions on each medium.

a. Preliminary Subculturing of Different Bacterial Cultures (Key Step 1)

Make subcultures on agar slants from the stock cultures of the following microorganisms using yeast-mannitol agar (YMA) slants for the rhizobia, and nutrient agar slants (Appendix 3) for the other bacteria: *Rhizobium meliloti*, *R. leguminosarum* bv. phaseoli, *Rhizobium* sp. [chickpea (*Cicer arietinum*)], *Bradyrhizobium* sp. (*Lotononis*), *B. japonicum*, *Bacillus subtilis*, *Escherichia coli*, *Staphylococcus aureus*, and *Pseudomonas* sp.

Also needed are:

1. Surface-sterilized nodules (saved from Chapter 1).

2. A homogenate of nonsurface-sterilized nodules of slow-growing and fast-growing rhizobia.

3. A broth culture of rhizobia mixed with other species of bacteria.

b. Comparing Cell Morphology and Gram Stain Reactions of Rhizobia with Those of Other Microorganisms (Key Steps 2 and 3)

Make wet mounts of the cultures provided and examine under the phase-contrast microscope. Note the motility, size, and shape of the rhizobia compared with other bacteria. Place a loopful of sterile distilled water onto a clean, pre-flamed and cooled microscope slide. Flame the loop and transfer a small sample of the bacterial growth from the slant culture to the water on the slide. Mix thoroughly and make a thin smear approximately 1 cm^2 in diameter. For broth cultures, transfer a loopful and make smear directly on the dry slide. Air dry, heat fix, and allow to cool. Flood the smear with diluted carbolfuchsin for 60 s. Rinse carefully in a gentle stream of water and blot dry. Locate smears under low power (10×, 25×, or 40×) objective. Apply a drop of oil to the smear and observe with the 100× oil immersion objective using bright field illumination.

The carbolfuchsin stain makes the bacteria easily visible (cells appear pink). Note the characteristic rod shape of the cultured cells of rhizobia and compare the size and shape of these to that of bacteroids seen in the nodule preparation. Also compare rhizobia with the other bacteria and note the difference in size and form. Refer to Figure 3.1 for the morphology of the microorganisms.

c. Determining Gram Stain Reactions of Various Bacteria (Key Step 3)

1. Make thin smears of the various bacteria provided and heat fix.

2. Stain the smears with solution I (crystal violet) for 1 min.

3. Wash lightly with water and flood with solution II (iodine).

4. Drain immediately and flood again with solution II for 1 min.

5. Drain solution II and decolorize with solution III (95% alcohol) for 5–15 s in the case of a thin smear and 30 s if the smear is thick.

6. Wash with water and blot dry carefully.

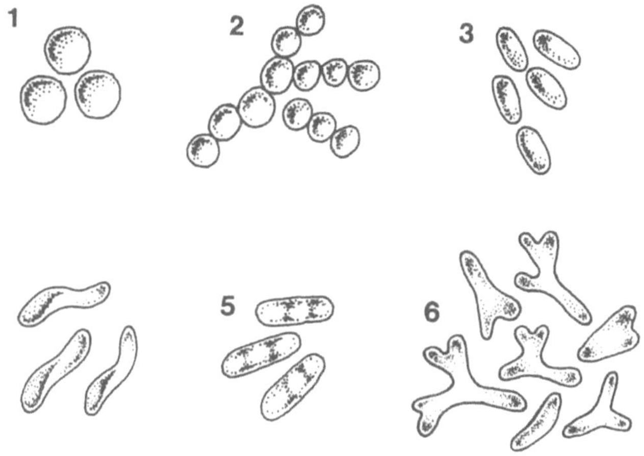

FIGURE 3.1 Shapes of bacteria: 1) coccus, 2) Staphylococcus, 3) rod (e.g., *E. coli*), 4) *Spirillum*, 5) cultured cells of rhizobia, and 6) bacteroids of rhizobia (e.g., *Lens* sp.).

7. Counterstain with solution IV (safranin) for 1 min.
8. Wash with water and air dry.

Observe the preparation under oil immersion.

The Gram stain procedure separates bacteria into two groups: Gram-positive and Gram-negative organisms. *Gram-positive* organisms retain the crystal violet stain after treating with iodine and washing with alcohol, and appear dark violet after staining (e.g., *Bacillus subtilis* and *Staphylococcus aureus*). *Gram-negative* organisms lose the violet stain after treating with iodine and washing with alcohol but retain the red coloration of the counterstain, safranin (e.g., *Rhizobium*, *Pseudomonas*, and *E. coli*).

d. Characterizing Growth of Rhizobia Using a Range of Media (Key Steps 4 and 5)

Rhizobia can be described according to their growth in solid and in liquid media. The size, shape, color, and texture of colonies and the ability to alter the pH of the medium are generally stable characteristics useful in defining strains or isolates. Typical colony characteristics, when grown on standard YM medium, are described as follows.

Shape

Usually discrete, round colonies varying from flat () to domed () and even conical (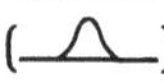) shape on agar surfaces. Colonies usually have a smooth margin. When growing subsurface in the agar, colonies are typically lens shaped.

Color and Texture

Colonies may be white-opaque or they may be milky- to watery-translucent. The opaque colony growth is usually firm with little gum, whereas the less dense colonies are often gummy and soft. Colonies may be glistening or dull, evenly opaque or translucent, but many colonies develop darker centers of rib-like markings with age. Red or pink (e.g., from *Lotononis*) and yellowish (e.g., from *Stylosanthes*) occur, but are not common.

Growth Rate

Growth rate generally ranges from 3–5 days for fast growers (e.g., from *Leucaena, Psoralea,* and *Sesbania*), 5–7 days for slow growers (e.g., from *Macroptilium, Desmodium,* and *Galactia*), to 7–12 days (e.g., some *Stylosanthes* and *Lupinus*) to achieve maximum colony size on agar or growth in liquid medium. Growth rate varies according to the temperature of incubation (optima 25–30°C), origin (culture or nodule), aeration (in liquid cultures), and composition of medium.

Size

When well separated on agar plates, colony size may vary from 1 mm for many slow-growing strains (e.g., from *Stylosanthes, Zornia, Aeschynomene,* and *Lupinus*) to 4–5 mm for faster-growing strains (e.g., from *Leucaena, Psoralea,* and *Sesbania*). In crowded plates, colonies remain smaller and discrete, but coalesce to confluent growth when colonies join.

Select two preparations from the pure cultures, the nodule homogenates or the mixed cultures of rhizobia and other bacteria. Streak out on plates containing each of the following media: YMA, YMA + bromthymol blue, YMA + congo red (CR), and peptone glucose agar + bromcresol purple.

These indicators and selective media are used as presumptive tests for purity of cultures. Their interpretation is as follows.

Rhizobia generally do not absorb CR when plates are incubated in the dark. Colonies remain white, opaque, or occasionally pink. Contaminating organisms usually absorb the red dye. However, reactions depend on the concentration of CR and the age of the culture. Rhizobia will absorb the red dye if plates are exposed to light during the incubation or exposed to light for an hour or more after growth has occurred. As stated in Chapter 1, some strains of *Rhizobium meliloti* are an exception and absorb CR strongly.

Freshly prepared YMA plates containing bromthymol blue have a pH of 6.8 and are green. Slow-growing rhizobia show an alkaline reaction in this medium, turning the dye blue. Fast-growing rhizobia show an acid reaction, turning the medium yellow.

Rhizobia grow poorly, if at all, on peptone glucose agar and cause little change in pH, when incubated at 25–30°C. Heavy growth indicates contamination. Note that growth and color reactions described here are dependent on the strain metabolism on the standard YMA media. Reactions may differ when other types of media are used. There are media that are used for special purposes. For instance, if a faster growth rate is desired, the arabinose gluconate (AG) medium (Appendix 3) may be used. Other special purpose media are also listed in Appendix 3.

e. Observing Growth Reactions on Modified Media (Key Steps 6, 7, and 8)

If mannitol and yeast extract are not available, the basic growth medium may be modified, since rhizobia can utilize C and N from various sources (Table 3.1).

Make up the media as follows:

1. Prepare 1500 ml of a mineral salts solution containing the inorganic constituents of YMA (Appendix 3). Add 28 g of agar and heat in the autoclave or water bath. Dispense the melted mineral salts agar solution in 12-ml portions into test tubes. Sterilize in the autoclave and keep melted in the water bath at 50°C.

2. Prepare C source stock solutions (10 g/100 ml) of mannitol (M), sucrose (Su), arabinose (A), and glycerol (G). Sterilize by autoclaving, except for A, which should be filter sterilized.

3. Prepare stock solutions of the N sources (Appendix 3), yeast-water from baker's yeast (B), yeast extract (Y) 0.5 g/100 ml, soybean extract (S), and ammonium chloride (NH_4Cl) (N) 0.4 g/100 ml. Sterilize by autoclaving.

Pipette into separate, sterile Petri dishes 1.5 ml of (2) and 1.5 ml of (3). Pour one test tube (12 ml) of the melted (50°C) mineral salts agar preparation (1) to each plate, so as to provide the combinations shown in Table 3.1 in duplicate.

Mix immediately after adding the agar by rotating each dish gently three times clockwise and counterclockwise, and three times to the right and to the left, as well as forward and backwards. Allow the plates to cool overnight for the agar to solidify. Remove any contaminated plates.

Streak each of the five rhizobial strains onto each medium as you would streak for isolation (chapter 1). Streaking two or more cultures onto one plate may be necessary

TABLE 3.1 Combinations of C and N Sources for Growth Media of Rhizobia

	N Source[1]			
	B	Y	S	N
C Source[2]		Media		
M	MB	MY	MS	MN
Su	SuB	SuY	SuS	SuN
A	AB	AY	AS	AN
G	GB	GY	GS	GN

[1]B, baker's yeast; Y, yeast extract; S, soybean extract; and N, NH_4Cl.

[2]M, mannitol; Su, sucrose; A, arabinose; and G, glycerol.

if there is a shortage of plates. However, this practice should be avoided if possible because of an aerosol effect during the streaking process.

Incubate and compare growth on the various media at 3, 7, and 10 days after plating.

REQUIREMENTS

a. Preliminary Subculturing of Different Bacterial Cultures

Transfer chamber
Incubator
Inoculation loop, flame
Microorganisms on slants, surface-sterilized nodules, homogenate of nonsurface-sterilized nodules, and a mixed-broth culture as indicated in (a).

b. Comparing Cell Morphology and Gram Stain Reactions of Rhizobia with Those of Other Microorganisms

Microscope with bright field and phase-contrast equipment
Inoculation loop, flame
Microscope slides
Immersion oil, lens paper
Running water or wash bottle with water
Carbolfuchsin solution (Appendix 3)
Tissue paper or paper towels
Microorganisms from (a)

c. Determining Gram Stain Reactions of Various Bacteria

All requirements of (b), not including phase-contrast equipment
Gram stain solutions (Appendix 4)

d. Characterizing Growth of Rhizobia Using a Range of Media

Transfer chamber
Incubator
Inoculation loop, flame
Plates of YMA
Plates of YMA + bromthymol blue
Plates of YMA + CR
Plates of peptone glucose agar + bromcresol purple
Consult Appendix 3 for growth media preparation

e. Observing Growth Reactions on Modified Media

Incubator, autoclave, water bath, pH meter, suction pump, balance
Bacteriological filter unit (0.2-μm pore size)
Volumetric flasks (eight of 100 ml)
Sterile pipettes, 5 and 10 ml
Spatula, weighing paper
Large flask or beaker, 2–3 liter
Test tubes, test tube rack
Sterile Petri dishes
Inoculation loop, flame
Mineral salts: for YMB (Appendix 3)
Distilled water
Agar
Mannitol, sucrose, arabinose, glycerol
Brewer's yeast, yeast extract, soybean extract, ammonium chloride
Rhizobia from (a)

KEY REFERENCES

Allen, O.N., and E.D. Allen. 1950. Biochemical and symbiotic properties of the rhizobia. Bacteriol. Rev. 14:273–330.

Fred, E.B., I.L. Baldwin, and E. McCoy. 1932. Root Nodule Bacteria and Leguminous Plants. University of Wisconsin, Madison.

Graham, P.H., and C.A. Parker. 1964. Diagnostic features in the characterization of the root nodule bacteria of legumes. Plant Soil 20:383–396.

4

Demonstrating Genetic Diversity in Rhizobia Using Patterns of Carbohydrate Utilization and Intrinsic Antibiotic Resistance

A large range of biochemical and metabolic tests are frequently used to differentiate between rhizobial species. These tests include vitamin requirements, salt, acid and alkali tolerance, carbohydrate utilization, and resistance to antibiotics. Carbohydrate utilization properties are of taxonomic significance, while resistance to low levels of antibiotics can be used for rhizobial strain characterization and identification.

In this chapter, cultures of *Rhizobium* spp. and *Bradyrhizobium* spp. are inoculated on media containing various carbohydrates and antibiotics. The ability of *Rhizobium* spp. to utilize a wider range of carbohydrates than *Bradyrhizobium* spp. is demonstrated. Rhizobia are grouped according to their intrinsic antibiotic resistance (IAR) patterns. The potential for applying the two techniques in rhizobial classification and ecology is implied.

KEY STEPS/OBJECTIVES

1. Culture rhizobia.
2. Prepare test media containing carbohydrates.
3. Prepare test media containing antibiotics.
4. Design experiment.
5. Inoculate test media.
6. Monitor growth on plates.
7. Tabulate data.

a. Culturing Rhizobia and Examining Purity (Key Step 1)

The rhizobia that will be studied in this chapter are listed in Table 4.1. Check the purity of the cultures by Gram stain and growth on yeast-mannitol agar (YMA) containing bromthymol blue and congo red (CR) (Chapter 1). Discard contaminated cultures. Maintain the cultures on YMA slants stored at 4°C. Note that the 32 cultures listed in Table 4.1 represent *Rhizobium* spp. and *Bradyrhizobium* spp. Other rhizobial and bradyrhizobial strains (16 each) may be substituted if they were previously checked for purity and authenticated on the host legume.

Inoculate a loopful of each culture in 1.0 ml yeast-mannitol broth (YMB) in 16- × 125-mm screw-cap tubes. Aerate the cultures on a rotary shaker. Since the *Bradyrhizobium* spp. grow slower than the *Rhizobium* spp., inoculate the bradyrhizobia 3 days ahead of the *Rhizobium* spp. When the inoculation is staggered, both species of rhizobia will be ready for the experiment after 6 days.

b. Preparing Carbohydrate Solutions (Key Step 2)

Note that many carbohydrates are heat labile and these solutions are sterilized by membrane filtration. Carbohydrates are prepared most commonly as 10% (w/v) solutions in water. Heat-labile carbohydrate solutions are added to the autoclaved carbohydrate-free basal agar medium before pouring plates. In this chapter, 19 carbohydrates are tested for rhizobial utilization.

Prepare 10% solutions (1 g carbohydrate in 10 ml distilled or deionized water) of each of the following heat-labile carbohydrates (prepare the solutions in 50-ml beakers): arabinose, rhamnose, xylose, galactose, gluconate, mannose, maltose, trehalose, dextrin, inulin, raffinose, adonitol, dulcitol, and erythritol. Label the beakers. Also, make 10% solutions of the following heat-stable carbohydrates: fructose, glucose, lactose, sucrose, and mannitol. Label the beakers. Store all the carbohydrate solutions in a refrigerator. Beakers must be covered with aluminum foil or Parafilm during temporary storage.

c. Preparing Carbohydrate-Free Basal and Test Medium (Key Step 2)

The carbohydrate-free medium is essentially similar to YMA medium (Appendix 3) but with the following modifications that reduce the yeast extract to 0.05 g liter^{-1}. The carbohydrates for testing are added later.

Prepare 2.5 liters of the carbohydrate-free medium following the composition described in Appendix 3, but with low yeast extract (0.05 g liter^{-1}) and without carbohydrates. Dispense 90 ml of the medium in 250-ml Erlenmeyer flasks. Prepare at least 21 flasks of the medium and add 1.5 g of agar powder to each flask.

The heat-stable sugar solutions (fructose, glucose, lactose, sucrose, and mannitol) can be added to the basal medium at this point. Label the flasks appropriately. Autoclave the medium in all of the flasks at 15 lb/in^2 for 15 min. Transfer the flasks to a water bath equilibrated at 50°C. Pour plates with the media containing the heat-stable car-

Table 4.1 Suggested List of Rhizobial Strains for Studying IAR Patterns and Carbohydrate Utilization

Rhizobial Species	Host of Isolation	Designations
Rhizobium meliloti	*Medicago sativa*	TAL 380 (SU 47)
	Medicago sativa	TAL 1372 (POA 116)
	Medicago sativa	TAL 1373 (POA 135)
Rhizobium sp.	*Cicer arietinum*	TAL 620 (CB 1189)
	Cicer arietinum	TAL 480 (UASB 67)
	Cicer arietinum	TAL 1148 (Nit.27A3)
Rhizobium spp.	*Leucaena leucocephala*	TAL 82
	Leucaena leucocephala	TAL 1145 (CB 3060)
	Leucaena leucocephala	TAL 582 (CB 81)
R. leguminosarum bv. phaseoli	*Phaseolus vulgaris*	TAL 182
	Phaseolus vulgaris	TAL 1383 (CIAT 632)
	Phaseolus vulgaris	TAL 1797 (CIAT 899)
R. leguminosarum bv. viceae	*Lens culinaris*	TAL 634 (Nit. 92A3)
	Pisum, sativum	TAL 1236 (Allen 344)
	Vicia faba	TAL 1397 (Nit. 175F9)
	Vicia faba	TAL 1399 (Nit. 175F12)
Bradyrhizobium japonicum	*Glycine max*	TAL 102 (USDA 110)
	Glycine max	TAL 379 (CB 1809)
	Glycine max	TAL 377 (USDA 138)
Bradyrhizobium spp.	*Arachis hypogaea*	TAL 1000
	Arachis hypogaea	TAL 1371 (Nit. 8A11)
	Cajanus cajan	TAL 1127 (IHP 38)
	Cajanus cajan	TAL 1132 (IHP 195)
Bradyrhizobium spp.	*Calopogonium mucunoides*	TAL 652 (UMKL 46)
	Phaseolus lunatus	TAL 22
	Phaseolus acutifolius	TAL 644 (CIAT 257)
	Vigna unguiculata	TAL 169 (Nit. 176A22)
	Vigna unguiculata	TAL 209
	Vigna unguiculata	TAL 173 (Nit. 176A30)
	Vigna mungo	TAL 441 (UPLB M6)
	Vigna mungo	TAL 420 (THA 301)
	Macrotyloma africanum	TAL 309 (CB 756)

bohydrates. Pour four plates (approximately 25 ml of medium per Petri dish) for each test carbohydrate. (Square style 100- × 15-mm Petri dishes, if available, are preferred over the standard round 85-mm dishes. However, square style dishes take more media.) Label the plates.

d. Adding Membrane-Sterilized Carbohydrate Solutions and Pouring Plates (Key Step 2)

The heat-labile carbohydrate solutions are prepared ahead of time and sterilized using sterile 0.2-μm disposable membrane filters and plastic hypodermic syringes (10 ml) with the Luer-Lok system. One filter-syringe assembly is used for each solution. All 10 ml of the carbohydrate solution must be sterilized, then added to the basal medium.

Remove the carbohydrate solutions from the refrigerator. Work with one solution at a time in the laminar flow hood. Draw the contents of the beaker into a 10-ml syringe. Secure the prefilled syringe to the filter with a clockwise twist. Hold the prefilled syringe vertically with the attached filter at the mouth of a flask of media. Press the syringe plunger with your thumb to begin filtration. Swirl the contents of the flask and pour the plates. Store the plates in a refrigerator after the agar has solidified. If time is available, leave the plates in the laminar flow hood 24–48 h to check for contaminated plates.

e. Preparing Antibiotic Solutions (Key Step 3)

The 32 rhizobial strains listed in Table 4.1 will be tested on eight antibiotics to study IAR patterns. The antibiotics and concentrations (micrograms per milliliter final concentration in YMA) are as follows: neomycin sulfate, 1.25 and 2.5; kanamycin sulfate, 4.0 and 10.0; polymyxin B sulfate, 5.0 and 10.0; nalidixic acid, 5.0 and 10.0; streptomycin sulfate, 2.5 and 10.0; erythromycin, 2.5 and 5.0; vancomycin hydrochloride, 5.0; and carbenicillin, 10.0. Make all solutions of antibiotics in water except erythromycin (in ethanol) and nalidixic acid (in 1 M NaOH).

Prepare fresh or use previously frozen (−20°C) stock solutions of the previously mentioned antibiotics as follows: nalidixic acid, 5 mg ml^{-1} in 1 M NaOH stock; erythromycin, 5 mg ml^{-1} in ethanol; and all others as 10 mg ml^{-1} in sterile water. Sterilize the antibiotic solutions using 0.2-μm disposable membrane filters and plastic hypodermic syringes as described for sterilizing carbohydrate solutions in step (d). Store stock solutions of antibiotics in small presterilized McCartney bottles or other small screw-capped glass containers. Stock solutions must be kept frozen (−20°C) if intended for later use. Stock solutions are conveniently removed from the storage bottles/vials after thawing using micropipettes with disposable tips. Frozen stock solutions should not be used after a 2-week storage time.

f. Preparing Antibiotic Testing Media (Key Step 3)

The YMA containing the antibiotic must be prepared 24 h before use. Standardized practice of antibiotic media preparation is important in obtaining consistent results.

Prepare 15 Erlenmeyer flasks (250-ml capacity) each containing 100 ml of YMA medium. Sterilize the medium for 15 min at 121°C and 15 lb/in^2. Keep the sterilized medium in a water bath at 50°C.

Using a micropipette with sterile disposable tips, draw out the appropriate volume of the stock antibiotic solution from the storage vials and deliver into the YMA kept at 50°C. Swirl the contents gently to ensure uniform distribution of the antibiotic. Pour four plates for each antibiotic. Plain YMA without antibiotic serves as a control.

g. Designing the Experiment (Key Step 4)

A total of 32 cultures (16 *Rhizobium* spp. and 16 *Bradyrhizobium* spp.) are studied. The multiple inoculator (Appendix 19) selected is the type with 48 prongs. Therefore, 16 cultures, each replicated three times, can be inoculated per plate of the test medium. Study the numbering system used on the plastic microtiter plates. The 48-prong multiple inoculator requires the wells in the vertical rows, numbers 1–6. Because the two groups of rhizobia differ in their growth rates, inoculate them on separate plates of the test medium. Set up controls.

h. Inoculating the Test Media (Key Step 5)

Inoculating the rhizobia onto the test media (carbohydrates and antibiotics) can be accomplished the same day. Two presterilized multiple inoculators are needed.

Combine 1 ml of each culture with 9 ml of quarter-strength YMB to obtain a 10-fold diluted culture (approximately 10^8 cells ml^{-1}). With a sterile Pasteur pipette, transfer seven drops (approximately 0.025 ml drop^{-1}) of the diluted culture per well into three wells (i.e., wells A1, A2, and A3) of a sterile microtiter plate. Likewise, with a fresh Pasteur pipette, transfer the second culture into wells A4, A5, and A6. Transfer the third culture into wells B1, B2, and B3. Completely transfer all 16 cultures of a group (either *Bradyrhizobium* spp. or *Rhizobium* spp.) into the wells of the microtiter plate. Similarly, fill up the wells of a second microtiter plate with the 16 cultures of the second group of rhizobia.

Record the identity of each culture in the wells of the microtiter plate. Mark the lower-half of each Petri dish with an arrow to indicate orientation of the plate in relation to the location of cultures in the microtiter plates. Inoculate plates set up for studying carbohydrate utilization first.

Load one of the two multiple inoculators with the cultures in the wells of the microtiter plate. Carefully inoculate the control (YMA only) plate first. Spray the prongs of the first multiple inoculator with 95% alcohol/methanol. Flame and set aside to cool. Load the second multiple inoculator with cultures and inoculate a plate of test media containing a specified carbohydrate. Sterilize the second multiple inoculator.

Following the previously described sequence on the use of the two multiple inoculators, proceed to inoculate the various carbohydrate test media prepared for the experiment. Complete the experiment by similarly inoculating the various antibiotic media. Begin by inoculating the control plate. Inoculate duplicate plates of each test media × strain combination. Incubate the plates at 28°C in an incubator.

i. Observing and Recording (Key Step 6)

Make periodic observations to monitor growth on the test media. For *Rhizobium* spp., all plates should be read and the experiment terminated at 3 days. For the *Bradyrhizobium* spp., complete observations and recording after 6 days.

For the carbohydrate-utilization test media, two controls are needed to record the effects. The test medium containing mannitol is the standard or reference positive control that should indicate most growth stimulation. The basal medium control should indicate minimal to no growth. In the case of the growth of the rhizobia on antibiotic test media, resistance or inhibition is compared to maximal growth produced on the YMA control plates containing no antibiotic.

Record the growth of the rhizobia as follows: 4 = good, 1 = weak, and 0 = no growth (sensitive). Growth response of several rhizobial strains on carbohydrate (mannitol and sucrose) medium and IAR patterns on two concentrations of streptomycin are illustrated in Figures 4.1 and 4.2.

j. Tabulating Data (Key Step 7)

Treat the growth response data for carbohydrate utilization and IAR patterns separately. Recognize and group strains of rhizobia showing similar carbohydrate-utilization patterns. Distinct groups of strains may emerge. Similarly, group the data from IAR test to show the rhizobial groups with distinct IAR patterns. It is important to note that there could be interstrain variation in a larger collection of the same species.

REQUIREMENTS

a. Culturing Rhizobia and Examining Purity

Cultures of rhizobia representing all inoculation groups (see Table 4.1)
Transfer chamber
Inoculation loop, flame
YMB in 16- × 125-mm screw-cap tubes
Shaker
Microscope, microscope slides, immersion oil
Gram-stain reagents

b. Preparing Carbohydrate Solutions

Distilled or deionized water
Carbohydrates: arabinose, rhamnose, xylose, galactose, gluconate, mannose,

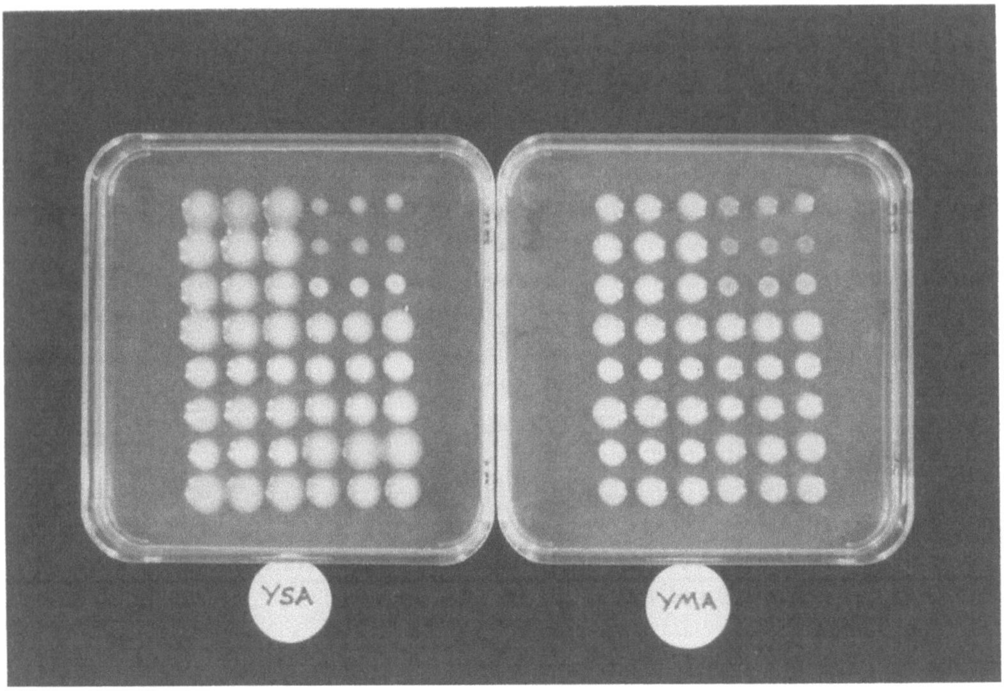

FIGURE 4.1 Sucrose (YSA) and mannitol (YMA) utilization patterns of 16 isolates of *Rhizobium* spp. (*Gliricidia sepium*).

maltose, trehalose, dextrin, inulin, raffinose, adonitol, dulcitol, erythritol, fructose, glucose, lactose, mannitol, and sucrose
Beakers, 50-ml; serum vials or small McCartney bottles
Aluminum foil or Parafilm
Transfer chamber, refrigerator

c. Preparing Carbohydrate-Free Basal and Test Medium

Basal medium, 21- × 90-ml in Erlenmeyer flasks, 250 ml
Agar (Difco)
Autoclave, water bath (50°C)
Heat-stable carbohydrates [see (c)]

d. Adding Membrane-Sterilized Carbohydrate Solutions and Pouring Plates

Heat-labile carbohydrates
Disposable, sterile plastic syringes, 10 ml with Luer-Lok system

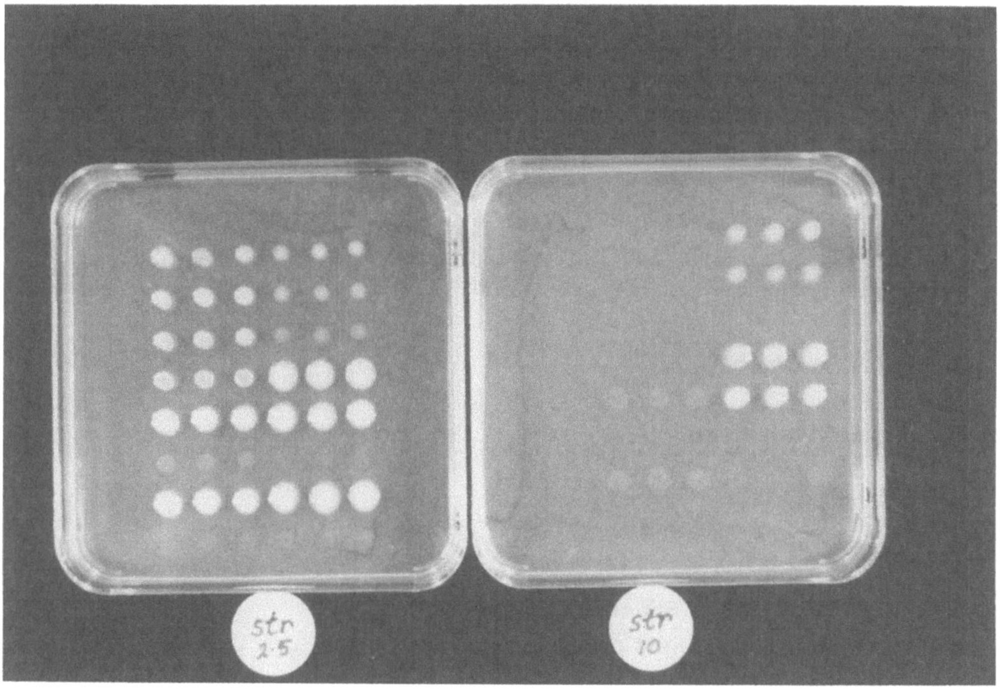

FIGURE 4.2 IAR patterns of 16 isolates of *Rhizobium* spp. (*Gliricidia sepium*) at 2.5 μg ml^{-1} and 10 μg ml^{-1} streptomycin (str).

Disposable, sterile membrane filters, 0.2 μm
Autoclaved basal medium [see (c)]
Water bath (50°C)
Petri dishes (square style, 100 × 15 mm or round 85 mm)
Transfer chamber
Refrigerator

e. Preparing Antibiotic Solutions

Distilled or deionized water
1 M NaOH, ethanol
Serum vials or small McCartney bottles
Antibiotics: neomycin sulfate, kanamycin sulfate, polymyxin B sulfate, nalidixic acid, streptomycin sulfate, erythromycin, vancomycin hydrochloride, and carbenicillin
Sterile, disposable hypodermic syringes, 1 ml
Sterile 22-gauge needles
Micropipettes, 0–200 μl, with disposable tips

f. Preparing Antibiotic Testing Media

Erlenmeyer flasks, 250-ml capacity
YMA, 100 ml, in Erlenmeyer flasks
Autoclave
Water bath (50°C)
Antibiotic stock solutions
Petri dishes (square style, 100 × 15 mm or round 85 mm)
Transfer chamber

g. Designing the Experiment

No special requirements

h. Inoculating the Test Media

Test media in Petri dishes
Microtiter plates, sterile Pasteur pipettes
Multiple inoculators (12 × 4 prongs)
Sterile pipettes, 10 ml
Sterile 25% (v/v) YMB
Alcohol in spray bottle
Transfer chamber
Cultures of *Rhizobium* spp. and *Bradyrhizobium* spp. (a)

i. Observing and Recording

No special requirements

j. Tabulating Data

No special requirements

KEY REFERENCES

Eaglesham, A.R.J. 1987. The use of intrinsic antibiotic resistance for *Rhizobium* study. pp. 185–204. *In* G.H. Elkan (ed.) Symbiotic Nitrogen Fixation Technology. Marcel Dekker, Inc., New York.

Elkan, G.H., and L.D. Kuykendall. 1982. Carbohydrate metabolism. pp. 147–166. *In* W.J. Broughton (ed.) Nitrogen Fixation. Vol. 2. Clarendon Press, Oxford.

3. Graham, P.H. 1964. Studies on the utilization of carbohydrates and Krebs Cycle intermediates by rhizobia. Antonie van Leeuwenhoek J. Microbiol. Serol. 30:68–72.

5

Quantifying the Growth of Rhizobia

This chapter deals with routine enumeration techniques for pure cultures of rhizobia. The total or direct count is performed using the microscope. Optical density measurements are used to estimate the number of cells in broth culture. The viable count is accomplished through plating methods. The mean generation times of a *Rhizobium* sp. and a *Bradyrhizobium* sp. in broth culture are computed.

KEY STEPS/OBJECTIVES

1. Inoculate yeast-mannitol broth (YMB) with rhizobia.
2. Calibrate Pasteur pipettes.
3. Determine the total count.
4. Measure the optical density of the broth cultures.
5. Make a serial dilution and plate by the pour-plate, spread-plate, and drop-plate methods.
6. Read and calculate the viable counts obtained by the three methods.
7. Compare results of the counting methods.
8. Inoculate flasks with diluted culture(s) for the generation time experiment.
9. Determine viable counts periodically.
10. Plot growth curve and determine the mean generation time.

a. Preliminary Culturing of Fast- and Slow-Growing Rhizobia (Key Step 1)

Inoculate two flasks, each containing 50 ml of YMB, with fast-growing *Rhizobium leguminosarum* bv. phaseoli strain TAL 182, and two other flasks with a slow-growing *Bradyrhizobium japonicum* strain TAL 379. Other strains of *Rhizobium* spp. and *Bradyrhizobium* spp. may be used instead. Incubate the flasks at 25–30°C on a rotary shaker at 100 rpm. TAL 182 should be started 4–5 days in advance of the exercise; TAL 379,

7–9 days in advance. Take the culture flask of TAL 182 from the shaker after 4–5 days and remove a 20-ml subsample for procedures described in (c) and (d).

b. Becoming Familiar with the Petroff-Hausser and Helber Counting Chambers (Key Step 3)

The Petroff-Hausser counting chamber (Figure 5.1) is a precision-machined glass plate that has a sunken platform at its center. The depth of this sunken platform is exactly 0.002 cm. A channel borders two sides of the platform. The surface of the platform is etched with a grid system that consists of 25 large squares, each of which is divided into 16 smaller squares. Because of the precisely machined gap between the grid surface and the overlying glass coverslip, it is possible to relate the number of cells observed in a field to the volume of fluid in which they are suspended. Knowing the volume above each square, the concentration of rhizobia (total cells per milliliter) can be calculated.

The data in Table 5.1 apply to the Petroff-Hausser counting chamber and also to the Helber counter, except for the total grid area. The Helber counter has only 16 large squares, resulting in a total grid area of 6.4×10^{-3} cm^2. Other differences are the absence of a frame and a circular "drainage moat" surrounding the grid area instead two channels.

c. Using the Petroff-Hausser and Helber Counting Chambers (Key Step 3)

Chamber and coverslip should be soaked in a mild liquid detergent, thoroughly rinsed with distilled water, and then air dried. This will ensure an even flow of the liquid into the chamber and prevent the formation of air bubbles.

A fully grown broth culture contains approximately 10^9 cells per milliliter. Make a 1:10 dilution with sterile water to bring the suspension within a countable range. From this dilution, make another dilution series in sterile water of 10, 20, 40, and 80%. Choose the dilution that you consider is within the best counting range. Trial runs with the various dilutions may be needed to select the best dilution for the count. This may require washing and drying the counting chamber several times, until the best dilution has been selected for counting.

Slide the clean Petroff-Hausser chamber into its frame, place the coverslip into position, and press it down lightly to ensure a firm seating on the supporting surface of the chamber. The frame of the Petroff-Hausser chamber has a small indentation on the inside of one of its long edges. To this area, deliver a small drop of the diluted culture suspension using a fine tipped Pasteur pipette. The culture suspension will quickly spread over the grid. Excess culture (if a large drop was added) will overflow into the two channels at the edges of the etched platform. If these channels flood completely, the coverslip may not rest flush on the surface of the glass plate. If this happens, clean the counting chamber and start again.

The Helber counting chamber is somewhat easier to use than the Petroff-Hausser

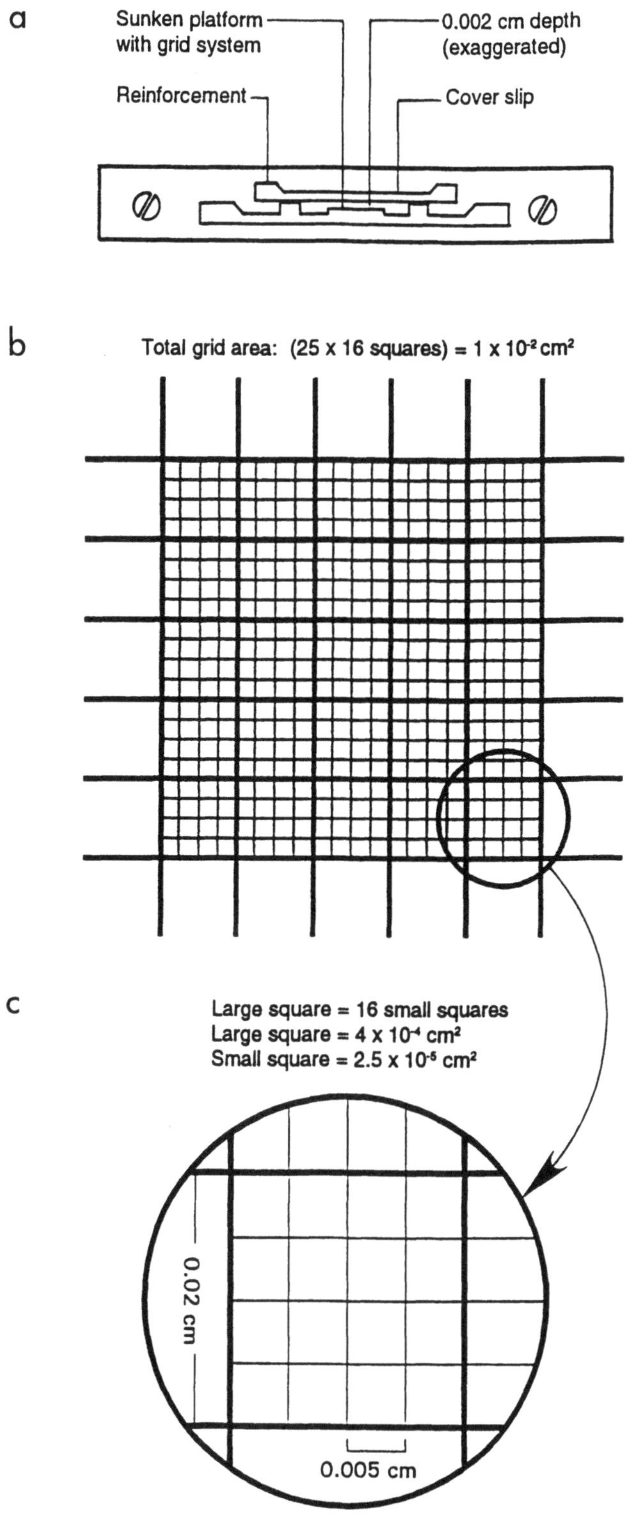

FIGURE 5.1 Petroff-Hausser counting chamber. (a) Cross section, (b) top view of entire grid, and (c) magnified view of single, large square containing 16 small squares.

TABLE 5.1 Brief Details of the Petroff-Hausser Counting Chamber

Area (cm^2)		Corresponding Volume (ml)	Factor
Total grid	1×10^{-2}	2×10^{-5}	5×10^{4}
Large square[1]	4×10^{-4}	8×10^{-7}	1.25×10^{6}
Small square[2]	2.5×10^{-5}	5×10^{-8}	2×10^{7}

[1]The large squares are most suitable for counting rhizobia. The countable range is 8–80 cells per large square.
[2]The countable range for small squares is 3–8 cells.

chamber. Place a loopful of the diluted sample directly on the grid system and apply the cover glass. The amount of sample must be such that the space between the platform and the cover glass is just filled, with no or minimal overflow into the moat. The absence of a frame allows freer movement of the microscope objective while counting. This is especially helpful if a larger objective (100× magnification) is chosen.

Place the chamber under the 40× objective of a phase-contrast microscope. Count cells in individual large squares. To avoid counting the same cells twice, omit bacteria on the upper- and left-borderlines of each square. Count at least 10 fields (8–80 cells per large square) to obtain coefficients of variability of 10%.

Use the following formula to calculate the number of cells: Bacteria in 1 ml of original cell suspension = dilution × cells per square × factor for square.

If 20 cells (mean of 10 squares) were counted in a large square and the original broth culture was diluted by a factor of 10, and then again by a factor of 2, the total number of rhizobia per milliliter of undiluted broth would be:

$$20 \times 10 \times 2 \times (1.25 \times 10^{6}) = 5 \times 10^{8} \text{ cells ml}^{-1}$$

Note that this direct count included dead as well as viable rhizobia and also the cells of contaminants, if present. Most direct total counts are of variable reliability in that they may overestimate the viable count by a factor of more than 2, as 50% or more of the cells counted may not be viable. This method is suitable only for counting log phase broth cultures in liquid media, and not in peat, soil, or other particulate materials.

d. Estimating Cell Concentration by Optical Density (Key Step 4)

The optical density of a bacterial suspension is generally correlated with the number of cells it contains. Optical density measurements are a simple and convenient estimate of cell numbers because they require little manipulation and aseptic conditions need not be observed. Dilute 5–10 ml of the TAL 182 broth culture to 10, 20, 40, and 80% of its original concentration. Measure the light absorbed by each concentration with a spectrophotometer at a wavelength of 540 nm. Use YMB to calibrate the instrument at zero. Relate the different concentrations to the actual cell count obtained with the Petroff-Hausser chamber by plotting the optical density (OD) against the total cell number. This method is useful only for cells grown in clear media. Measurements must be taken no later than the late-log phase of growth when most cells are still viable. Dead cells and gum, usually associated with an older culture, will render OD readings unreliable.

e. Determining the Number of Viable Cells in a Culture by Plating Methods (Key Steps 2, 5, 6, and 7)

Serial Dilution

Make serial dilutions of the TAL 182 broth culture. Based on the total count, the number of viable cells will be approximately 1.0×10^9 ml^{-1}. A countable range for plate counts is 30–300 cells ml^{-1}. To achieve this concentration, set out eight tubes, each containing 9 ml of sterile water. Most tap water is suitable for this purpose. Deionized water may be used where tap water is highly chlorinated and therefore toxic to rhizobia. Do not use saline for this purpose because some rhizobial strains are sensitive to sodium chloride. One milliliter of the broth culture is diluted in tenfold steps (10^{-1} through 10^{-8}). With a sterilized 1-ml serological pipette equipped with a rubber bulb of 1-ml capacity, suck up broth culture from tube 1 to the 1-ml mark. Immediately expel the broth culture back into the tube with sufficient vigor to effect a thorough mixing. Repeat sucking up and expelling five times and then transfer 1 ml to tube 2. Take a new sterile pipette, attach the rubber bulb, and remove 1 ml of cell suspension from tube 2. Mix five times as before and transfer 1 ml to tube 3. Repeat this procedure using a fresh sterile pipette each time, until the dilution series has been completed. Figure 5.2 illustrates the serial dilution procedure.

Pour-Plate Count

Aliquots of three dilutions from the dilution series are pipetted in duplicate onto three sterile Petri dishes. Melted agar is then added and mixed with the culture. Proceed as follows: Use a fresh pipette for each strain and for each dilution in the series. Begin with the highest dilution in the series. With the aid of the suction bulb, fill and empty the pipette by sucking in and out five times with the diluted culture, then aseptically transfer 1 ml to a sterile Petri dish. Open the Petri dish only enough to allow the pipette to enter, and deliver the sample. Flame the pipette briefly (but do not overheat it) by passing it through the Bunsen burner flame prior to each successive removal of aliquots for replication (two per dilution) from the same tube. Similarly, with the same pipette, remove 1-ml aliquots in duplicate from the 10^{-7} and 10^{-6} dilutions into more Petri dishes. Aseptically pour 15–20 ml yeast-mannitol agar (YMA) (kept melted at 50°C in a water bath) onto each of the cell suspensions in the Petri dishes. To disperse the cells evenly, gently move each Petri dish clockwise and counterclockwise, allowing an equal number of swirls in each direction. To further ensure uniform dispersion of the cells, move the

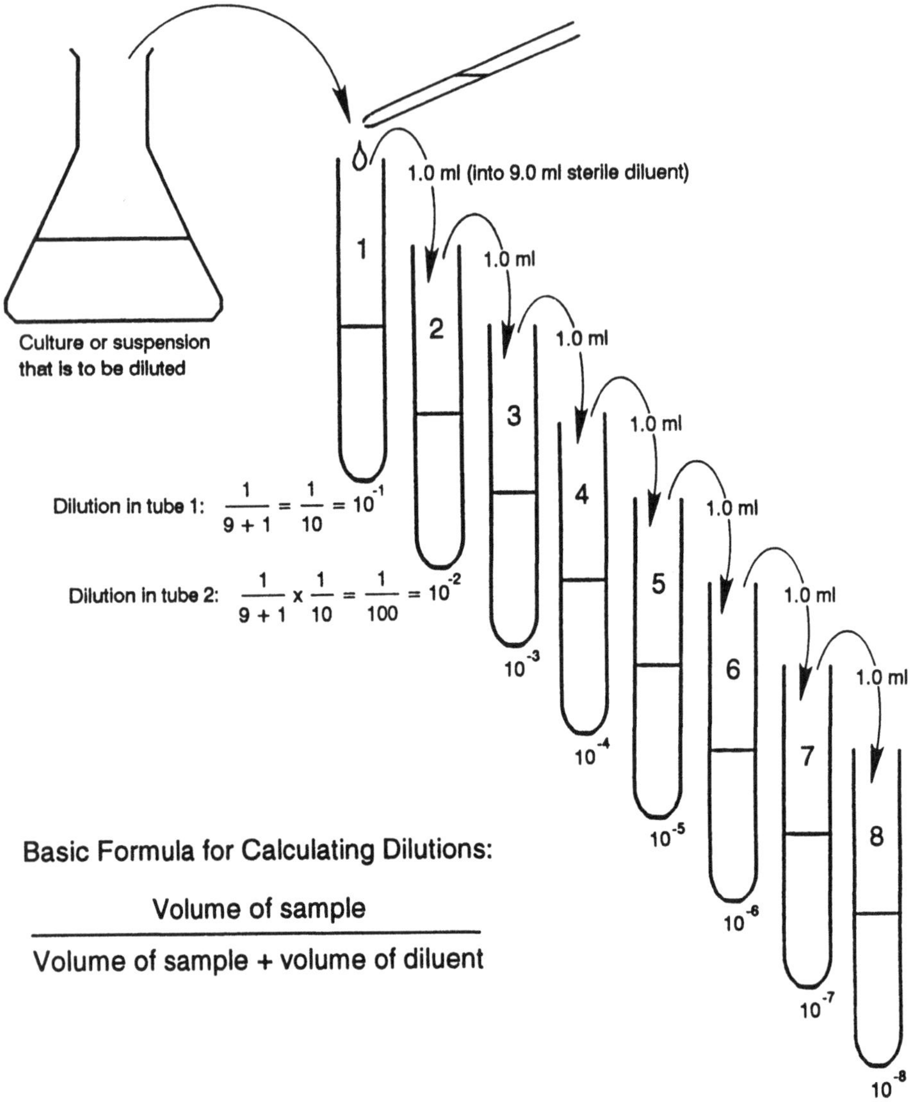

Figure 5.2 Procedure for serial dilution.

Petri dish three times forward and backward, then to the left and right. Allow the agar to set, invert the dishes, and incubate at 25–30°C. Read the plates after 3–5 days.

Prepare serial dilutions of TAL 379. Make pour plates with dilutions 10^{-8}, 10^{-7}, and 10^{-6} in duplicates. Incubate the plates for 7–9 days, checking them daily during the incubation. Lens-shaped colonies develop in the YMA and normal colonies develop on the surface. For counting, choose a plate that yields between 30 and 300 colonies. Plates resulting from dilutions that yield lower numbers of colonies tend to overestimate, while more crowded plates usually underestimate the actual numbers.

Multiply the average number of colonies by the dilution factor. If the average number of colonies at 10^{-7} dilution is 50, then the original broth culture had a concentration of:

$$50 \times 10^{7} = 5.0 \times 10^{8} \text{ cells ml}^{-1}$$

Spread-Plate Method

Use the same serially diluted samples of TAL 379 prepared for the previously described pour-plate method. Begin with the 10^{-7} dilution and deliver 0.1 ml of the sample into each of four plates of YMA, previously dried at 37°C for about 2 h. Using the same pipette, dispense 0.1-ml samples from the 10^{-6} and 10^{-5} dilutions, in that order. Prepare a glass spreader by bending a 20-cm glass rod of 4-mm diameter to the shape of a hockey stick, dip it into alcohol, and flame; then cool the spreader by touching it to the surface of a separate YMA plate. Lift the cover of each Petri dish just enough to introduce the spreader and place it in position on the agar surface. Spread the sample evenly over the agar surface, sterilizing and cooling the spreader between samples. Incubate as before. Calculate the number of viable cells as outlined for the pour-plate method, adjusting for the smaller volume that was plated (0.1 ml instead of 1.0 ml). For example, if 50 colonies were counted on a plate inoculated with 0.1 ml of a 10^{-7} dilution, the results should be:

$$50 \times 10 \times 10^{7} = 5 \times 10^{9} \text{ cells ml}^{-1}$$

Drop-Plate Method

Both the spread- and pour-plate methods are lengthy and require many Petri dishes. A variation, known as the Miles and Misra drop-plate method, is more rapid and consumes less materials. Use agar plates that are at least 3 days old or have been dried at 37°C for 2 h. Radially mark off eight equal sectors on the outside bottom of the Petri dish, as seen in Figure 5.3. Label four sectors for replications of one dilution and four for another, allowing two dilutions per plate.

For this technique, calibrated pipettes are required. Calibrate at least 10 pipettes by the following method. Determine the weight of 100 drops of water on a sensitive balance or the volume of 100 drops of water in a 10-ml measuring cylinder. Calculate the weight or the volume of a single drop by dividing the total weight or volume by 100.

Pipettes with the same tip diameter (e.g., external diameter of 1 mm) deliver drops of virtually the same volume. After the drop size of a calibrated pipette has been established, more pipettes of the same tip diameter may be selected using a wire gauge. Alternatively, any Pasteur pipette may be cut to the same tip diameter with a fine file after matching its tip with a wire gauge.

Use the dilution series of TAL 379, which had been prepared earlier for the pour plate method. Plate dilutions of 10^{-7}, 10^{-6}, and 10^{-5}. Using a calibrated Pasteur pipette fitted with a rubber bulb, begin with the highest dilution and deliver one drop to each of the appropriate four sectors of the plate. To do this, hold the pipette vertically, about

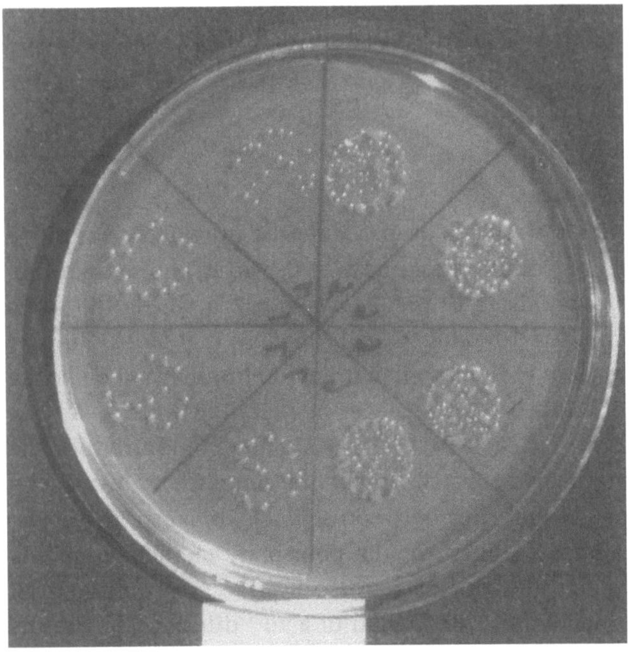

FIGURE 5.3 Growth of colonies of *Rhizobium* sp. from drops plated by the Miles and Misra drop-plate method.

2 cm above the agar surface; exert just enough pressure on the bulb to deliver one drop. Use the remaining four sectors of the plate for the next dilution. Allow the drops to dry by absorption into the agar; then invert and incubate at 25–30°C. Growth of colonies of *Rhizobium* spp. from drops plated by the drop-plate method will emerge as a pattern seen in Figure 5.3. The drop-plate method requires more practice than the other methods. Results may not match those of the pour-plate and spread-plate methods at first attempts. It is advisable to practice drop plating with water before using this method for the first time. Fewer colonies per drop require more drops to be counted in order to provide the same statistical precision.

After 3–5 days of incubation, with daily observations, count the colonies that TAL 182 formed. Open the Petri dish, invert it, and place it on the illuminator of a colony counter. With a fine-tipped felt pen, mark each colony counted while simultaneously operating a tally counter. Record your counts. The preferred counting range should be 10–30 colonies per drop.

If a pipette with a 14-gauge tip is used, one drop will be 0.03 ml. Divide 1 ml by 0.03 (which is 33), and multiply by the dilution factor and the average number of colonies per drop. Example: If the average number of colonies per drop is 30 at 10^{-5} dilution, the number of viable cells are:

$$33 \times 30 \times 10^{5} = 9.9 \times 10^{7} \text{ cells ml}^{-1}$$

Compare the viable count of TAL 182 with its total count and calculate the percentage viability in the original culture.

At the end of a 7–10-day incubation period, count the colonies of TAL 379 on plates prepared by the three methods. Calculate the number of viable cells per milliliter and compare the results obtained by the different methods. Discuss the advantages and disadvantages of the three plating methods.

It is important to note that plate counts, of whatever variety, are of value only for counting the viable rhizobia in pure culture. There is no selective medium that permits only the growth of rhizobia. Therefore, quantifying rhizobia in soil is difficult. Also, the plating methods do not distinguish between strains or species of rhizobia that have similar visual colony characteristics on YMA. When it is necessary to quantify the occurrence of viable cells of rhizobia in nonsterile materials, a plant infection method must be employed (Chapter 6).

f. Determining the Mean Generation (Doubling) Time of Rhizobia (Key Steps 8, 9, and 10)

The time required for a doubling of a given cell population or one cell to become two is referred to as the *generation time* or *doubling time*. The growth of rhizobia in broth culture is followed for a 7-day period. Viable counts are made each day throughout the duration of the experiment. A growth curve is obtained by plotting the log of the viable count versus time. From the curve, the mean generation (doubling) time is computed.

Bradyrhizobium japonicum strain TAL 379 and *R. leguminosarum* bv. phaseoli strain TAL 182 are used in this experiment. A total of 16 250-ml Erlenmeyer flasks, each containing 100 ml of full strength YMB, will be needed for each strain. Prepare 32 flasks for the two strains. Measure accurately 100 ml of YMB into each flask and sterilize. Obtain 1 ml each of the fully grown cultures of TAL 182 and TAL 379 from broth cultures prepared previously in this exercise. By the serial dilution procedure, dilute each culture to give 1×10^6 cell ml^{-1}. (It is approximated that when fully grown, each strain will have at least 1×10^9 cells ml^{-1}.)

Inoculate each flask with one drop (0.03 ml per drop) of the diluted broth culture. Use a calibrated Pasteur pipette for the inoculation. Inoculate 16 flasks with TAL 182 and another 16 with TAL 379. Two flasks will be sampled each day for each strain. Incubate flasks on a rotary shaker (100 rpm) at room temperature (25–30°C). Based on the presence of at least 1×10^6 cells ml^{-1} in the diluted broth, and by inoculating 0.03 ml or 3.0×10^4 cells of this sample into 100 ml of the broth, the starting number of cells at zero time should be 3.0×10^2 cells ml^{-1}.

Perform a zero time viable count for both strains. Remove 1 ml and dilute in 9 ml of quarter-strength YMB to give a 10^{-1} dilution. Use the spread-plate method to plate this dilution in duplicate. Perform viable counts for each culture every day for 7 days, taking care to allow the full 24 h between counts. The extent of dilution of a culture, the choice of dilutions to be plated, and the volume (0.1 ml, spread-plate method or 0.03 ml, by drop-plate method) to be plated will depend on the rate at which turbidity develops during growth.

Obtain the mean viable count for each day and transform the values to $\log_{10}$. Plot

$\log_{10}$ of the viable count (Y axis) versus time (X axis). Draw a smooth curve through the points.

The mean generation time is computed using values from the exponential phase. From the exponential phase, choose a straight line portion of the curve and note the values for viable count and time. Obtain the number of generations by transforming the value for viable count from $\log_{10}$ to $\log_2$ using the relationship:

$$\log_a x = \log_b x/\log_b a$$

when $a = 2$ and $b = 10$

then $\log_2 x = \log_{10} x/\log_{10} 2$

since $\log_{10} 2 = 0.3010$

therefore $\log_2 x = \log_{10} x/0.3010$

Divide the time (hours) by the number of generations to obtain the mean generation time. Compare the mean generation time of TAL 182 with that of TAL 379.

REQUIREMENTS

a. Preliminary Culturing of Fast- and Slow-Growing Rhizobia

Transfer chamber
Rotary shaker
Flasks (four) containing 50 ml of YMBroth each
Sterile pipettes, 10 ml
Slant cultures of TAL 182 and TAL 379

b. Becoming Familiar with the Petroff-Hausser and Helber Counting Chambers

No requirements

c. Using the Petroff-Hausser and Helber Counting Chambers

Phase-contrast microscope
Petroff-Hausser counting chamber and covers
Pasteur pipettes, rubber bulb
Pipettes, 10 ml
Wash bottle with distilled water
Small beaker with diluted liquid soap
Test tubes and rack
Tally counter
Broth cultures of TAL 182 and TAL 379

d. Estimating Cell Concentration by Optical Density

Spectrophotometer, cuvettes
Pipettes, 10 ml
Test tubes, rack
Broth cultures of TAL 182 and TAL 379 from (a)

e. Determining the Number of Viable Cells in a Culture by Plating Methods

Incubator, balance, water bath, colony counter, tally counter
Wire gauge (available through Scientific Products, Evanston, IL)
Dilution tubes with 9 ml of sterile quarter-strength YMB
Test tube rack
Sterile pipettes, 1 ml
Suction bulb
Liquid YMA in flask
Pasteur pipettes
Glass rod or spreader, beaker with alcohol, flame
Small beaker with water, small beaker (empty)
Sterile Petri dishes
YMA plates
Broth cultures of TAL 182 and TAL 379 from (a)

f. Determining the Mean Generation (Doubling) Time of Rhizobia

Rotary shaker, colony counter, autoclave
Spreader, small beaker of alcohol, flame
Sterile pipettes, 1 ml
Erlenmeyer flasks (32) with 100 ml of YMB each
Dilution tubes with 9 ml of sterile quarter-strength broth each
Plates of YMA

KEY REFERENCES

Hoben, H.J., and P. Somasegaran. 1982. Comparison of the pour, spread and drop-plate methods for the enumeration of *Rhizobium* spp. in inoculants made from presterilized peat. Appl. Environ. Microbiol. 44:1246–1247.

Vincent, J.M. 1970. A Manual for the Practical Study of Root Nodule Bacteria. IBP Handbook no. 15. Blackwell Scientific Publications, Oxford.

6

Counting Rhizobia by a Plant Infection Method

The plant infection count, also known as the most-probable-number (MPN) count, is used to determine the number of viable and infective rhizobia in the presence of other microorganisms. The trap legume selected in the MPN method must belong to the same cross-inoculation group of legumes nodulated by the rhizobia under investigation. This indirect method is commonly used to determine the quality of inoculants produced from nonsterile carrier materials. It is also used to determine the number of rhizobia in the soil.

In this chapter, the quality of soybean (*Glycine max*) inoculants prepared with *Bradyrhizobium japonicum* incorporated into presterilized and nonsterilized peat is determined by the plate and MPN count methods. The MPN of *B. japonicum* in both types of inoculants is determined using soybean plants grown in plastic growth pouches. The results of both determinations are compared.

KEY STEPS/OBJECTIVES

1. Prepare peat inoculants.
2. Prepare growth pouches.
3. Surface sterilize and pregerminate seeds.
4. Transfer pregerminated seeds from seedling agar to growth pouches.
5. Prepare serial dilutions of peat sample(s); initiate MPN and plate counts.
6. Make periodic observations of plants and water if needed.
7. Count colonies on plates.
8. Harvest and record nodulation.
9. Determine the MPN.
10. Compare results of plant infection and plate counts.

a. Preparing Inoculants (Key Step 1)

Start in duplicate, 100-ml cultures of a strain of slow-growing soybean rhizobia e.g., *B. japonicum* (TAL 102) in 250-ml flasks. Aerate on a rotary shaker for 7 days. Test the purity of the fully grown broth culture by Gram stain (Chapter 3), and pH measurement and agglutination with its specific antiserum (Chapter 9).

Prepare or obtain two sealed polyethylene bags of 50 g neutralized peat sterilized by gamma irradiation, or neutralized peat packaged and sealed in autoclavable, polypropylene bags that were autoclave & sterilized (Chapter 27). Also needed are two sealed polyethylene bags, each containing 50g of neutralized nonsterile peat.

Following the methods described in Chapter 27, inject 40 ml per bag of the fully grown (1×10^9 cells ml^{-1}) broth cultures. Prepare two bags of TAL 102 in sterile peat and two bags of TAL 102 in nonsterile peat to produce the peat-based inoculants. The bags of sterile peat are injected first. Allow the inoculants to mature at 25–30°C for at least 2 weeks.

b. Setting Up the Plant Infection Count in Plastic Growth Pouches (Key Step 2)

In this experiment, the sterile pouches used are made of polypropylene (16×18 cm) with paper wick liners. Growth pouches serve well as inexpensive, space-saving substitutes for Leonard jars. They are susceptible to contamination, which air and insects introduce. Also, the growth pouches are not shielded against radiated heat. Therefore, their use is restricted to growth chambers or growth rooms where contamination is controlled. As in growth tubes, plants cannot be grown to maturity in growth pouches.

Leonard jars and growth tubes are also frequently used for MPN counts. Leonard jars (Appendix 11) are convenient growth units for large-seeded legumes and are primarily used in the greenhouse. Growth tubes are used on growth shelves or in growth chambers where space is limited. As in authentication (Chapter 1), a large-seeded legume of the same cross-inoculation group may be substituted by a small-seeded one for the MPN count. Growth tubes (seedling-agar slants or NifTAL tubes) may then be used to save space and labor (Appendix 7).

Add 30 ml of sterile plant nutrient solution (Appendix 3) into each growth pouch. (The growth pouches purchased are sterile. However, if contamination is suspected, the pouches may be sterilized by autoclaving after including the plant nutrient solution.) Arrange the pouches in a rack (Figure 6.1). Set up one rack of 60–70 pouches for each bag of inoculant to be tested. Suggestions for building a growth pouch rack are given in Appendix 8. Detection of any contamination is of great importance and requires sufficiently replicated uninoculated controls. Contamination will also be indicated if sporadic nodulation occurs at the higher dilutions.

Figure 6.1 Soybean plants growing in growth pouches.

c. Planting Seeds in Growth Pouches (Key Steps 3 and 4)

Surface sterilize and pregerminate 100 soybean seeds as explained in Appendix 10. Select seeds of uniform size and high viability (95–100%). Use more seeds if the viability rate is lower.

Select 60–70 well-germinated seeds of similar size and radical length (1–1.5 cm). Transfer one seed to each pouch aseptically. Place each seed in the trough of the paper wick liner.

To prevent the growing radical from pushing the seed out of the pouch, a hole is made in the trough of the wick and the radical is inserted into the hole during planting. Holes are easily made in the trough with fine-tipped, sterile forceps when the wick is wet. Two forceps are needed: one for holding the wick and the other for making the hole.

When the plants are 5–7 days old, reorganize the growth pouches on the rack. Discard plants of poor growth and select 50 healthy plants. Forty pouches are needed to count dilutions for 10^{-1}–10^{-10} in quadruplicate plus one control pouch following each group of four inoculated pounches. This brings the number of pouches to 50. Repeat this set-up in separate racks for each inoculant to be tested.

d. Inoculating for the MPN Count (Key Steps 5–8)

Make a tenfold dilution of each inoculant by transferring the content of each bag (100 g) into separate 2.0-liter flasks containing 900 ml of sterile water. Remove the peat inoculant through a 2–3-cm opening made by cutting off one corner of each bag. Close each flask with a sterile rubber stopper and shake vigorously for 5 min by hand. This gives 10^{-1} dilution of the peat inoculant in sterile water. Continue the serial dilutions using 1.0-ml aliquots of inoculant suspension and 9.0-ml sterile water in tubes to obtain dilutions 10^{-2} to 10^{-10} (Figure 5.2). Alternatively, 1 g of peat inoculant may be weighed and aseptically transferred into 99 ml of sterile water. The first dilution is then 10^{-2}.

Plate the 10^{-5}, 10^{-6}, and 10^{-7} dilutions of each inoculant by the spread-plate method. The drop-plate method (Chapter 5) may be used for inoculants prepared from sterile peat. Plate in quadruplicate on yeast-mannitol agar containing congo red (YMA + CR). Inoculants prepared from nonsterile carriers should be plated by the spread-plate method on YMA containing a fungicide such as brilliant green (1.25 $\mu g\ ml^{-1}$) or pentachloronitrobenzene (PCNB) (0.5 g in 100 ml acetone plus one drop of Tween 80 added to 400 ml of medium). Plate in duplicates and, if possible, include an additional YMA plate containing CR for each dilution. Incubate at 25–30°C for 5–8 days.

Inoculate the soybean plants that have been set up for the MPN count. Pipette 1 ml of each dilution (from 10^{-1}–10^{-10}) to each one of the four replicates in each set. Begin by inoculating with aliquots from the highest dilution and proceed down the series with the same pipette.

Observe the plants periodically and replenish the nutrient solution if necessary. Nodulation may be evident after 2 weeks. Make the final observation after 3 weeks and record presence (+) or absence (−) of nodules. (For tree legumes make final observations after 5–7 weeks.) Count rhizobia on plates as described in Chapter 5.

e. Determining the MPN (Key Steps 9 and 10)

For each set, write down the dilutions used and record the nodulation. The actual number of nodules on each plant and the number of plants in each replication have no bearing on the MPN count. If replications are in quadruplicate, the reading may be 4, 3, 2, 1, or 0 nodulated units. The highest dilution used should show no nodulation in each replication, indicating the absence of rhizobia.

Refer to tables in Appendix 14, indicating tenfold dilutions (Table A14.7) for the estimation of the number of rhizobia by the plant infection method. If twofold or fourfold dilutions are used, refer to Tables A14.5 and A14.6, respectively. The letter *n* indicates the number of replications, and *s* signifies the number of dilution steps. Dilutions may be made in duplicate or quadruplicate. Each series should end with a dilution at which no nodules are formed.

The MPN is calculated from the most likely number (*m*) found in the MPN tables. To find this number, use the procedure shown in the following example:

1. Record nodulation (+ or −) as shown in Table 6.1.
2. Take note of the number of replications used ($n = 4$).
3. Count the number of dilution steps used ($s = 8$.)
4. Add up the total number of (+) units (+ = 18).
5. Find this number 18 in Table A14.7 (calculated for tenfold dilutions).
6. Locate the most likely number (m) column $s = 8$, on the same line as 18, which is 5.8×10^3.

The MPN may now be calculated from m by using the following formula:

m = Likely number from the MPN table for the lowest dilution of the series

d = Lowest dilution (first unit used in the tabulation)

v = Volume of aliquot applied to plant

The MPN per gram of inoculant is:

$$X = \frac{m \times d}{v} = \frac{(5.8 \times 10^3) \times 10^2}{1} = 5.8 \times 10^5 \text{ rhizobia g}^{-1} \text{ inoculant}$$

For additional information, refer to Appendix 14, which is essential to this chapter for the evaluating and understanding the plant infection count. Compare results obtained by the plant infection (MPN) and plate-count methods.

TABLE 6.1 Example for Recording Nodulation for the MPN Count

	Nodulation				
	Replications				
Dilution	I	II	III	IV	No. of Nodulated Units
10^{-2}	+	+	+	+	4
10^{-3}	+	+	+	+	4
10^{-4}	+	+	+	+	4
10^{-5}	+	+	+	+	4
10^{-6}	+	−	+	−	2
10^{-7}	−	−	−	−	0
10^{-8}	−	−	−	−	0
10^{-9}	−	−	−	−	0
				Total	18

REQUIREMENTS

a. Preparing Inoculants

Platform shaker, incubator, water bath
Agglutination tubes and rack, test tubes, rack
Pipettes, 1 and 10 ml
Saline, flame, alcohol in spray bottle, adhesive tape
Sterile 50-ml syringe, 18-gauge needles
Requirements for Gram stain (Appendix 3)
Solution of bromthymol blue (BTB) (0.5% in alcohol)
Erlenmeyer flasks, 250 ml (four) containing 100 ml broth each
Sterile peat, 50 g polyethylene per bag (two)
Nonsterile peat, 50 g polyethylene per bag (two)
Culture of TAL 102, antiserum for TAL 102

b. Setting Up the Plant Infection Count in Plastic Growth Pouches

Growth chamber, autoclave
Forceps, flame
Measuring cylinder (50 ml) or adjustable filling unit
Growth pouches 16 × 18 cm with paper wick liners (available from Scientific Products, Evanston, IL)
Plant nutrient solution (Appendix 3)

c. Planting Seeds in Growth Pouches

Requirements for seed sterilization (Appendix 10)
Water agar plates
Soybean seeds

d. Inoculating for the MPN Count

Incubator
Sterile pipettes, 1 and 10 ml
Pasteur pipettes, calibrated, sterile; rubber bulbs for Pasteur pipettes
Flame, spray bottle with alcohol
Erlenmeyer flasks of 2-liter capacity (four) containing 900 ml of sterile water each
Rubber stoppers, sterile, to fit 2-liter flasks
Dilution tubes with 9 ml of sterile water, racks
Plant nutrient solution
YMA containing CR (YMA-CR)

YMA containing brilliant green (YMA-BG)
Plants in growth pouches from (b)
Peat inoculant from (a)

e. Determining the MPN

Records of observations
MPN tables (Appendix 14)

KEY REFERENCES

Vincent, J.M. 1970. A Manual for the Practical Study of Root Nodule Bacteria. IBP Handbook no. 15. Blackwell Scientific Publications, Oxford.

Weaver, R.W., and L.R. Frederick. 1972. A new technique for the most-probable-number counts. Plant Soil 36:219–222.

Woomer, P., J. Bennett, and R. Yost. 1990. Overcoming the inflexibility of most-probable-number procedures. Agron. J. 82:349–353.

7

Counting Serologically Specific Rhizobia in Soil and Peat Inoculants Using Membrane Filters and Immunofluorescence

In the plate-count method, it is assumed that each viable cell forms a colony. Therefore, the number of colonies growing on the medium is a measure of the bacterial content of the material analyzed. Bacterial numbers can also be determined by direct counting procedures, provided the cells are made visible by specific staining. Rhizobia in soils and peat inoculants can be stained with specific fluorochrome-conjugated antisera and counted using fluorescent microscopy.

In this chapter, a serologically specific rhizobial strain is inoculated into soil or peat carrier for inoculants. A suspension of the inoculated material in water is processed to facilitate extractability and release the strain from soil or peat particles. The processed soil suspension and peat inoculant suspension are then passed through membrane filters to trap the rhizobia on the surface of the filters. The trapped cells of the rhizobia are stained with the strain-specific fluorescent antibody and counted using a microscope equipped for fluorescence microscopy. This direct counting procedure is suitable for counting populations in excess of 10^5 cells g^{-1} soil or peat inoculant.

KEY STEPS/OBJECTIVES

1. Select a suitable rhizobial strain for enumeration.
2. Culture the rhizobial strain.
3. Inoculate soil and peat with the selected strain.
4. Prepare extracting solution and flocculating mixture.
5. Stain membrane filters.
6. Calibrate microscope objectives.

7. Process inoculated soil and peat inoculants for counting.
8. Perform membrane filtration.
9. Stain the specific rhizobial strain.
10. Count rhizobia.
11. Calculate results.

a. Selecting a Suitable Rhizobial Strain for Enumeration (Key Steps 1 and 2)

Select a rhizobial strain for which good quality fluorescent antibody (FA) is available. Since most soils will normally contain indigenous rhizobia, it is important to ensure that the rhizobial strain selected does not cross-react serologically with the indigenous rhizobia in the test soil. Information on the serological characteristics of the indigenous rhizobia in one or several soils is needed to select a noncross-reacting rhizobial test strain. When the desired serological information is not available, there are other possible approaches that illustrate the principles of the technique.

One way to avoid cross-reactions is to pick rhizobia of a legume not indigenous to the locality. For example, a chickpea (*Cicer arietinum*) rhizobial strain is suitable for most soils of Southeast Asia and Hawaii since chickpea is not indigenous or grown in these places. Chickpea rhizobia are serologically specific and cross-reactions with other rhizobial groups are remote. Unlike in soils, any rhizobial strain may be used in peat, especially when the peat has been processed by flash drying and milling. Peats that have not been flash dried may carry rhizobia. Therefore, prior testing of nonsterile peat may be necessary to rule out cross-reactions.

In this experiment, chickpea rhizobial strain TAL 620 (Table A24.1) will be used to count rhizobia in soils and peat inoculants. Other rhizobia and their fluorescent antibodies may also be used.

Prepare 200–250 ml of yeast-mannitol broth (YMB) in a 500-ml Erlenmeyer flask. Inoculate the broth with TAL 620 and shake (aerate) the culture on a rotary shaker. After 4 days of growth, the populations will be close to $1\text{–}2 \times 10^9$ cells ml^{-1}.

b. Inoculating Soil and Peat with the Selected Strain (Key Step 3)

Collect approximately 1 kg of soil from a selected site. Break up clumps, sieve (5-mm wire sieve) the soil, and air dry it for at least 2 days. Accurately weigh out 100 g of the air-dried field soil and oven dry it at 110°C for 48 h to determine the moisture content. Weigh the soil again to determine the amount of moisture lost due to oven drying. Calculate the amount of moisture that must be added to the air-dried soil to bring the soil to field capacity (Appendix 21). Assuming a 10% soil moisture content (dry weight basis) for 100 g of a particular soil, the addition of 20 ml of moisture is needed to bring the soil to field capacity if field capacity is 30%. Adding 20 ml of the broth culture of TAL 620 (ca. 10^9 cells ml^{-1}) to the soil will result in approximately 2×10^8 cells g^{-1} inoculated soil. Take 200 g of air-dried soil in a 500-ml beaker. Pipette the appropriate volume of the broth culture of TAL 620 to bring the soil to field capacity. Mix thoroughly with a spatula or glass rod. Cover the mouth of the beaker with aluminum foil to prevent moisture loss and store in a refrigerator until needed.

To a second 200-g batch of air-dried soil in a 500-ml beaker, add sterile water to bring the soil to field capacity. Mix thoroughly with a clean, sterile glass rod or spatula. Cover the mouth of the beaker with aluminum foil and store in a refrigerator. This uninoculated soil treatment serves as a control. (To study the limitations of the method, add appropriately diluted broth culture to other batches of soil to obtain cell densities ranging from 10^4 to 10^7 cells g^{-1} soil and enumerate.) Inoculate two packages of sterile (autoclaved or gamma irradiated) peat or nonsterile peat with broth culture of TAL 620 as described in Chapter 27. Store the peat inoculants at room temperature or in a refrigerator.

c. Preparing Extracting Solution and Soil Flocculating Mixture (Key Step 4)

For soil, an extracting solution and flocculating mixture are necessary to release rhizobia bound to soil particles. The extracting solution is obtained by mixing partially hydrolyzed gelatin and dibasic ammonium phosphate $[(NH_4)_2HPO_4]$ solution.

Prepare 1% gelatin by placing 1.0 g of powdered gelatin in 100 ml of distilled water. Adjust the pH to 10.3 with 1 N sodium hydroxide solution. Autoclave the pH adjusted gelatin for 10 min at 15 lb/in.^{2} Allow the partially hydrolyzed gelatin to cool.

Prepare 0.1 M ammonium phosphate solution by dissolving 13.2 g of the salt in 1000 ml of distilled water. Obtain the extracting solution by diluting 1 part of partially hydrolyzed gelatin with 10 parts of ammonium phosphate solution. Store the extracting solution in the refrigerator and warm it to room temperature before use. Extracting solutions should not be stored for extended periods because bacterial growth can easily set in. To obtain the flocculent mixture, mix the following reagents: 1 g powdered magnesium carbonate ($4\ MgCO_3\ Mg(OH)_2 \cdot 4H_2O$) with 1.6 g calcium chloride ($CaCl_2 \cdot 2H_2O$).

d. Staining Membrane Filters (Key Step 5)

Polycarbonate membrane filters are preferred to cellulose or other types of filters. Polycarbonate filters are available stained (black) or unstained (white). White filters need to be stained black to obtain a good contrasting background to facilitate counting fluorescing cells. The filters are stained by soaking in irgalan black (chemical dye) solution.

Prepare the dye solution by mixing 1 g of irgalan black with 500 ml of 2% (v/v) acetic acid. Place approximately 100 ml of the dye solution in a 250-ml beaker. Using blunt forceps, drop the filter membranes into the dye solution. Soak 20–30 filters overnight at room temperature and keep the beaker closed with a piece of aluminum foil. Carefully drain off the dye solution into an Erlenmeyer flask and store in a refrigerator for future use. Rinse the membranes in the beaker with several changes of distilled or deionized water. Remove the rinsed filters and lay them on paper towels to air dry in a still-air environment. Stack the dry filters back into the storage container.

e. Calibrating Microscope Objectives (Key Step 6)

When microscopy is employed in direct counting of microorganisms, the area of the microscope field under a given magnification needs to be known to calculate the number of microorganisms per unit weight or volume of the sample.

With a stage micrometer, measure the diameter of the microscope fields covered by the 40× and 100× oil immersion objective fields. Calculate and record the areas of the fields covered by the 40× and 100× objectives.

f. Processing Soil and Peat Inoculant for Counting (Key Step 7)

Weigh 10 g of inoculated soil containing TAL 620 [prepared in (c)] into 90 ml of extracting solution in a 160-ml dilution bottle. Similarly, weigh 10 g of control (uninoculated soil) into 90 ml of extracting solution in another bottle. Dispense the soil by shaking with a wrist-action shaker for 15 min. Add 0.7 g of flocculent mixture to each soil suspension. Return the bottles to the wrist-action shaker and shake for another 15 min. Allow the soil suspensions to settle for 1 h. At the end of this time the soil particles will settle, leaving a supernatant containing the rhizobial cells. Rhizobial cells will also settle during the flocculation, but a high proportion will remain in the supernatant. In tenfold dilution steps, dilute the supernatant of both inoculated and noninoculated soils to 10^{-4}.

The extracting solution and flocculent mixture are not needed to release the rhizobia from peat particles. Weigh 10 g of peat inoculant into 90 ml of sterile water in a milk dilution bottle. Shake for 15 min on a wrist-action shaker. As done earlier with the soil suspension, dilute the peat suspension up to 10^{-4} in tenfold dilution steps. Refrigerate the tubes containing 10^{-4} dilutions of the soil and peat suspensions until needed.

It is important to note that for direct counting, the number of rhizobia expected per gram of soil or peat inoculant will determine the choice of dilution level (e.g., 10^{-3}, 10^{-4}, 10^{-5}, etc.) and the volume to be filtered. When the expected numbers are close to 10^{8}–10^{9} cells g^{-1}, as in this exercise, filtering 5–10 ml of 10^{-4} dilutions of either supernatant (soil) or the peat inoculant suspension will give a countable range.

When investigating the autoecology of a known rhizobial strain introduced into a soil previously, process the intended soil sample for direct counting after a subsample has been taken for determining the moisture content.

g. Performing Membrane Filtration (Key Step 8)

The soil and peat inoculant suspensions need to be filtered to trap the rhizobial cells on the surface of the membrane filter. The membrane filtration apparatus and accessories are illustrated in Figure 7.1. A clamp (not shown in the figure) holds together the funnel

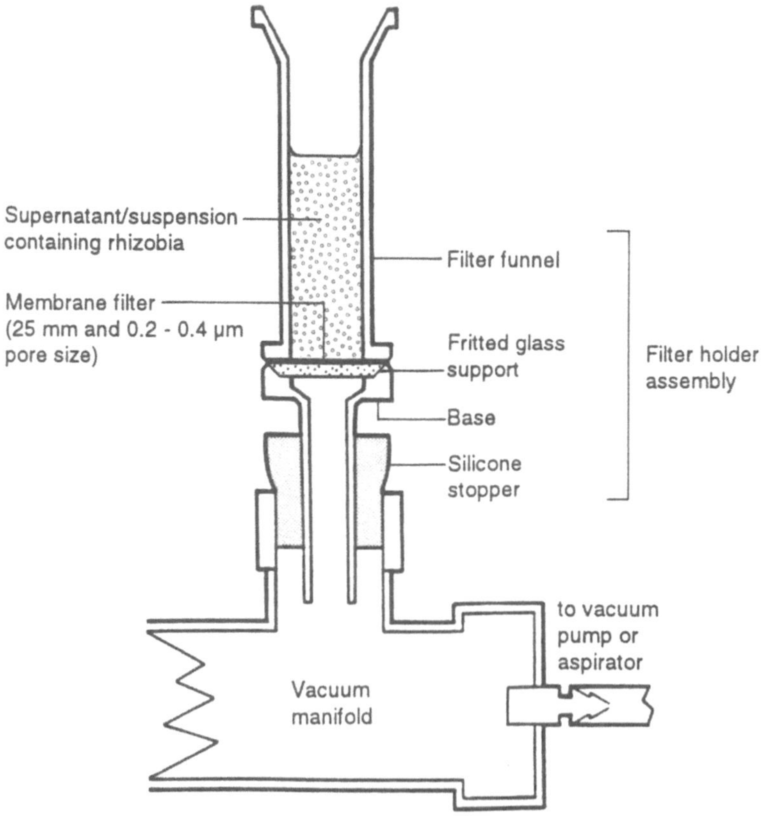

FIGURE 7.1 Position of the membrane filter ready for filtration (clamp not shown).

to the base. The construction of an inexpensive vacuum manifold, which can be used to filter six different samples simultaneously, is shown in Figure 7.2.

Unclamp the funnel from the base. Place one to two drops of water on the fritted glass surface of the base. With filter forceps (broad, unserrated tips to prevent damage to the filters), place one irgalan black-treated filter on the base. The drops of water aid in the adhesion of the filter to the base. Replace the funnel in position on the base and clamp.

For the processed soil suspensions (inoculated and noninoculated soil), pipette 5 ml of each of the diluted (10^{-4}) supernatants into two separate filter funnels. Similarly, pipette 5 ml of the 10^{-4} diluted peat inoculant suspension into a third filter funnel. Apply negative pressure (suction) using an aspirator pump connected to a faucet. An electrically operated vacuum pump, if available, is preferred. After the samples have passed through the filter, rinse the inside of each filter funnel with 10–20 ml of saline. After rinsing, gently reduce the negative pressure (suction) before closing it completely.

FIGURE 7.2 Fully assembled vacuum manifold with filtration units and vacuum pump (moisture trap not included).

h. Staining the Rhizobial Strain Trapped on the Membrane Filter (Key Step 9)

The cells of TAL 620 trapped on the filter are stained with the homologous specific FA. Before specific staining is achieved, nonspecific staining is suppressed by treating the filters with rhodamine gel. Preparation and method of application of rhodamine gel is described in Appendix 4.

Carefully unclamp and remove the filter funnel. With filter forceps, transfer the filters to clean microscope slides. Label each slide to indicate the treatment. With a Pasteur pipette, add rhodamine gel to each filter to cover the entire filter surface. Put the filter in an oven at 60°C until it is almost dry. (Note that the filter adheres to the glass slide. The slide and filter will separate during the staining process later.) Let the filter cool before initiating the staining.

Stain the entire surface of the filter with 1:4 saline (0.85% NaCl) diluted FA of TAL 620 for 20 min. Ensure that the slides (filters) are kept level in a moisture-saturated environment during the staining period. At the end of the staining period, bring the slide close to and level with the base of the filter holder. With a filter forceps, slide the filter onto the base (fritted glass surface). Reposition the filter funnel and clamp. Rinse remaining material from the slide into the funnel with saline.

Fill each filter funnel with 15–20 ml of filtered saline. Apply suction. Rinse each filter with at least another 80 ml of filtered saline. Prolonged filter rinsing will help remove unreacted FA and also result in a dark background that is highly desirable for direct counting.

i. Counting FA-Stained Cells on the Membrane Filter (Key Step 10)

Upon completion of the rinsing with filtered saline, unclamp and remove the filter funnel. Carefully transfer the filters to their respective slides. Add one drop of mounting fluid (buffered glycerol, pH 9.3) to the center of the filter and place a clean coverslip over it. Allow the pressure (weight) of the coverslip to spread the buffered glycerol over the filter.

Counting is most reliable when a 100× oil immersion objective is used. At this magnification, the typical rod-shaped fluorescing cells of rhizobia are clear and unmistakable (Figure 7.3). Lower magnifications (40×) yield less satisfactory results. A compound microscope equipped with an epifluorescence condenser is required. Count at least 20 fields per filter.

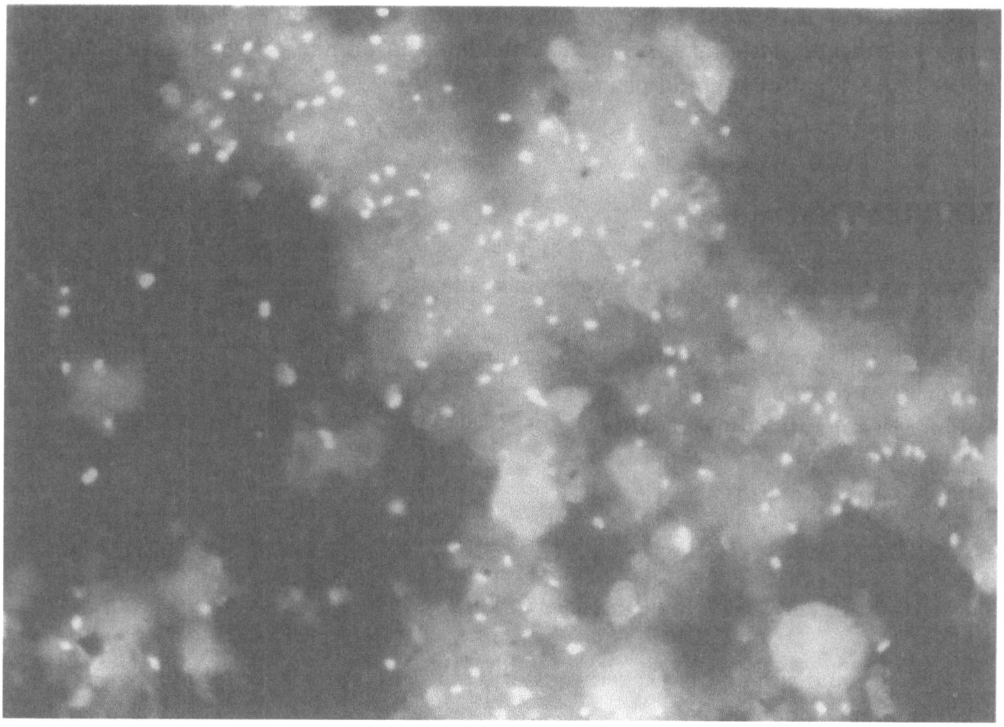

Figure 7.3 Fluorescent antibody stained cells of a specific strain of rhizobia (seen as white round or short rods) in a tropical soil as processed and visualized by the membrane filter immunofluorescence (MFIF) technique. Note that most of the cells are bound to the soil matrix. (Photograph contributed by B.B. Bohlool.)

j. Calculating the Results (Key Step 11)

The number of rhizobia per gram of soil (or inoculant) $= N \times \frac{A}{a} \times \frac{D}{v}$

where N = Mean number per microscope field
A = Effective filtering area (mm^2)
a = Area of microscope field (mm^2)
D = Dilution factor
v = Volume (ml) of supernatant (soil) or inoculant suspension filtered

Determine a for the 100× objective from the calibrations done in (f).

The effective filtering area (A) is that part of the membrane filter which is defined by the internal circumference of the lower-end of the filter funnel. Measure the diameter and calculate A.

The A/a ratio can be calculated as a constant for each specific objective. Alternatively, the R^2/r^2 ratio (where R and r are the radii of the membrane filter and microscope field, respectively) may be used as a constant instead of the A/a ratio.

The number of rhizobia per gram may be expressed on a dry-weight basis for soils or on a wet-weight basis for inoculants. Calculate the number of rhizobial strain TAL 620 per unit weight of soil and peat inoculant.

REQUIREMENTS

a. Selecting a Suitable Rhizobial Strain for Enumeration

Transfer chamber, inoculation loop, flame
Chickpea rhizobial strain TAL 620
Fluorescent antibody TAL 620
250 ml of YMB
Slant culture of TAL 620
Shaker

b. Inoculating Soil and Peat with the Selected Strain

500 g selected soil and packaged presterilized peat (autoclaved/irradiated)
Sieve and wooden mallet
Balance, newspapers
500-ml beakers and aluminum foil
Pipettes and 50-ml syringes (sterile)
Refrigerator
Materials for measuring field capacity (Appendix 21)

c. Preparing Extracting Solution and Soil Flocculating Mixture

Gelatin powder (Difco Laboratories, Detroit, MI)
Dibasic ammonium phosphate: $(NH_4)_2HPO_4$
Sodium hydroxide (1 N)
pH meter, beakers, glass rod
Autoclave
Powdered magnesium carbonate: 4 $MgCO_3$ $Mg(OH)_2 \cdot 4H_2O$
Calcium chloride: $CaCl_2 \cdot 2H_2O$

d. Staining Membrane Filters

Polycarbonate membrane filters (0.4-μm pore size, 25-mm diameter)(Nucleopore Corporation, Pleasanton, CA)
Irgalan black (acid black no. 107, CIBA-GEIGY, Greensboro, NC)
Erlenmeyer flask
Acetic acid (2%, v/v)
Beakers, paper towels, blunt forceps
Deionized/distilled water
Aluminum foil

e. Calibrating Microscope Objectives

Microscope with epifluorescence condenser
40× and 100× oil immersion objective
Stage micrometer

f. Processing Soil and Peat Inoculant for Counting

Peat inoculant and soil inoculated with TAL 620 from (c)
Noninoculated soil
Milk dilution bottles (160 ml) with 90 ml extracting solution or water
Wrist-action shaker
Balance
Tubes (one rack) containing 9 ml sterile water
Pipettes, sterile (1 ml and 10 ml)
Refrigerator

g. Performing Membrane Filtration

Vacuum manifold
Filter-holder assemblies
Irgalan black stained membrane filters, filter forceps

Membrane-filtered saline
Pipettes, 5 or 10 ml
Aspirator or vacuum pump

h. Staining the Rhizobial Strain Trapped on the Membrane Filter

Fluorescent antibody of TAL 620, saline
Rhodamine gel, filter forceps
Oven (60°C)
Vacuum manifold, filter-holder assemblies, membrane filters with samples from (g), aspirator vacuum pump
Filter forceps, microscope slides, coverslip (24 × 30 mm)
Pasteur pipettes

i. Counting FA-Stained Cells on the Membrane Filter

Membrane filters with samples from (h)
Microscope with epifluorescence condenser and 100× objective
Buffered glycerol (mounting fluid)
Microscope slides/coverslips
Tally counter, filter forceps

j. Calculating the Results

No special requirements

KEY REFERENCES

Demezas, D.H., and P.J. Bottomley. 1986. Autoecology in rhizospheres and nodulating behavior of indigenous *Rhizobium trifolii*. Appl. Environ. Microbiol. 52:1014–1019.

Kingsley, M.T., and B.B. Bohlool. 1981. Release of *Rhizobium* spp. from tropical soils and recovery for immunofluorescence enumeration. Appl. Environ. Microbiol. 42:241–248.

Schmidt, E.L. 1974. Quantitative autoecological study of microorganisms in soils by immunofluorescence. Soil Sci. 118:141–149.

Additional References and Recommended Reading

Allen, O.N., and E.K. Allen. 1981. The Leguminosae. A Source Book of Characteristics, Uses, and Nodulation. The University of Wisconsin Press, Madison.

Annear, D.I. 1964. Recoveries of bacteria after drying in glutamate and other substances. Aust. J. Exp. Biol. Med. Sci. 42:717–722.

Baldwin, I.L., and E.B. Fred. 1929. Nomenclature of the root nodule bacteria of the Leguminosae. J. Bacteriol. 17:141–150.

Bergersen, F.J. 1961. The growth of rhizobia in synthetic media. Aust. J. Biol. Sci. 14:349–360.

Bushby, H.V.A., and K.C. Marshall. 1977. Some factors affecting the survival of root nodule bacteria on desiccation. Soil Biol. Biochem. 9:143–147.

Chen, W.X., G.S. Li, E.T. Wang, H.L. Huang, and J.L. Li. 1991. *Rhizobium huakuii* sp. nov. isolated from the root nodules of *Astragalus sinicus*. Int. J. Syst. Bacteriol. 41:275–280.

Dart, P.J. 1977. Infection and development of leguminous nodules. pp. 367–472. *In* R.W.F. Hardy and W.S. Silver (eds.) A Treatise of Dinitrogen Fixation. Section III, Biology. John Wiley & Sons, New York.

Date, R.A., and J. Halliday. 1979a. Selecting *Rhizobium* for acid, infertile soils of the tropics. Nature (London) 277:62–64.

Date, R.A., and J. Halliday. 1979b. Collection of strains of *Rhizobium*. pp. 21–26. *In* G.O. Mott and A. Jiménez (ed.) Handbook for the Collection, Preservation and Characterization of Tropical Forage Germplasm Resources, CIAT, Colombia.

Dreyfus, B., J.L. Garcia, and M. Gillis. 1988. Characterization of *Azorhizobium caulinodans* gen. nov., sp. nov., a stem-nodulating nitrogen-fixing bacterium isolated from *Sesbania rostrata*. Int. J. Syst. Bacteriol. 38:89–98.

Dye, M. 1982. A note on some factors affecting the survival of *Rhizobium* cultures during freeze drying and subsequent storage. J. Appl. Bacteriol. 52:461–464.

Eaglesham, A.R.J., J.M. Ellis, W.R. Evans, D.E. Fleischman, M. Hungria, and R.W.F. Hardy. 1990. The first photosynthetic N_2-fixing *Rhizobium*: Characteristics. pp. 805–811. *In* P.M. Greshoff et al. (ed.) Nitrogen Fixation: Achievements and Objectives. Chapman and Hall, Ltd., London.

El Essawi, T.M., and A.S. Abdel Ghaffar. 1967. Cultural and symbiotic properties of rhizobia from Egyptian clover (*Trifolium alexandrinum*). J. Appl. Bacteriol. 30:354–361.

Elkan, G.H. 1981. The taxonomy of the *Rhizobiaceae*. pp. 1–12. *In* K.L. Giles and A.G. Atherly (eds.) International Review of Cytology, Suppl. 13, Biology of the *Rhizobiaceae*, Academic Press, New York.

Graham, P.H., M.J. Sadowsky, H.H. Keyser, Y.M. Barnet, R.S. Bradley, J.E. Cooper, D.J. De Ley, B.D.W. Jarvis, E.B. Roslycky, B.W. Strijdom, and J.P.W. Young. 1991. Proposed minimal standards for the description of new genera and species of root- and stem-nodulating bacteria. Int. J. Syst. Bacteriol. 41:582–587.

Hahn, N.J. 1966. The Congo Red reaction in bacteria and its usefulness in the identification of rhizobia. Can. J. Microbiol. 12:725–733.

Herridge, D.F., and R.J. Roughley. 1975. Variation

in colony characteristics and symbiotic effectiveness of *Rhizobium*. J. Appl. Bacteriol. 38:19–27.

Hubbell, D.H. 1970. Studies on the root hair "curling factor" of *Rhizobium*. Bot. Gaz. (Chicago) 4:337–342.

Jarvis, B.D.W., Pankhurst, E.E., and Patel, J.J. 1982. *Rhizobium loti*, a new species of legume root nodule bacteria. Int. J. Syst. Bacteriol. 32:378–380.

Jordan, D.C. 1982. Transfer of *Rhizobium japonicum* Buchanan 1980 to *Bradyrhizobium* gen. nov., a genus of slow-growing, root nodule bacteria from leguminous plants. Int. J. Syst. Bacteriol. 32:136–139.

Jordan, D.C. 1984. Family III. *Rhizobiaceae* Conn. 1938, 321. pp. 234–256. *In* N.R. Krieg and J.G. Holt (eds.) Bergey's Manual of Systematic Bacteriology, Vol. I. The Williams & Wilkins Co., Baltimore.

Jordan, D.C., and Allen, O.N. 1974. Family III. *Rhizobiaceae* Conn. 1938. pp. 261–164. *In* R.E. Buchanan and N.E. Gibbons (eds.) Bergey's Manual of Determinative Bacteriology, 8th ed. The Williams & Wilkins Co., Baltimore.

Keyser, H.H., B.B. Bohlool, T.S. Hu, and D.F. Weber. 1982. Fast growing rhizobia isolated from root nodules of soybean. Science 215:1631–1632.

Kneen, B.E., and LaRue, T.A. 1983. Congo Red absorption by *Rhizobium leguminosarum*. Appl. Environ. Microbiol. 45:340–342.

Kuykendall, L.D. 1987. Isolation and Identification of Genetically Marked Strains of Nitrogen-Fixing Microsymbionts of Soybean in Symbiotic Nitrogen Fixation Technology, G.H. Elkan (ed.) Marcel Dekker, Inc., New York.

Kuykendall, L.D., and G.H. Elkan. 1976. *Rhizobium japonicum* derivatives differing in nitrogen fixing efficiency and carbohydrate utilization. Appl. Environ. Microbiol. 32:511–519.

Li, D., and D.H. Hubbell. 1969. Infection thread formation as a basis of nodulation specificity in *Rhizobium*—strawberry clover associations. Can. J. Microbiol. 15:1133–1136.

Lindstrom, K. 1989. *Rhizobium galegae*, a new species of legume root-nodule bacteria. Int. J. Syst. Bacteriol. 39:365–367.

Martinez-Romero, E., L. Segovia, F. Martins, A.A. Franco, P. Graham, and M.A. Pardo. 1991. *Rhizobium tropici*, a novel species nodulating *Phaseolus vulgaris* L. beans and *Leucaena* sp. trees. Int. J. Syst. Bacteriol. 41:417–426.

Norris, D.O. 1958. A red strain of *Rhizobium* for *Lotononis bainesii*. Aust. J. Exp. Agric. 9:629–632.

Norris, D.O. 1963. A porcelain bead method for storing *Rhizobium*. Empire J. of Exp. Agric. 31:255–258.

Norris, D.O. 1965. Acid production by *Rhizobium*. A unifying concept. Plant Soil 22:143–166.

Norris, D.O., and R.A. Date. 1976. Legume bacteriology. pp. 134–174. *In* N.H. Shaw and W.W. Bryan (eds.) Tropical Pastures Research; Principles and Methods. Bull. 51. Commonwealth Bureau of Pastures and Field Crops, Hurley, England.

Okon, Y., Y. Eshel, and Y. Henis. 1972. Cultural and symbiotic properties of *Rhizobium* strains isolated from nodules of *Cicer arientinum* L. Soil Biol. Biochem. 4:165–170.

Scott, J.M., and F.E. Porter. 1986. An analysis of the accuracy of a plant infection technique for counting rhizobia. Soil Biol. Biochem. 18:355–362.

Sherwood, M.T. 1970. Improved synthetic medium for growth of *Rhizobium*. J. Appl. Bacteriol. 33:708-713.

Stevens, W.L. 1958. Dilution series: A statistical test of technique. J.R. Stat. Soc. Ser. B 20:205–214.

Sutton, W.D., C.E. Pankhurst, and A.S. Craig. 1981. The *Rhizobium* bacteroid state. pp. 149–171. *In* K.L. Giles and A.G. Atherly (eds.) International Review of Cytology, Suppl. 13, Biology of the *Rhizobiaceae*, Academic Press, New York.

Toomsan, B., O.P. Rupela, S. Mittal, P.J. Dart, and K.W. Clark. 1984. Counting *Cicer-Rhizobium* using a plant infection technique. Soil Biol. Biochem. 16:503–507.

Trinick, M.J. 1973. Symbiosis between *Rhizobium* and the non-legume *Trema aspera*. Nature (London) 244:459–460.

Tsien, H.C., P.S. Cain, and E.L. Schmidt. 1977. Viability of *Rhizobium* bacteroids. Appl. Environ. Microbiol. 34:854–856.

Turk, D., and H.H. Keyser. 1993. Accuracy of most-probable-number estimates of rhizobia for tree legumes. Soil Biol. Biochem. 25:69–74.

Vincent, J.M. 1974. Root-nodule symbiosis with *Rhizobium*. pp. 265–341. *In* A. Quispel (ed.) The Biology of Nitrogen Fixation, North Holland Publishing Company, Amsterdam.

Woomer, P.L., P.W. Singleton, and B.B. Bohlool. 1988. Reliability of the most-probable-number technique for enumerating rhizobia in tropical soils. Appl. Environ. Microbiol. 54:1494–1497.

SECTION II

Identification of Rhizobia

Rhizobia that have dramatic differences in such important traits as host specificity, infectiveness (invasiveness), and effectiveness are indistinguishable from each other under the microscope. However, there are many circumstances in which recognition of a particular rhizobial strain and monitoring its occurrence following introduction to a soil environment is important in ecological studies. Indirect procedures are available for this purpose.

SEROLOGICAL MARKERS

Any substance that provokes an immune response when introduced into the tissue of an animal or human is referred to as an *antigen*. In work with rhizobia, rabbits are commonly used for immunization and the antigens are rhizobial cell preparations. As a result of antigen injections, complex immunological reactions result in the rabbit producing special proteins called *globular antibodies (immunoglobulins)*. These antibodies are found in the serum portion of the blood. The study of the reactions of the immune serum with the antigens outside the animal is known as *serology*. Antigen-antibody reactions are highly specific in that the antibody reacts only with the antigen that elicited its formation.

The classes of immunoglobulins (Ig) in humans and most mammals including rabbits are IgA, IgG, IgM, IgD, and IgE. They mediate different immunological functions. For rhizobial research, the most important of these are the IgG and IgM antibodies because of their abundance in the rabbit immune serum and also because of their involvement in serological reactions and immunoassays. However, it is important to note that the ratio between IgG and IgM can vary with time, and probably immunization route and form of antigen. The IgG antibody is a symmetrical molecule and consists of four peptide chains, two heavy and two light, held together by disulfide (-s-s-) bridges as shown in Figure II.1.

Each antibody molecule has two antigen-binding (F_{ab}) units. The F_{ab} units contain the variable region that is responsible for the specificity of the IgG molecule. Variations in the polypeptide sequence of the F_{ab} region are complementary to the antigenic determinant (i.e., specific molecular group of the rhizobial cell), and provide the basis for the highly specific antigen-antibody binding.

The F_c region of the IgG antibody has secondary immunological functions. From the application point of view, the F_c region is highly significant because all the "tag" or

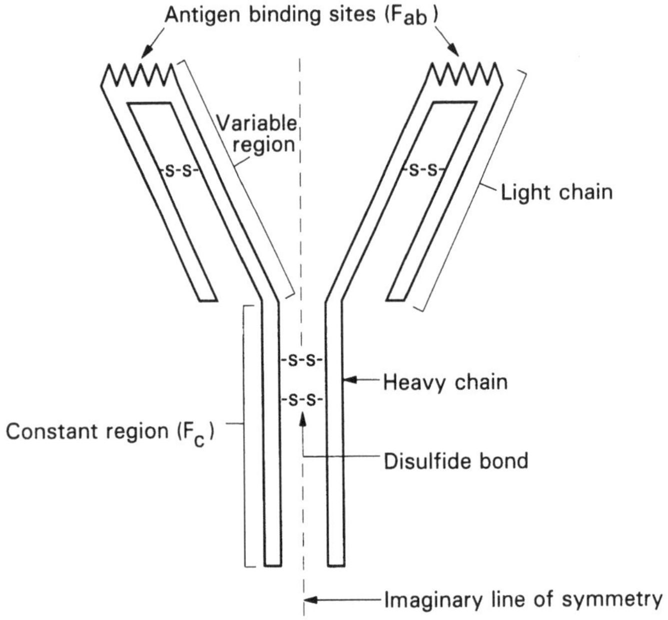

FIGURE II.1 The IgG antibody molecule.

"label" molecules [e.g., fluorescein isothiocyanate (FITC) and alkaline phosphatase] are generally attached to the F_c region for the visualization of the antigen-antibody reactions as in immunofluorescence and enzyme immunoassays.

As for other bacteria, antigens of rhizobia can be categorized into somatic, flagellar, and capsular, depending on their derivation. Somatic antigens are closely related to the rhizobial cell wall and are usually designated by the letter O. Some somatic antigens may be tightly bound to the cell wall, in which case they are not removed by washing of the cells. Therefore, these antigens are only detected when whole cells of rhizobia react with the antibody, as in agglutination or immunofluorescence. The somatic antigens that are soluble and easily removed by washing are detected by precipitation in gel, as in the Ouchterlony double-diffusion process. Somatic antigens are also heat stable. They are the most specific of the three groups of antigens.

The precipitating "internal antigens" are more widely shared and taxonomically significant. These are released from cells having fragile or broken walls. Because internal antigens are widely cross-reactive within and between species, they require recognition and interpretation in gel immunodiffusion.

The tiny whip-like appendages (flagella) of the rhizobia are also antigenic and appropriately called *flagellar* or *H antigens*. They are heat labile and are commonly detected by agglutination or immunofluorescence. The capsular (extracellular) antigens are surface antigens and are found outside the cell itself. They are usually designated by the letter K.

In rhizobial serology, both cultured cells and nodule antigens (bacteroids) are used for strain identification. Basic concepts on some serological methods for identification of rhizobia are described as follows.

AGGLUTINATION

The process in which the antigens are linked together by their corresponding antibodies is called *agglutination*. The linked antigens may be microscopically or macroscopically visible as clumps, agglutinates or aggregates. The agglutination reaction depends on a firm structural relationship between an exposed bacterial antigen and the antibody. Linus Pauling's lattice hypothesis (Figure II.2) is the widely accepted concept for explaining the agglutination reaction. Pauling postulated that the antibody is bivalent and the antigen is multivalent, and that the antigen-antibody complexes are molded into a lattice or framework of alternating antigen-antibody particles.

PRECIPITIN REACTION

In recent years, the precipitation reactions of somatic antigens have been used extensively for work with rhizobia. The precipitation reaction occurs when certain soluble antigens are brought into contact with the corresponding antibody.

Precipitation differs from agglutination in that the precipitating antigens are not whole bacterial cells (cellular), but are proteins or polysaccharide molecules in solution. In the double-diffusion technique, gels, usually clarified agar, are used as matrices for combining diffusion with precipitation. The reactants simply diffuse through the gel towards each other and precipitation results when the equivalence points have been reached. Antigen preparation of a rhizobial strain will give rise to one or more lines of precipitation in the presence of the homologous antibody. When two antigens are present in a system, they behave independently of one another. The different types of precipitation reactions are illustrated in Figure II.3. Rhizobial strains that share some of or all their antigens will cross-react with respective antisera. These cross-reactions may be encountered in both agglutinations and precipitations.

IMMUNOFLUORESCENCE

Certain chemical dyes (FITC and lissamine rhodamine) have the property of fluorescing when excited by near UV light. Rhizobial antibodies developed in rabbits can be conjugated to these fluorescing chemical dyes or fluorochromes. In work with rhizobia, the chemical dye commonly used for labeling the specific antibody is FITC, which has an apple-green fluorescence upon irradiation with blue light. In practice, a smear of rhi-

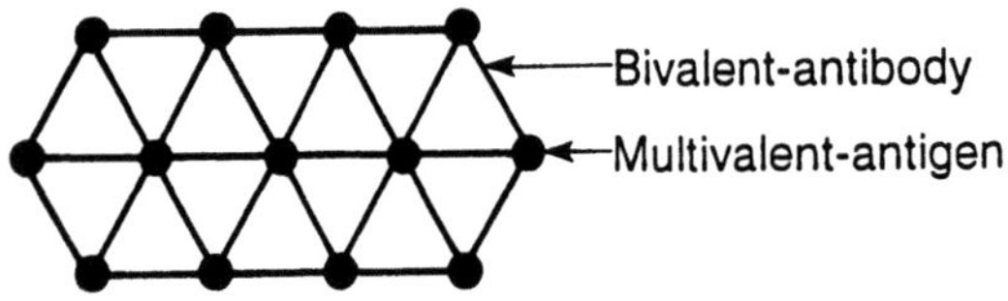

FIGURE II.2 Lattice formation in an antigen-antibody reaction.

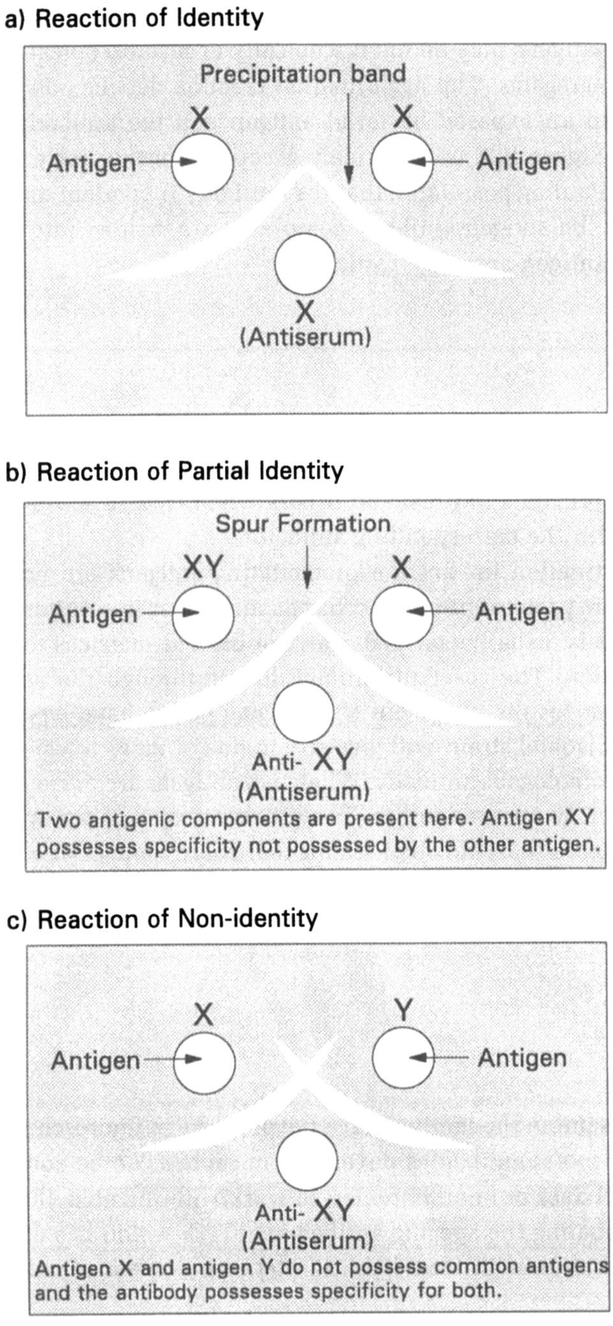

FIGURE II.3 Precipitation reactions: (a) identity, (b) partial identity, and (c) nonidentity.

zobial cells (cultured, or from a nodule) is made on a microscope slide, and this smear is allowed to react or stain with the specific antibody labeled with FITC. After appropriate washing to remove uncombined and excess-labeled antibody, the smear may be viewed through a UV microscope fitted with appropriately complementary filters. An apple-green fluorescence of the bacterial (rhizobial) cells would mean that the antigen smear has reacted with the FITC-labeled antibody.

There are two types of fluorescent antibody techniques, namely the direct- and indirect-immunofluorescence. In the *direct* method, the specific antiserum is conjugated and is used as a stain in the procedure. This is different from the *indirect* method, where the unconjugated (unlabeled) specific or primary antibody is first reacted with the antigen smear, and after sufficient time is allowed for antigen-antibody reaction, the smear is then washed free of excess antiserum. This step is followed by staining with the FITC-labeled secondary antibody.

In serological work with rhizobia, the specific or primary antibody against the rhizobial strain is most often developed in rabbits. The secondary antibody is developed by immunization of goats or sheep with purified rabbit immunoglobulins from a previously unimmunized rabbit. Thus, the rabbit immunoglobulin serves as an antigen for immunization of the goat or sheep. Therefore, the antibody produced in the goat or sheep will not only react with the rabbit antiserum, but will also react with rhizobial antigen with specific unlabeled rabbit antibody attached when the indirect procedure is employed. Though the results are the same, the indirect method is considered more sensitive. The indirect method requires the labeling of only the immune serum from the goat or sheep, and involves two reaction steps; the indirect method is also known to give more nonspecific staining reactions. In the direct method, each rabbit antiserum developed against each rhizobial strain must be conjugated. The two methods are illustrated diagrammatically in Figures II.4 and II.5.

ENZYME-LINKED IMMUNOSORBENT ASSAY (ELISA)

Enzyme-linked immunosorbent assay (ELISA) is one of the several enzyme immunoassays used in detecting antigens and antibodies. Basically, in an enzyme immunoassay, either the antibody or antigen is tagged with an enzyme (e.g., alkaline phosphatase). In completing the assay, the presence or absence of the enzyme-labeled component is detected by the addition of an appropriate substrate (e.g., paranitrophenylphosphate) resulting in a colored product. ELISA falls in the category of heterogeneous enzyme immunoassays in which the unbound reactants are separated from the bound reactants by washing/rinsing steps before the positive reaction is generated. The application of the ELISA technique for identifying rhizobia in nodules, culture, and peat inoculants is well documented. The direct and indirect ELISA techniques are summarized in Figure II.6.

In both ELISA approaches, a 96-well plastic microtiter plate (solid support) is used

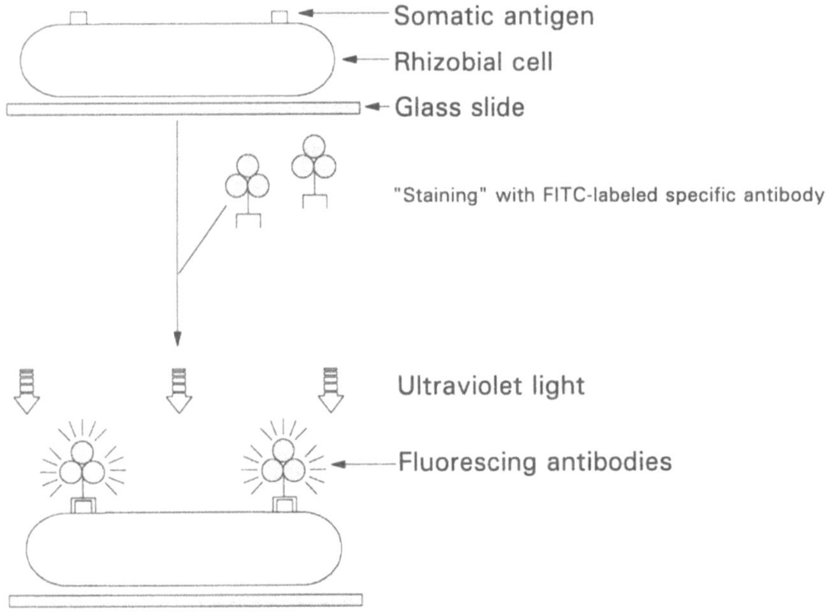

FIGURE II.4 Direct immunofluorescence.

to immobilize the antigen or antibody. In direct ELISA, the specific antibody (Ab1) developed for a particular strain of rhizobia is immobilized in the wells of the plate. Excess unreacted Ab1 is washed off. The rhizobial antigen is then added to the Ab1-coated wells. After an incubation period, excess unreacted antigen is removed by washing. This is followed by the addition of an enzyme-Ab1 conjugate, which binds to its specific antigen. Excess enzyme Ab1 is washed off. The substrate is then added and the reaction is stopped following incubation; the colored product is measured colorimetrically. In work with rhizobia, Ab1 is developed in rabbits.

In indirect ELISA, which is more popular with rhizobial workers, the antigen is immobilized first in the wells. This is followed by the addition of Ab1, incubation, and washing. The next reactant added is enzyme-Ab2 conjugate. Ab2 is usually sheep or goat antibody against Ab1. The enzyme-Ab2 conjugate specifically binds to Ab1. After addition of the substrate, the reaction is completed as with direct ELISA.

MEMBRANE IMMUNOBLOT

The membrane immunoblot procedure is another enzyme immunoassay that has been developed to detect antigen or antibodies (proteins) immobilized (bound) onto a membrane support. This technique has been applied in inoculant quality control and eco-

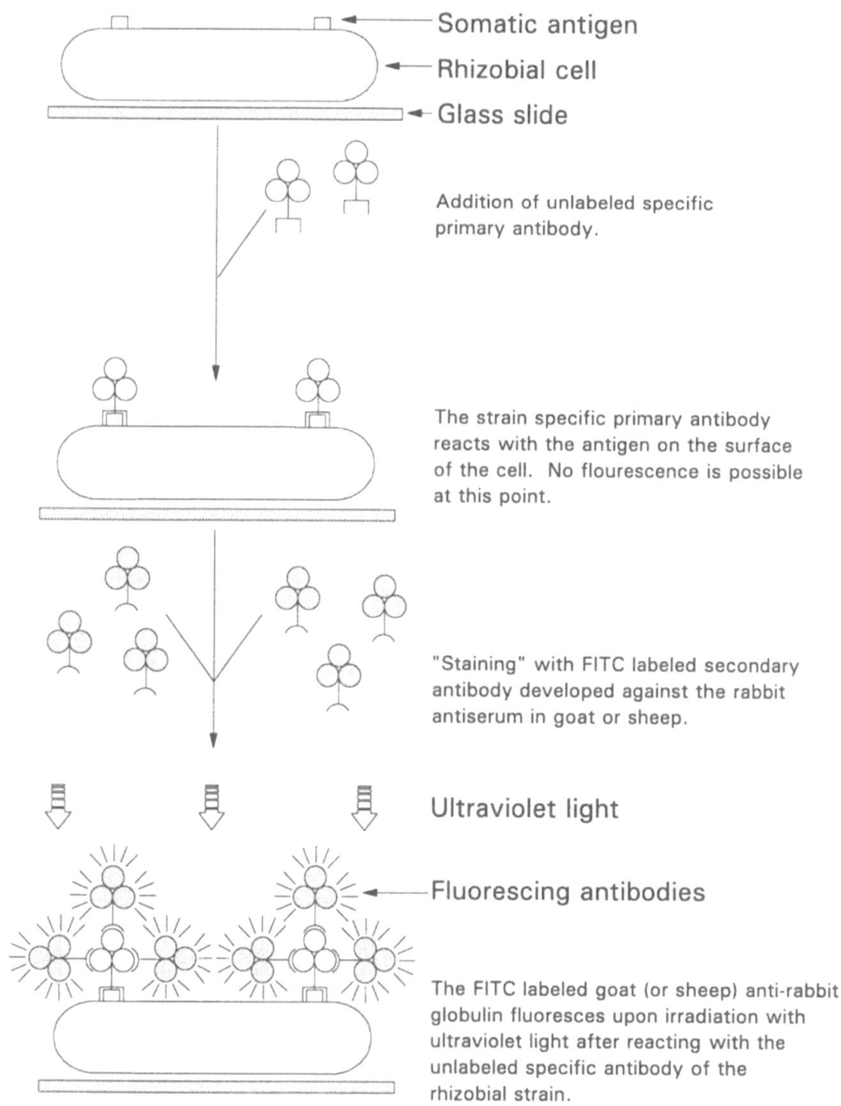

FIGURE II.5 Indirect immunofluorescence.

logical studies of rhizobia. The rhizobial cells (antigens) are blotted or applied onto membranes made of nitrocellulose or nylon. After incubating the membrane-bound antigens with the homologous antibody (Ab1) solution, and washing to remove excess unbound Ab1, the membrane is immersed in a solution containing enzyme Ab2. As with ELISA, Ab2 is usually sheep or goat antibody against Ab1. Ab2 has been conjugated with alkaline phosphatase enzyme and binds specifically to Ab1. The assay is completed by the addition of substrate reagents. These reagents are a mixture of 5-bromo-4-chloro-3-

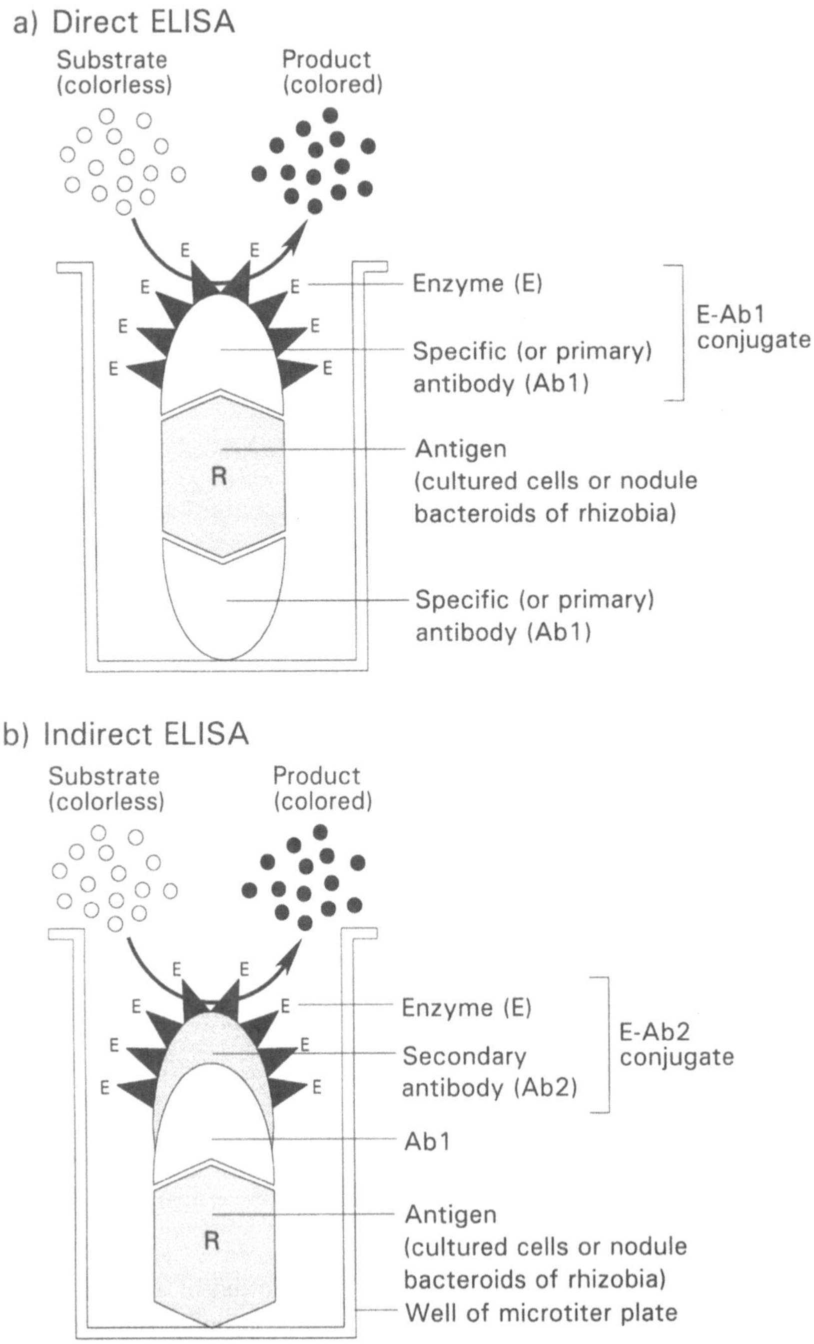

FIGURE II.6 (a) Direct and (b) indirect ELISA techniques applied in rhizobial strain identification.

indolyl-phosphate (BCIP) and Nitro Blue Tetrazolium (NBT). Purple dots develop where a positive reaction has taken place.

ANTIBIOTIC RESISTANCE MARKERS

When high-density inocula of a rhizobial strain are inoculated into media containing an antibiotic, a few cells may exhibit resistance as a result of spontaneous genetic changes or mutations. The resistance of a rhizobial strain to a particular antibiotic is a useful marker. If the mutant strain is used to inoculate a legume, then nodules occupied by that strain may be identified by plating nodule isolates on media containing the respective antibiotic. The mutant rhizobial strain will grow on the antibiotic media and other bacteria will be suppressed. It is important that antibiotic-resistant mutants that are selected for inoculation experiments have not lost their infectiveness (ability to form nodules) nor their effectiveness (ability to fix N_2) in the symbiosis with the host plant. The symbiotic capacity of the mutant should be compared with its parent culture from time to time. The mutant should be stable throughout the steps of infection, nodulation, N_2-fixation, and subsequent reisolation.

Streptomycin resistance is frequently used as a marker for rhizobia. Mutants resistant to this aminoglycoside are stable, have a low incidence of cross-resistance, and infrequently lose their symbiotic capacity. Besides streptomycin, spectinomycin and rifampicin have also been used. Highly resistant mutants with single- or double-markers (streptomycin-spectinomycin or streptomycin-rifampicin) can be obtained with one exposure of the rhizobia to low concentrations of these antibiotics or by successive selection for resistance.

Cross-resistance is a phenomenon whereby a bacterium develops resistance to a second antibiotic as a result of resistance to the first. This may happen if the antibiotics are closely related. The parallel use of antibiotic and serological markers, both relatively stable in themselves, provides a means of confirming the stability of each marker independently in ecological research with rhizobia. Compared to the serological marker techniques (fluorescent antibody, enzyme immunoassays, gel diffusion, and agglutination), the development and use of antibiotic resistant markers is relatively inexpensive and does not require sophisticated equipment.

BACTERIOPHAGE MARKERS: PHAGE TYPING

Viruses that infect bacteria (*bacteriophage*) were independently discovered by Twort and by d'Herelle in 1917 and 1919, respectively. Since then, the processes of infection and multiplication have been well defined. The first step involves the adsorption of the virus to specific receptors on the bacterial cell (somatic), flagella or pili. This is followed by the injection of the viral nucleic acid into the bacteria. The nucleic acids utilize the machinery of the host cell to replicate, leading to the accumulation of several copies of the viral nucleic acid. These nucleic acids are packaged as newly synthesized viral-coat protein and are then released by lysis of the host cell, liberating many infective viruses.

Susceptibility of a certain bacterial strain to a particular bacteriophage forms the basis for phage typing. One approach in a phage-typing scheme is the use of a group of phages with different host specificities. The bacteria can then be placed into groups (lysotypes) if they are susceptible to some of the phages and not others. Through this means, bacteriophage-marked rhizobia can be indirectly traced in soil, in isolates from nodules, and in laboratory experimentation.

8

Developing Antisera

Serological methods are important for rhizobial strain identification. This chapter covers the development of antisera against several strains of rhizobia. Antigens are prepared and then injected into rabbits by three different routes. Antisera are developed for serological techniques that require soluble and insoluble somatic antigens. Injection and bleeding techniques are practiced.

KEY STEPS/OBJECTIVES

1. Culture rhizobia on yeast-mannitol agar (YMA) bottle flats.
2. Check for culture purity by Gram stain.
3. Harvest and prepare antigens for immunization.
4. Begin immunization; inject antigen intramuscularly.
5. Inject antigen intravenously.
6. Give intraperitoneal injections.
7. Trial bleed.
8. Determine the antiserum titers.
9. Harvest blood through cardiac puncture.
10. Give subcutaneous booster injections.

a. Culturing Rhizobia for Antigens (Key Steps 1 and 2)

Inoculate selected strains of rhizobia on two 500-ml YMA flats (Appendix 3) and incubate at 25–30°C. Broth culture in 250-ml Erlenmeyer flasks may also be used, but the medium must be fully defined to avoid complications with antigenic components from the yeast in yeast-mannitol broth (YMB) (Appendix 3).

Check for purity (by Gram stain) at the end of the specified time for growth, e.g., 3–5 days for fast growers and 7–10 days for slow growers. Strains that produce a lot of gum should be harvested earlier.

b. Preparing Antigens for Immunodiffusion (Key Step 3)

When the cultures are ready for harvest, aseptically add about 10 ml of sterile, filtered saline and 20 sterile glass beads to the YMA slants. Close the culture vessel and hold it level so that the saline irrigates the entire surface. Tilt back and forth so that the glass beads dislodge the rhizobial cells into the suspension.

Transfer the suspension (but not the glass beads) to sterile centrifuge tubes and spin-down the cells at approximately 5000 $\times$ g for 15–20 min. Discard the supernatant and resuspend the precipitate again in sterile saline. The gummy substance in the supernatant consists of polysaccharides and is found especially in older cultures. It should be discarded at this point. Do not repeat the centrifugation because excessive washing would remove the soluble antigens essential to the immunodiffusion reaction. Resuspend the precipitate by dropwise addition of sterile saline and with frequent agitation to obtain a thick suspension of 1×10^{10} cells ml^{-1}.

Store about one-half of the thick suspension in the refrigerator, for reference. Dilute the remainder to 1×10^9 cells ml^{-1} using the McFarland standards (Appendix 6). Dispense the diluted suspension into small (5 ml), sterile serum vials in 2-ml portions to be used for injections. Add a preservative (1% Merthiolate) to each 2-ml sample and also to the thick suspension. Merthiolate is used extensively in serology as a preservative. When used in liquids at a final concentration of 1:10,000, it does not interfere with serological reactions. The vials may be stored at 4°C for several weeks, or kept frozen for several months.

c. Preparing Somatic Antigens for the Agglutination and the Fluorescent Antibody Techniques (Key Step 3)

The insoluble somatic antigens found on the surface of the cells are required. Frequent washing eliminates soluble and most of the flagellar antigens.

Harvest a fully grown culture from YMA flats as before. Cells should be centrifuged, the supernatant discarded, and the pellet resuspended in filter-sterilized saline, using a vortex mixer. This sequence of centrifugation and resuspension is repeated three times and the cell concentration is adjusted to approximately 1×10^9 cells ml^{-1}. Transfer the suspension to a sterile serum bottle and close with a rubber septum. Insert a small-gauge (about 23 gauge) needle through the septum to act as an air and steam vent. Heat the antigen for 1 h at 100°C to inactivate any remaining flagellar antigens. This is accomplished by partly immersing the serum bottle in boiling water or by subjecting it to heat in a steam bath. Add Merthiolate solution to the antigen suspension after heating.

d. Immunizing the Rabbit (Key Steps 4, 5, and 6)

A variety of injection schedules have been used to produce antisera of sufficiently high titers. Three examples are given in Appendix 12. The schedule used in this chapter employs three different routes of injection.

Pipette 2 ml of antigen and 2 ml of Freund's complete adjuvant into a 50-ml beaker. Emulsify it by repeatedly drawing the mixture into a glass or plastic syringe (no needle attached) and expelling it through the orifice. The right consistency is reached when a drop of this emulsion does not disperse immediately in water. Freund's complete adjuvant is made from mineral oil and killed cells of *Mycobacterium tuberculosis* or *M. butyricum*. It is used to enhance the effect of the antigen.

Inject 1 ml of the antigen-adjuvant emulsion into the thigh muscle on each hind leg of the rabbit. After 2 weeks, give an intravenous injection of 1 ml of antigen without adjuvant. After 4 weeks, give an intraperitoneal injection of 1 ml of antigen without adjuvant.

e. Trial Bleeding for Titer Determination (Key Steps 7 and 8)

Seven days after the last injection, conduct a test bleed through the marginal ear vein (Appendix 12). Transfer the blood into a sterile screw-cap test tube. Allow the blood to clot at room temperature for approximately 2 h. Detach the blood clot from the test tube wall by moving a wooden applicator stick around the clot. Refrigerate overnight to separate the serum from the clot.

Decant clear serum into a test tube, minimizing carry-over of red blood cells. Since only a very small amount of blood is obtained in the trial bleeding, centrifugation may not be practical. Determine the agglutination titer (Chapter 9).

f. Collecting Blood and Giving Booster Injections (Key Steps 9 and 10)

If the titer is satisfactory (not less than 1:1600), bleed the rabbit from the heart by cardiac puncture using a bleeding rack (Figure A12.1, Appendix 12). Obtain 30–50 ml of blood. Transfer the blood into a sterile, screw-cap test tube of 50-ml capacity.

After the blood has been clotted and refrigerated, decant the serum and centrifuge at 5000 × g for 15 min (under refrigeration, if possible) to clear the serum of red blood cells. Transfer the clear serum supernatant into an appropriate container for storage by freezing. Serum should be stored in 1–2-ml portions in suitable-sized vials. This may not be necessary if the blood is to be processed for conjugation with fluorescein isothiocyanate (FITC). Sera from different rabbits receiving the same antigen may be pooled.

If the titer was too low in the trial bleeding (less than 1:1600), give a booster injection of 1 ml of antigen subcutaneously immediately after the titer determination. Bleed the rabbit 1 week later by cardiac puncture. If more antiserum is desired, the level of immunoglobulins in the rabbit can be maintained by booster injections 3 weeks after each bleeding. However, in such a case, it would be advisable to make intraperitoneal injections of sterile saline each time after the blood has been taken to replenish the liquid level in the animal. The volume of saline injected should be equal to the volume of blood taken from the rabbit.

Note: Storage vials should also be adequately labeled to indicate rhizobial strain, serum batch number, and date. Records of injection schedules and agglutination titer determinations should be noted. Weight, age, sex, and other relevant information on the animals are also usually recorded.

REQUIREMENTS

a. Culturing Rhizobia for Antigens

Microscope
Incubator
Microscope slides, immersion oil
Inoculation loop, flame
Gram-stain solutions (Appendix 3)
Wash bottle with distilled water
YMA flats in 500-ml medicine bottle or Erlenmeyer flasks (250 ml) containing 100 ml of a defined medium (Appendix 3)
Cultures of rhizobia

b. Preparing Antigens for Immunodiffusion

Centrifuge, balance, vortex mixer (optional)
Sterile glass beads (4-mm diameter, sterilized and stored in test tubes at 20 beads per tube)
Pipettes, 10 ml
Centrifuge tubes (30–50-ml capacity) with caps, rack
Saline, sterile (membrane-filtered 0.85% NaCl, w/v)

c. Preparing Somatic Antigens for the Agglutination and the Flourescent Antibody Techniques

Centrifuge, balance, vortex mixer (optional)
Stove or Bunsen burner with tripod and gauze screen, or steam bath
Centrifuge tubes 30–50 ml with caps, rack
Pipette, 1 ml
Glass beads as in (b)
Saline, sterile
Serum bottle with rubber septum, syringe needle (small gauge)
1% membrane-filtered Merthiolate (also called Thimerosal) solution (Gallard-Schlesinger Chemical Manufacturing Corp., Carle Place, New York, or Sigma Chemical Corp., St. Louis, MO)
Inoculated YMA flats from (b)

d. Immunizing the Rabbit

Large towel (approximately 100 × 75 cm)
Small, sterile beaker (50-ml capacity)
Glass syringes (10-ml capacity)
Plastic syringes (1–5-ml capacity)
Syringe needles, 20, 22, and 26 gauge (sterile)
Freund's complete adjuvant (Difco Laboratories, Detroit, MI)
Rhizobial antigens from (c)

e. Trial-Bleeding for Titer Determination

Refrigerator
Large towel, scalpel, petrolatum, razor blade
Cotton wool or tissue paper, alcohol (70%)
Test tubes with caps, rack
Wooden applicator sticks or thin glass rods
Requirements for titer determination (Chapter 9)

f. Collecting Blood and Giving Booster Injections

Centrifuge, balance, freezer
Bleeding rack (Appendix 12)
Syringes (with 18-gauge needles)
Screw-cap test tubes (50 ml), rack
Centrifuge tubes (50 ml with caps), rack
Vials, for storing serum, 5 ml
Merthiolate solution (Sigma Chemical Corp., St. Louis, MO) (Appendix 4)
Alcohol (70%), cotton wipes or soft tissue paper
Glass vials for bulk storage, 20 ml

KEY REFERENCES

Schmidt, E.L., R.O. Bankole, and B.B. Bohlool. 1968. Fluorescent antibody approach to the study of rhizobia in soil. J. Bacteriol. 95:1987–1997.

Vincent, J.M. 1970. A Manual for the Practical Study of Root Nodule Bacteria. IBP Handbook no. 15. Blackwell Scientific Publications, Oxford.

9

Somatic Agglutination Reactions with Pure Cultures of Rhizobia

This chapter will describe relatively simple serological procedures for estimating the concentration of antibodies in a serum by agglutination. The rhizobial antigen is heat treated to inactivate the flagellar antigens and to illustrate the granular clumps characteristic of the insoluble somatic antigen reaction. The highest dilution of the serum at which positive agglutination occurs is used to define the titer.

KEY STEPS/OBJECTIVES

1. Culture rhizobia.
2. Harvest culture for antigen preparation.
3. Prepare serial dilutions of antiserum and perform titration in trays, tubes, and on microscope slides.
4. Read and record titers.

a. Preparation of Somatic Antigens from Cultured Cells (Key Steps 1 and 2)

Obtain a young broth- or agar-slant culture of a strain from Chapter 8. Inoculate in duplicate, two yeast-mannitol agar (YMA) slopes in 500-ml flat culture bottles with inoculum from the broth or slant culture. If a broth culture is used, 1–2 ml of the broth can be squirted onto the agar surface and spread with a loop.

Under aseptic conditions, harvest the culture in saline after 3–5 days for fast-growing and 7–10 days for slow-growing rhizobia. Wash the cells three times in filter-sterilized saline by repeated resuspension and centrifugation (5000–8000 rpm or 4420–11,300 $\times$ g). To inactivate the flagellar antigens, heat treat the antigen preparation as in Chapter 8. Finally, visually adjust the concentration of cells to approximately 1×10^9 cells ml^{-1} with sterile saline, and use the McFarland barium-sulfate standards (Appendix 6). If a photoelectric nephelometer or spectrophotometer is available, the turbidity may be adjusted more precisely.

b. Dilution of Stock Antiserum (Key Step 3)

Prepare the two fold dilutions of the antiserum as follows: Arrange 10 test tubes (16 × 125 mm) in a row on a test-tube rack. Label them 1 through 10. Pipette 9.6 ml of saline into tube 1. Pipette 2.5 ml of saline into tubes 2–10. Accurately pipette 0.4 ml of the stock antiserum into tube 1. Mix the saline and serum thoroughly by sucking the serum-saline mixture into the pipette and then expelling the contents. Repeat this process five times. Expelling should be done gently to avoid frothing. This tube now contains antiserum of a 1/25 dilution.

Using a fresh pipette, remove 2.5 ml of diluted serum from tube 1 and transfer to tube 2. Mix well. (The dilution of the serum in tube 2 will be $1/25 \times 1/2 = 1/50$.) Using a fresh pipette each time, repeat the dilution down the series by transferring 2.5 ml of the diluted serum successively from the previous tube to the next until reaching tube 10. (Tube 10 should have a serum dilution of 1/12,800). Familiarize yourself with the identification system used for wells in the plastic agglutination tray.

c. Performing Agglutinations in Microtiter Trays (Key Step 3 and 4)

Start with the highest dilution (tube 10) and a clean Pasteur pipette (calibrated to deliver 0.03 ml drop^{-1}, Chapter 5). Place two drops of the diluted antiserum into well A10 of the plastic agglutination tray. Next, using the same Pasteur pipette (after blotting the tip dry), place two drops of the antiserum of the next highest dilution (tube 9) into well A9 of the agglutination tray. Repeat until all the dilutions of the antiserum have been dispensed into the respective wells of row A of the agglutination tray.

Next, with a clean, calibrated Pasteur pipette, dispense two drops of the homologous antigen (approximately 1×10^9 cells ml^{-1}) into each of the wells from well A1 through A10. Avoid touching the antiserum in the well or the walls of the well with the tip of the antigen pipette. Discard the antigen pipette after use.

Work from the well containing the most to the least dilute antiserum. Using a clean glass applicator (a fine capillary tube sealed at both ends or a fine solid-glass rod rounded and smooth at both ends), carefully stir the antigen-antiserum mixture in each well. Avoid spillage into neighboring wells. Rinse the stirring rod in a beaker of water and wipe dry with tissue paper between each well. Change the rinsing water frequently. The same stirring rod may be used for each new well.

Place two drops of serum of 1/25 dilution into well A11. Add two drops of saline with another calibrated Pasteur pipette. This serves as the serum-saline control. Place two drops of saline into well A12. Add two drops of antigen. This serves as the antigen-saline control.

Seal all wells (A1 through A12) with a strip of cellophane tape. Float the agglutination tray in a water bath at 52°C for 4 h and then hold overnight in the refrigerator. Alternatively, the reaction mixture may be incubated in an incubator at 37°C for 2 h and then transferred to a refrigerator before reading the reactions. Figure 9.1 shows the steps for the antiserum titer determination in wells. Read and record positive agglutinations

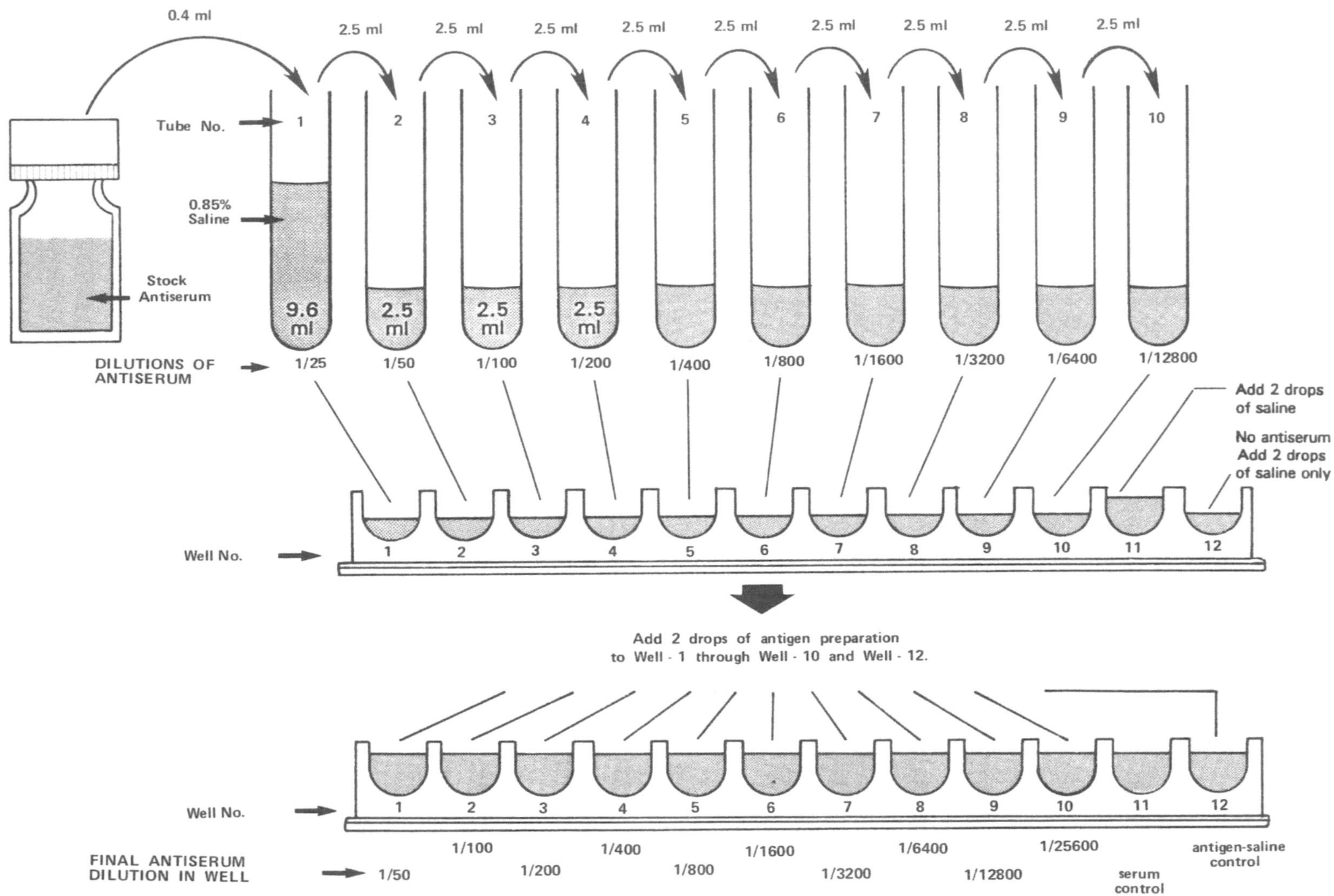

FIGURE 9.1 Scheme for antiserum titer determination in wells.

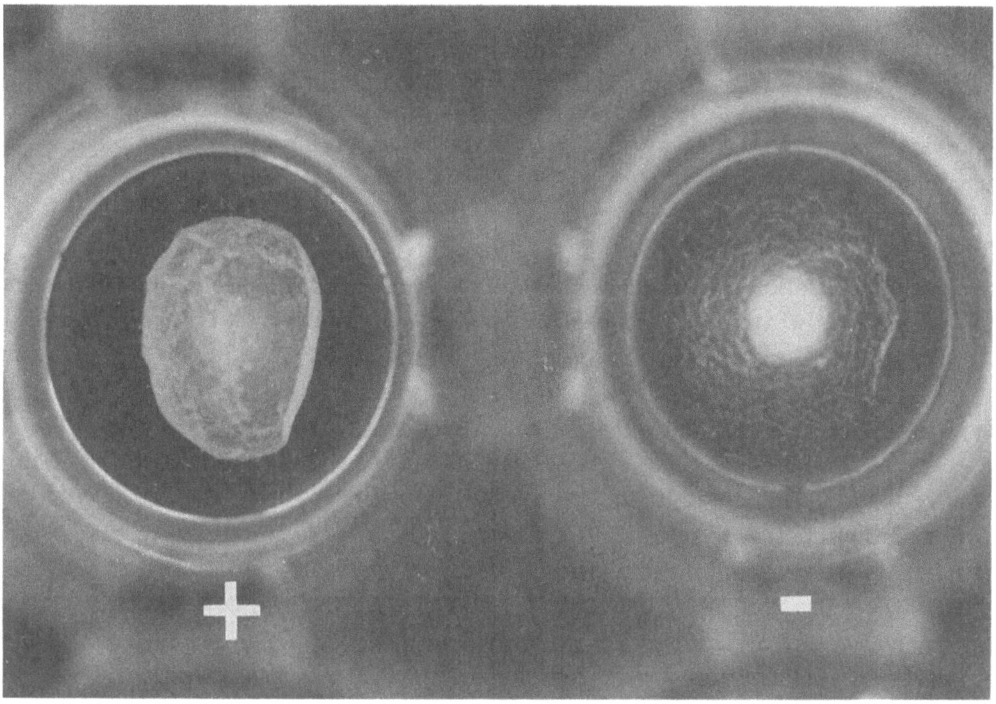

FIGURE 9.2 Positive (+) and negative (−) agglutination reactions in wells of agglutination trays.

at the highest dilution of the serum (Figure 9.2). Positive agglutination will appear as granular clumps with clear supernatant. Negative agglutinations are indicated by cells settling on the bottom of the well and turbid supernatant.

To calculate the titer (serum titer is the reciprocal of the highest serum dilution at which positive agglutination occurs), multiply the highest dilution of the serum at which positive agglutination occurs by two. This is because equal volumes of the diluted serum and antigen were titrated in the well. Example: If positive agglutination was detected at 1/3200 dilution of the serum, the true titer will be $3200 \times 2 = 6400$.

Further confirmation of a positive reaction can be made by gently stirring the reactants in the well with a sterile inoculating needle with a 2-mm loop. Stirring will cause the granular clumps to float in suspension. Observe with a stereo-microscope or magnifying glass and note the granular clumps suspended in a clear suspending solution. Stir the antigen-saline control and observe the turbid appearance showing no separation into granular clumps and no clear suspending solution. Flame the inoculating needle for reuse in other wells. Flaming will remove contaminating reactants and thus prevent carry-over. The antigen-saline control will help in distinguishing between positive and negative agglutinations. Record titer of antiserum and date of experiment.

d. Performing Agglutinations in Tubes (Key Steps 3 and 4)

Prepare a twofold dilution series of the antiserum as described for tray agglutinations (10 dilutions ranging from 1/25 through 1/12,800). Remaining diluted serum prepared previously for the tray method may be used, provided both methods are done on the same day.

Arrange 24 tubes in agglutination tube racks. Tubes with an internal diameter of 5 mm and a length of 60 mm are suitable. Special tubes called Dreyer tubes, if available, are preferred. Label them adequately to facilitate reading the antiserum dilution in each tube. Alternatively, arrange the tubes systematically in a tube rack to avoid labeling.

Dispense 10 drops (0.03 ml $drop^{-1}$) of each dilution into the series of agglutination tubes set up for the titration. Set up duplicate tubes for each antiserum dilution. Add 10 drops of each antigen (approximately 1×10^9 cells ml^{-1}) to each of the agglutination tubes with a clean Pasteur pipette. Avoid contact of the antigen pipette with the mouth or walls of the tubes containing the serum. Do not attempt to stir or mix the reaction mixture in the tubes. Also set up antigen-saline and antiserum-saline tubes as controls.

Incubate the reaction mixtures in the tubes in a water bath at 52°C for 4 h. Alternatively, the tubes may be incubated in a water bath at 37°C for 24 h. Read the tubes after the initial incubation. Read the tubes again after keeping them overnight in the refrigerator (4°C). Place glass beads of suitable size in the mouth of the tubes to prevent evaporation. Parafilm can be substituted if glass beads are not available. Covering the mouths of tubes is not necessary if the water bath is equipped with a lid/cover. The filled portion of the tubes should be immersed half-way in the water bath, thereby facilitating mixing of the antigen and antiserum through convection.

Record positive agglutinations (Figure 9.3). These appear as granular clumps in a

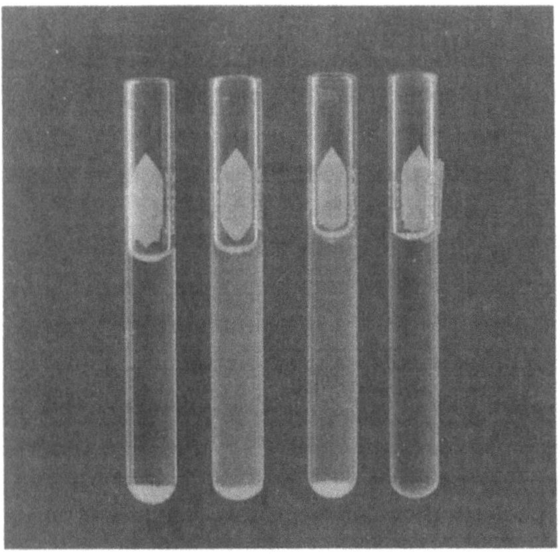

FIGURE 9.3 Antigen-antibody reactions by agglutination in tubes. From left to right: First tube indicates positive agglutination shown by a clear supernatant and agglutinate on the bottom; second tube (antigen control) shows turbidity indicating no agglutination; third tube shows a partial reaction; and the fourth tube (serum control) indicates no reaction.

clear supernatant. When spun gently, negative tubes will produce a wisp-of-smoke effect arising from the bottom of the tube, indicating the sediment is not granular. Calculate the titer as outlined for tray agglutination.

e. Performing Agglutinations on Microscope Slides (Key Steps 3 and 4)

This method has the advantages of using only small amounts of serum and giving results within minutes. There is, however, some loss of accuracy and a degree of subjectivity in interpretations. The method should not be depended on without some objective evidence (e.g., tube agglutination) to support interpretation.

Partition a microscope slide into two sections with melted petrolatum. Prepare 10 slides and lay them in a row on black paper. One section is used for one dilution of the test serum and the other section for the antigen control. Label one section T for test serum and the other C for antigen control.

Using the remaining diluted serum prepared for the tray methods, with a calibrated Pasteur pipette (starting from the highest dilution) place a drop (0.03 ml) of the serum in the test-serum section of each slide. To each drop of serum add a drop (0.03 ml) of the antigen. Also place a drop of antigen on each control section. Add a drop (0.03 ml) of saline to the antigen in the control section of the slides. Stir the antigen-antiserum mixtures with the loop of an inoculation needle. Flame and cool the needle after each mixing to avoid carry-over between tests.

Immediately after mixing, slowly rock the slide back and forth for 1–2 min. With a high-titer antiserum, the agglutination should occur quickly and should be detectable with the naked eye or a dissecting microscope. Record positive agglutinations and calculate the titer. Compare the results of this method with those from the tray and tube agglutination methods.

REQUIREMENTS

a. Preparation of Somatic Antigens from Cultured Cells

Centrifuge
McFarland barium-sulfate standards (or photoelectric nephelometer)
Boiling water or steam bath
Centrifuge tubes
Test tubes
Serum vials with rubber stoppers
Sterile pipettes, 1 ml, and Pasteur pipettes
Sterile glass beads
Sterile saline, 100 ml
YMA slopes in 500-ml flat culture bottles
Broth or agar slant culture of rhizobia

b. Dilution of Stock Antiserum

Antiserum (1 ml) from Chapter 8
Sterile saline (200 ml)
Sterile pipettes, 1 ml (two)
Sterile pipettes, 5 ml (10)
Test tubes (10) and rack

c. Performing Agglutinations in Microtiter Trays

Plastic agglutination tray (rigid polystyrene Uplate, Cooke Laboratory Products, Alexandria, VA)
Calibrated Pasteur pipettes (0.03 ml $drop^{-1}$)
Glass applicator
Rubber bulb (1–2-ml capacity)
Cellophane tape
Tissue paper
Diluted antiserum from (b)
Antigen suspension (1×10^9 rhizobia ml^{-1})
Binocular dissecting microscope
Magnifying glass (hand lens)
Water bath (52°C), incubator (37°C), refrigerator (4°C)

d. Performing Agglutinations in Tubes

Calibrated Pasteur pipettes
Agglutination tubes or other small tubes
Tube rack
Parafilm
Diluted antiserum from (b)
Antigen suspension from (c)
Water bath (52°C), incubator (37°C), refrigerator (4°C)

e. Performing Agglutinations on Microscope Slides

Clean microscope slides
Calibrated Pasteur pipettes
Rubber bulb
Petrolatum
Diluted antiserum from (b)
Antigen suspension from (c)

KEY REFERENCES

Vincent, J.M. 1982. Serology. pp. 235–273. *In* W.J. Broughton (ed.) Nitrogen Fixation, Vol. 2. *Rhizobium*. Clarendon Press, Oxford.

Wollum II, A.G. 1987. Serological techniques for *Bradyrhizobium* and *Rhizobium* identification. pp. 149–155. *In* G.H. Elkan (ed.) Symbiotic Nitrogen Fixation Technology. Marcel Dekker, Inc., New York.

10

Agglutinating Antigens from Root Nodules

This experiment is designed to demonstrate that the *Bradyrhizobium japonicum* bacteroids in soybean (*Glycine max*) nodules share common antigenic properties with its bacterial genotype in culture. Therefore, bacteroids from fresh, desiccated, or oven-dried nodules can be used directly for identifying the occupant strain by simple agglutination. This direct method eliminates the time-consuming steps of isolating the strain in pure culture prior to its use as an antigen in an agglutination reaction.

KEY STEPS/OBJECTIVES

1. Develop antiserum of *B. japonicum*.
2. Prepare Leonard jars.
3. Prepare broth inoculum.
4. Pregerminate soybean seeds.
5. Plant and inoculate pregerminated soybean seeds.
6. Harvest plants for nodules.
7. Perform agglutinations with bacteroid antigens.
8. Read and record agglutinations.

a. Developing Antisera (Key Step 1)

Inoculate *B. japonicum* strain TAL 379 onto yeast-mannitol agar (YMA) flats. Harvest culture and make antigen preparation for the development of antibodies for agglutination as described in Chapter 8. Other strains of *B. japonicum*, for which antisera are available, may be substituted in place of TAL 379.

b. Culturing Soybean Plants Nodulated with a Serologically Marked Strain of *B. japonicum* (Key Steps 2–6)

Prepare Leonard jars. This should be done well ahead of the experiment. Inoculate 100 ml of yeast-mannitol broth (YMB) in a 250-ml Erlenmeyer flask with a loopful of TAL 379 from an agar slant. This should be initiated at least 7 days before planting the Leonard jars to give sufficient time for culture growth.

Surface sterilize soybean seeds as described in Appendix 10. Plate seeds on water-agar plates for germination. Do not invert plates for soybean and other large-seeded legumes. Pregerminate seeds 1–2 days before planting and inoculating Leonard jars. Plant three germinated soybean seeds in each Leonard jar. Inoculate each seed with 1 ml of TAL 379 broth culture. Plant four jars.

Harvest and wash the soybean nodules after 30–35 days of plant growth. Separate the nodules from the roots. Pack and seal the washed nodules in small polyethylene bags (100- × 100-mm size and 0.04-mm thickness). Use one bag per plant and label. Bags of the specified size, or slightly smaller, can be purchased commercially or can be made in the laboratory if the polyethylene material and a bag sealer are available.

c. Separating Bacteroid Antigens from Nodules for Agglutination (Key Step 7)

Fill a 1-liter beaker with approximately 500 ml of water and bring it to a boil, then control the heating source to produce gentle boiling. Immerse one bag of nodules in the boiling water for 3–5 min, then remove the bag and cool. Save the remaining bags of nodules for Chapter 11.

Cut open the plastic bag with scissors. Using forceps, transfer the nodules to the agglutination tray, one nodule per well. Have one nodule in each well, beginning at well A1 through A10 (refer to the well identification system on the agglutination tray, see Figure 10.1). Leave wells B1 through B10 empty; these wells will be used for agglutinations with antigens separated in the series of wells in row A.

With a Pasteur pipette, place six drops (0.03 ml $drop^{-1}$) of saline into each well containing a nodule. (Excess heat-treated nodules can be stored frozen and thawed later for use in agglutination without losing the ability of the bacteroid antigens to agglutinate.)

Gently press out (do not homogenize) the nodule contents into the saline in the wells with fine forceps or a round-ended glass rod (4-mm diameter). Gently stir the exuding nodular contents into the saline and push the resulting nodule tissue against the wall of the well. Rinse the rod and wipe dry for each nodule. Suitable flat toothpicks can be substituted for forceps or glass rods. Toothpicks are used once and discarded.

With a fresh Pasteur pipette, transfer three drops of the antigen from well A1 to B1. Rinse the pipette thoroughly with hot water by sucking the hot water into the pipette and emptying the pipette contents in another beaker. Next, transfer (with the same pipette) three drops from well A2 to B2. With alternate rinsing of the pipettes between transfers, transfer the antigen A3 to B3, A4 to B4, and so on until reaching A10 to B10.

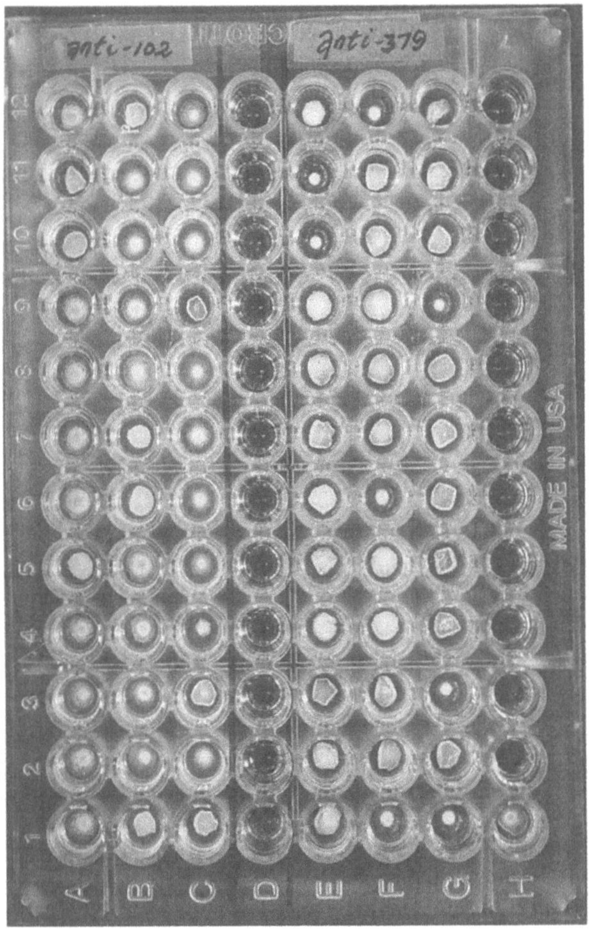

FIGURE 10.1 Identifying bacteroid antigen from soybean nodules by agglutination in a microtiter tray.

(In these transfers, the same pipette is used each time as the nodules were formed by one strain, TAL 379. If the nodules were formed by strains from a mixed inoculum, each antigen transfer must be done with a fresh Pasteur pipette.) Variable Finn pipettes with disposable tips are excellent substitutes for Pasteur pipettes, especially when large numbers of nodules need to be identified, since a fresh tip can be used for each nodule.

d. Agglutinating the Bacteroid Antigens with Homologous Antiserum (Key Steps 7 and 8)

Prepare a 1/25 dilution of antiserum TAL 379 by diluting 0.4 ml of the stock antiserum in 9.6 ml of saline. Dispense the diluted antiserum by placing three drops in each of wells B1 through B10. Place three drops of the 1/25 diluted antiserum and three drops of saline in well B11 (serum control). Set up the antigen-saline control in well B12 by placing three drops of antigen from well A1 followed by adding three drops of saline. Mix the reactants in the wells with a round-ended glass applicator, starting at well B10 and proceeding towards well B1. Use the same applicator for mixing the contents in the wells. Rinse and wipe dry the applicator between each mixing. The contents in the wells may also be mixed by holding the plate loosely in a level position with both hands and tapping the side of the plate with a free forefinger. Avoid spilling during this operation.

In the remaining wells, prepare antigens using nodules formed by *B. japonicum* strain TAL 378. Test the bacteroid antigen of TAL 378 against antiserum TAL 379. Cover the tops of the wells containing the reactants with a strip of cellophane tape. This will prevent evaporating during incubation. Leave a tab of tape to assist removal.

Place the trays at 37°C for 2 h in an incubator. At the end of this time, transfer the trays to 4°C in a refrigerator and leave overnight. Record the appearance of the positive agglutinations by comparison with the antigen-saline control. (The titer of the antiserum at which the agglutinations occurred would be $1/25 \times 3/6 = 1/50$.)

This method has been used with good results on nodules of *Glycine max*, *Centrosema pubescens*, *Vigna unguiculata*, and *Phaseolus lunatus*. These legumes have nodules of similar size. With other legume species, the volume of saline to be used in squashing the nodule to extract the bacteroid antigen has to be determined by trial and error. The volumes of the bacteroid antigen and the serum in the wells also have to be determined before large numbers of nodules are identified by agglutination. Avoid using thick suspensions of the bacteroid antigen because unreacted antigen produces turbidity, resulting in ambiguity in recognizing positive agglutinations. The ultimate purpose of manipulating the volumes of saline and serum to be used during the agglutination is to regulate the density of the antigen close to 1×10^9 bacteroids ml^{-1}.

REQUIREMENTS

a. Developing Antisera

See Chapter 8

b. Culturing Soybean Plants Nodulated with a Serologically Marked Strain of *B. japonicum*

Sterilizing solutions (Appendix 10)
Sterile water, 500 ml
Sterile wide-mouthed flask, 250 ml
Soybean seeds
Water-agar plates (three)
100-ml of broth culture of *B. japonicum* (TAL 379)
Sterile pipettes, 5 ml

Sterilized gravel (mulch for Leonard jars)
Leonard jars (four)
Isopropyl alcohol in spray bottle
Spirit lamp, matches
Forceps
Scissors, polyethylene bags

c. Separating Bacteroid Antigens from Nodules for Agglutination

Beaker, 1 liter
Scissors, fine forceps
Sterile Pasteur pipettes (or variable Finn pipettes with disposable tips, if available)
Bunsen burner
Round-ended glass rod
Sterile saline, 250 ml
Tissue paper
Agglutination tray (rigid polystyrene U plate)
Nodules containing bacteroids of *B. japonicum* TAL 379
Nodules containing bacteroids of *B. japonicum* TAL 378

d. Agglutinating the Bacteroid Antigens with Homologous Antiserum

Antiserum (TAL 379)
Calibrated Pasteur pipettes
Antigen from (c)
Glass applicator
Cellophane tape
Incubator (37°C), refrigerator (4°C)

KEY REFERENCES

Means, U.M., H.W. Johnson, and R.A. Date. 1964. Quick serological method of classifying strains of *Rhizobium japonicum* in nodules. J. Bacteriol. 87:547–533.

Somasegaran, P., R. Woolfenden, and J. Halliday. 1983. Suitability of oven-dried root nodules for *Rhizobium* strain identification by immunofluorescence and agglutination. J. Appl. Bacteriol. 55:253–261.

11

Performing Rhizobial Antigen-Antibody Reactions by Gel Immunodiffusion

Immunodiffusion in gel allows for the recognition of antigenically identical strains and for differentiating between closely related nonidentical strains. In this chapter, soluble and diffusible heat-stable somatic antigens of varying molecular sizes are studied. The antigen and test antiserum are filled in separate wells and allowed to diffuse toward each other slowly through agar, which provides the support medium. Line(s) of precipitation would form where optimal proportions of antigen and antibody molecules meet. Immunodiffusion is useful in analyzing the number of soluble somatic antigens present on the cells of rhizobia.

KEY STEPS/OBJECTIVES

1. Inoculate yeast-mannitol agar (YMA) flats for antigen preparation.
2. Prepare gel in plastic Petri dishes.
3. Harvest cultures for antigen preparation.
4. Perform immunodiffusion.
5. Observe and record diffusion patterns.

a. Preparing Gel for Diffusion (Key Step 2)

Place 100 ml of saline into a 250-ml Erlenmeyer flask. Add 0.75 g of Noble agar (Difco Laboratories, Detroit, MI) to the flask and melt by steaming, autoclaving, or heating in a microwave oven. If direct heat is applied to melt the agar, prevent charring of the agar on the bottom of the flask by constant stirring and controlling the heat. To the melted agar, add 1 ml of a 2.5% (w/v) solution of sodium azide (a preservative), and swirl the flask to ensure proper distribution of the sodium azide. Pipette 25 ml of the hot gel into Petri dishes kept on a level surface. Allow the agar to solidify. A total of four plates with a gel layer 4-mm thick should result.

Trace the outline of a Petri dish bottom on a sheet of white paper. Draw a hexagonal pattern of six circles (4-mm diameters) equidistant (5 mm from edge to edge) from one another in the center of the plate outline. Draw a seventh well in the center of the hexagonal pattern and shade in the circles (Figure 11.1). This paper pattern serves as a template for cutting out wells from the gel. Place a Petri dish (containing gel) on the template. The pattern of circles should be visible through the gel. Cut wells into the gel using a 4-mm cork-borer. The cork-borer should be held vertically when cutting the wells, otherwise wells with oblique walls will result. Carefully remove the gel plugs with pins or other suitable implements, or remove the plugs by suction using a Pasteur pipette (with a slightly bent tip) attached to a suction apparatus. (A Pasteur pipette attached to an aspirator or vacuum pump, with a trap in between for the gel plugs, is a suitable suction apparatus.)

It will take some practice to produce plates with seven intact wells. A drop of molten agar may be necessary to seal off the bottom of the well. Sealing the well is usually not necessary with the plastic Petri dishes, but it is essential for glass Petri dishes. The gel plates may be refrigerated if they are not required for immediate use. Make four sets (one set per Petri dish) of the hexagonal pattern of wells. With sufficient experience and care, three or seven sets of wells can be made per Petri dish.

b. Preparing Antigens (Key Steps 1 and 3)

Culture the following strains of *Bradyrhizobium* spp. on YMA flats:

TAL 651 (from *Calopogonium mucunoides*)

TAL 653, 655, and 855 (from *Centrosema pubescens*)

TAL 642 (from *Lablab purpureus*)

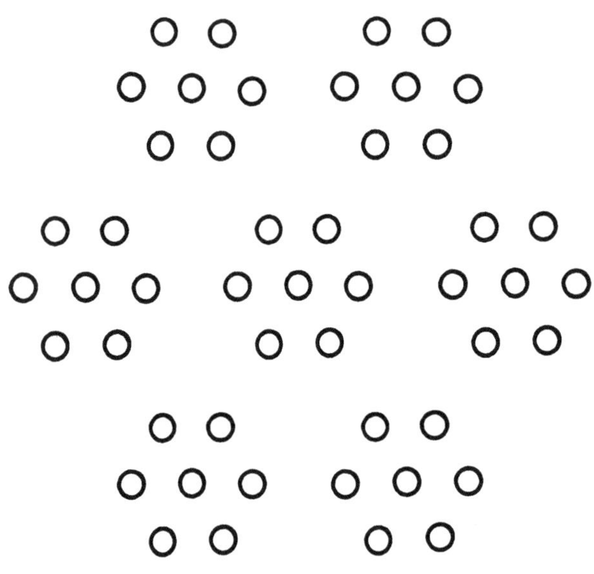

FIGURE 11.1 Hexagonal pattern template for Petri dishes with seven well sets.

Other rhizobia for which antisera are available may be used in place of the recommended strains.

Harvest the cultures after 7 days of growth (Chapter 8) and prepare antigen suspensions for immunodiffusion. A final volume of 1.0–1.5 ml of a dense antigen suspension containing approximately 1×10^{10} cells ml^{-1} is desirable.

Divide the antigen suspension of each strain into two small screw-capped tubes. Small McCartney bottles are better substitutes if available. Heat treat one sample for 1 h at 100°C by immersing the tube in boiling water. Leave the other sample unheated (untreated).

c. Setting Up Immunodiffusion Reactions (Key Steps 4 and 5)

Place two drops (0.03 ml $drop^{-1}$) of each heat-treated antigen in their respective wells. The positions of the different antigens for the diffusion are shown in Figure 11.2. Place the undiluted antiserum of TAL 655 in well 7.

Similarly, set up another set of wells for immunodiffusion with untreated (unheated) antigens.

Labeling the bottom of the Petri dishes is essential to facilitate identifying the antigens in the wells. Orientation of the dish can be established with a single line at the 12-o'clock position and a diagrammatic record of the location of each well.

Incubate the Petri dishes at room temperature in a water-saturated atmosphere. A saturated atmosphere is necessary to prevent moisture loss from the gel. Air-tight plastic boxes can be improvised to provide this environment by placeing wet paper towels on the inside before closing the boxes. Precipitation bands (Figure 11.3) will start forming after 1–2 days.

Make observations at 24 and 48 h. Record your observations in the form of drawings. Compare the diffusion patterns of the heated and unheated antigens. Interpret the diffusion patterns for reactions of identity, partial identity, and nonidentity. Heating can significantly alter the reactivity, concentration, and diffusibility of the somatic antigens leading to stronger and well-separated precipitin bands.

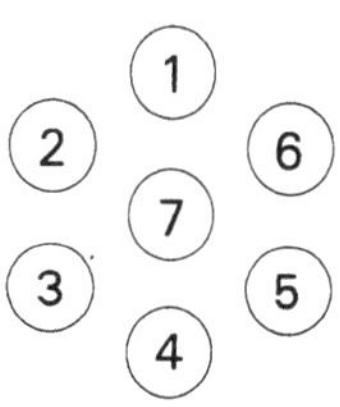

Well number	Antigen
1 and 4	TAL 655
2	TAL 653
3	TAL 855
5	TAL 651
6	TAL 642

FIGURE 11.2 Position of antigens and antiserum in immunodiffusion well set.

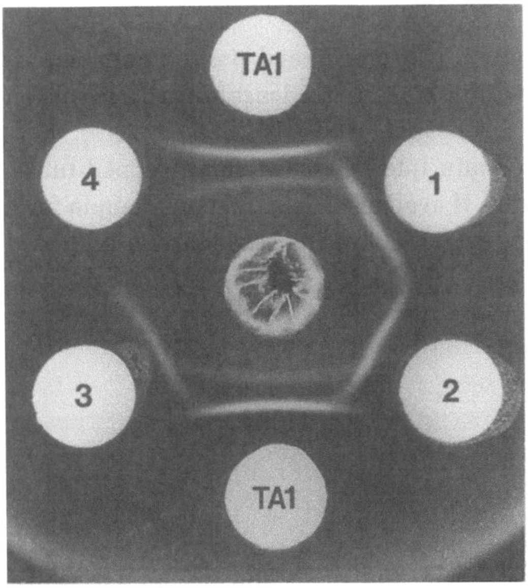

FIGURE 11.3 Identifying field isolates from clover (*Trifolium* spp.) nodules. The isolates are compared by immunodiffusion with a standard strain of *R. leguminosarum* bv. trifolii TA1. None of the four field isolates are identical with TA1. Isolates 1, 2, and 3 show partial identity (spur formation); isolate 4 has no antigen in common with TA1 (no precipitin bands); isolates 1 and 2 share identical antigens (joining of bands). (Photo contributed by W.F. Dudman.)

REQUIREMENTS

a. Preparing Gel for Diffusion

Autoclave, stove, or microwave oven
Saline, 100 ml
Erlenmeyer flask, 250 ml
Sodium azide
Noble agar (purified agar) (from Difco Laboratories, Detroit, MI)
Plastic Petri dishes (four)
Hexagonal pattern template
Cork-borer, 4 mm

b. Preparing Antigens

Agar slant cultures of bradyrhizobia (TAL 642, 651, 653, 655, and 855) or other rhizobia
YMA slopes (five) in 500-ml flat medicine bottles
Screw-capped tubes (or small McCartney bottles)
Steam or water bath

c. Setting Up Immunodiffusion Reactions

Pasteur pipettes
Rubber bulbs (1–2-ml capacity)
Air-tight plastic boxes (or substitute of similar function)

KEY REFERENCES

Dudman, W.F. 1964. Immune diffusion analysis of the extracellular soluble antigens of two strains of *Rhizobium meliloti*. J. Bacteriol. 88:782–794.

Dudman, W.F. 1971. Antigenic analysis of *Rhizobium japonicum* by immunodiffusion. Appl. Microbiol. 21:973–985.

Vincent, J.M. 1970. A Manual for the Practical Study of Root Nodule Bacteria. IBP Handbook no. 15. Blackwell Scientific Publications, Oxford.

12

Determining Strain Occupancy in Soybean Nodules by Gel Immunodiffusion

In this experiment, two serologically distinct strains of *Bradyrhizobium japonicum* are mixed in equal numbers to obtain mixed-strain inocula for studying interstrain competition for nodulation on soybean (*Glycine max*). The immunodiffusion technique, which is commonly employed for the detection of soluble antigens, is adopted for determining the identities of the strains occupying the nodules. The soluble somatic antigen characteristics of a particular strain and its bacteroid form are demonstrated as useful markers in identifying strains in competition experiments.

KEY STEPS/OBJECTIVES

1. Prepare Leonard jars.
2. Culture strains for stock broth cultures and for antigen preparation.
3. Pregerminate soybean seeds.
4. Sample stock broth cultures for viable counts and prepare mixed inoculum.
5. Plant and inoculate pregerminated seeds.
6. Harvest nodules.
7. Set up the Gelman immunodiffusion apparatus. Prepare gel in Petri dishes.
8. Prepare antigen from nodules and cultured cells of inoculum strains.
9. Perform gel diffusion in microscope slides and Petri dishes.
10. Record precipitin bands by drawing; analyze immunodiffusion patterns.

a. Preparing Mixed-Strain Inocula of *B. japonicum* (Key Steps 2 and 4)

Prepare two flasks, each containing 150 ml of yeast-mannitol broth (YMB). Inoculate one flask with *B. japonicum* strain TAL 379 str[1] and the other flask with *B. japonicum* strain TAL 378 spc. These two flasks will provide stock cultures of each strain. Allow 7 days for maximum growth of the strains. The nodules formed by these two strains will be identified by gel immunodiffusion here, and also by their antibiotic resistance properties, fluorescent antibody (FA) technique, immunoblot, and enzyme-linked immunosorbent assay (ELISA) techniques in other chapters. After 6 days of growth at 25–30°C on a rotary shaker, both strains of *B. japonicum* will have attained maximum growth.

Prepare mixed-strain inocula containing the two strains competing in equal numbers and also at different levels of inoculation as shown in Table 12.1. Pipette 5 ml of TAL 378 and 5 ml of TAL 379 into 90 ml of sterile water in a dilution bottle. Label this bottle A. Mix well and transfer 10 ml from bottle A to 90 ml of sterile water in bottle B. Carry out the tenfold dilution steps to obtain 10^{-7} dilution level of the mixed-strain inocula. Refrigerate all inocula until needed.

Use the drop- or spread-plate methods (Chapter 5) to obtain viable counts of TAL 378 spc and TAL 379 str. When the viable counts become available later, the actual ratios of the competing strains in the mixed inocula can be more accurately computed.

b. Culturing Soybean Plants Inoculated with a Single and a Mixture of *B. japonicum* strains (Key Steps 1, 3, 5, and 6)

The experimental treatments in this chapter are as follows:

1. Four mixed-strain inoculations (Table 12.1)

2. Two single-strain inoculations

3. Noninoculated controls

Eight replications of each treatment are set up.

Prepare 56 Leonard jars (Appendix 11). Surface sterilize and pregerminate 200–250 soybean seeds of good viability as described in Appendix 10. Allow 2 days for the pregermination of the seeds. Select well-germinated seeds and plant three per jar. Inoculate each seed at sowing with 1 ml of the broth inoculum of the appropriate treatment. Plant eight jars for each treatment and label adequately. Finally, plant the remaining three jars and leave them uninoculated. These three jars serve as uninoculated controls.

Seven days after planting, thin down to two uniform plants per jar. Harvest all treatment after 30–35 days. Carefully remove and wash the root system of each plant.

---

[1]The two strains used in this experiment are antibiotic resistant. TAL 379 is resistant to streptomycin (str) and TAL 378 is resistant to spectinomycin (spc). Other strains of *B. japonicum*, which are antigenically distinct and have been selected for resistance to different antibiotics, may be used in place of TAL 378 and TAL 379.

TABLE 12.1 Mixed-Strain Inocula containing TAL 378 spc and TAL 379 str at Different Inoculation Levels[1]

Bottle (Treatment)	Dilution	Inoculum Level (cells ml^{-1})	Inoculate[2]
A	10^{-1}	10^{8}	X
B	10^{-2}	10^{7}	—
C	10^{-3}	10^{6}	X
D	10^{-4}	10^{5}	—
E	10^{-5}	10^{4}	X
F	10^{-6}	10^{3}	—
G	10^{-7}	10^{2}	X

[1]TAL 378 is resistant to spectinomycin (spc) and TAL 379 is resistant to streptomycin (str).
[2]An X indicates treatments selected for inoculation.

Count and record the number of nodules on the roots of each plant. Detach the nodules from plants in replications I and II of all inoculation treatments and pack them in small plastic bags as described in Chapter 10. Label the bags adequately for later identification of the treatments. Nodules from replications I and II will be processed and analyzed by immunodiffusion.

Refrigerate nodules from replications III and IV. These nodules will be analyzed using antibiotic resistance. Nodules from replications V, VI, VII, and VIII will be oven dried and stored in glass vials. Replications V and VI will be analyzed by the FA technique while replications VII and VIII will be analyzed by immunoblot or ELISA techniques.

c. Preparing Nodule Bacteroid Antigens (Key Step 8)

Proceed as explained in Chapter 10. Prepare nodule bacteroid antigens from nodules of the treatments that received the mixed-broth inoculum as well as single-strain inoculation treatments. Examine at least 20 nodules from each replication.

d. Preparing Soluble Antigen from Cultured Cells (Key Step 8)

Inoculate one yeast-mannitol agar (YMA) flat each with TAL 379 str and TAL 378 spc. Harvest these strains after 7 days and prepare soluble antigen for immunodiffusion as described in Chapter 11.

e. Setting Up the Immunodiffusion System (Key Steps 7, 9, and 10)

The gel (Chapter 11) for immunodiffusion is prepared on microscope slides. Thin (1 mm) microscope slides are especially suitable for this method using the Gelman immunodiffusion apparatus (Figure 12.1.) described in this experiment. The various components of the apparatus include the immunoframes, immunoframe holders, rinsing tanks, and the immunodiffusion punch set. Familiarize yourself with their construction and use(s). The Gelman product numbers for the various components are given in the list of requirements.

Study the immunoframe that has been especially constructed to hold microscope slides. Each immunoframe holds six standard microscope slides, three in each of its two compartments. Each compartment is divided into three windows and one slide is centered over each window. All three slides must be placed in close contact with one another. Complete the arrangement of slides in each immunoframe and place the immunoframes on a clean and level surface. (A level surface is important to obtain gel of uniform thickness.)

With a Pasteur pipette, dispense minimal amounts of the molten gel around the

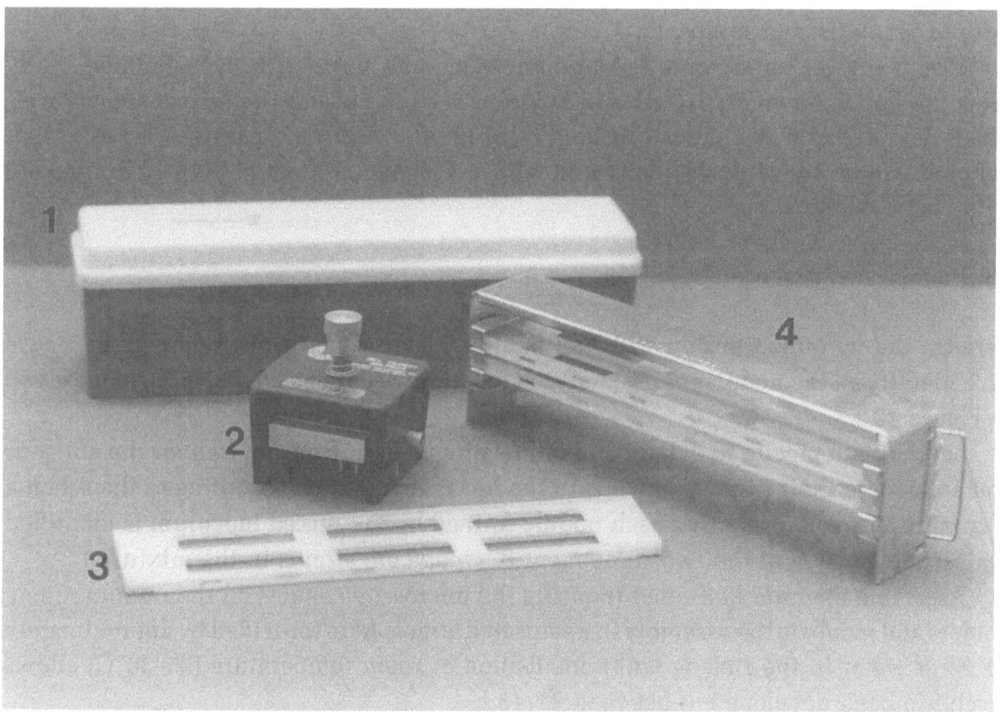

FIGURE 12.1 Gelman immunodiffusion apparatus. (1) Rinsing tanks, (2) immunodiffusion punch set, (3) immunoframe, (4) immunoframe holder.

edges of each slide to seal off the fine gaps at all points of contact between slides, and between slides and compartment walls. (Sealing is necessary to prevent leakage of the gel to the bottom when the melted gel is poured.) Allow the gel to cool to obtain a proper seal.

Pipette 10 ml of the molten gel into each compartment of the immunoframe. Empty the pipette beginning at one end of the compartment and proceeding to the other end, moving the pipette in a zig-zag motion, to evenly spread the gel over the slides. Complete layering the gel over all the slides in the immunoframes.

Allow 1 h for the gel to cool and set, in a dust-free environment. Improvise suitable covers to protect the gel from dust particles settling on its surface during the cooling process. When the gel is cool and set, mount the immunoframes onto the immunoframe holder. It can accommodate a maximum of three immunoframes. Place the whole assembly into a rinsing tank containing approximately 80 ml of water and replace the lid. Store the rinsing tank and its contents overnight at 4°C (refrigerator) or at room temperature (25–30°C) to improve gel setting.

Examine the immunodiffusion gel-punch set. The gel punch consists of a die and an attached system of cutters. The arrangement of cutters on the die allows the production of two sets of the hexagonal pattern of wells (used in this technique) on one slide at any one time. The gel punch is designed to fit the sides of the immunoframe and when the punch supports are properly mounted, the punch can be slid back and forth to the desired positions.

Mount the gel punch onto the immunoframe and position it over a slide. Gently press the punch down on the gel and hold for 3–4 s to cut out the hexagonal patterns. The 3-mm wells on the slides can hold 8–10 μl of antiserum or antigen. These small volumes can be conveniently delivered with a variable volume (5–50 μl) Finn pipette with disposable tips.

f. Performing the Immunodiffusion (Key Step 9)

Perform the immunodiffusion with the nodule bacteroid antigens, referring to Figure 12.2. Identify all the nodules being analyzed for strain occupancy using the scheme given in Figure 12.2.

Set the Finn pipette to deliver 8 μl of the antigen or antiserum. Deliver the antigens and antisera to their respective wells in the hexagonal system according to the scheme (Figure 12.2) given. (Note that each nodule formed by the mixed inoculum is identified against the antisera of the two-component *B. japonicum* strains in the mixture.)

Assemble the immunoframes (housing the microscope slides) on the immunoframe holders and incubate the assembly in a saturated atmosphere (provided by approximately 80 ml of water in the rinsing tank). Incubation at room temperature (25–30°C) allows precipitin band development between 24–48 h.

In Table 12.2, record the number of nodules giving positive precipitation bands against each antiserum. Nodules giving reactions of identity with both the antisera in-

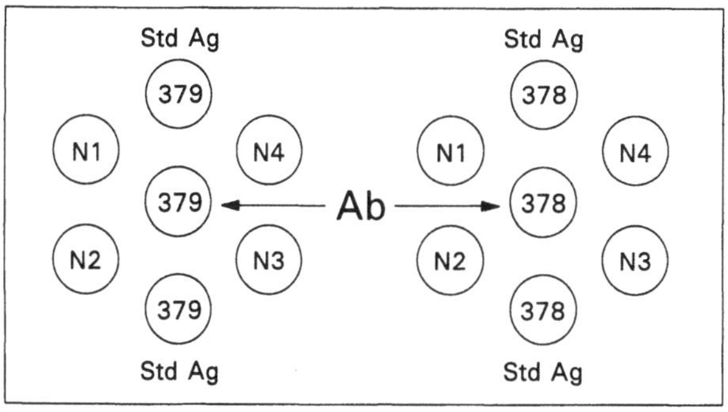

Microscope slide

Key: N1 - antigen preparation from nodule no. 1
N2 - antigen preparation from nodule no. 2
N3 - antigen preparation from nodule no. 3
N4 - antigen preparation from nodule no. 4
Std Ag - antigen preparation from cultured cells
Ab - antiserum (undiluted)

FIGURE 12.2 Scheme for determining the identity of nodules from soybean plants inoculated with a mixture of two strains of *B. japonicum*.

TABLE 12.2 Immunodiffusion Analysis of the Competition for Nodulation between Two Strains of *B. japonicum* Applied as Mixed-Strain Inocula

Mixed Inocula	No. of Nodules Examined	Nodule Occupancy (%)			Chi-Square Deviation (1 df)
		TAL 378	TAL 379	TAL 378 + TAL 379	
A					
C					
E					
G					

dicate mixed infections, i.e., the nodule contains both the strains from the mixed-broth inoculum. Use the nodule analysis data to examine whether the proportion of nodules formed by each strain was according to its representation in the mixture using chi-square analysis. If sufficient nodules and antisera are available, perform a parallel immunodiffusion exercise with gel prepared in Petri dishes (Chapter 11). Follow a similar scheme of nodule identification as detailed for the microscope slide method in this experiment.

REQUIREMENTS

a. Preparing the Mixed-Strain Inocula of *B. japonicum*

Transfer chamber
Agar slant cultures of *B. japonicum* strains TAL 379 str and TAL 378 spc
YMB 150 ml (two flasks)
Sterile Erlenmeyer flasks, 125 ml (two)
Sterile pipettes, 10 ml (five)
Sterile pipettes, 1 ml (15)
Sterile calibrated Pasteur pipettes
90 ml of sterile water in each milk dilution bottle
Quarter-strength YMB or sterile water (9 ml in 30-ml capacity screw-cap tubes)
YMA plates

b. Culturing Soybean Plants Inoculated with a Single and a Mixture of *B. japonicum* Strains

Leonard jars
Soybean seeds
Sterilizing solutions (Appendix 10)
Sterile water
Water agar plates
Pipettes, 10 ml (three to five)
Spirit lamp, alcohol in spray bottle, matches
Forceps
Inoculant broth of TAL 379 str and TAL 378 spc

c. Preparing Nodule Bacteroid Antigens

See Chapter 10

d. Preparing Soluble Antigens from Cultured Cells

Slant cultures of TAL 378 spc and TAL 379 str
YMA flats (two)
Other requirements as in Chapter 11

e. Setting Up Immunodiffusion Systems

Immunoframes (Gelman Product no. 51447, Ann Arbor, MI)
Immunoframe holders (Gelman Product no. 51448)

Rinsing tanks (Gelman Product no. 51457)
Immunodiffusion punch set (Gelman Product no. 51450)
Microscope slides without frosted ends (approximately 1-mm thick)
Finn pipettes (Variable Volumetrics Inc., Woburn, MA)
Pasteur pipettes

f. Performing the Immunodiffusion

All supplies as in (e)
Immunodiffusion wells on microscope slides
Nodule antigens in microtiter wells
Antisera for *B. japonicum* TAL 379 and TAL 378

KEY REFERENCES

Dudman, W.F., and J. Brockwell. 1965. Ecological studies of root-nodule bacteria introduced into field environments. I. A survey of field performance of clover inoculants by gel immune diffusion serology. Aust. J. Agric. Res. 19:739–747.

Skrdleta, V. 1969. Application of immunoprecipitation in agar gel for the serological typing of soybean root-nodules. Folia Microbiologica (Praha) 14:32–345.

13

Producing and Applying Fluorescent Antibodies

With the fluorescent antibody (FA) technique, the antigen-antibody complex can be viewed directly under the microscope. This allows the researcher to detect and identify microorganisms simultaneously. FAs are useful for rhizobial strain identification in ecological research. In this chapter, antisera are purified by ammonium sulfate precipitations and dialysis. The protein content of the dialysate is determined. The immunoglobulin fraction is then conjugated with fluorescein isothiocyanate (FITC). The FITC-antibody conjugate is separated from the unreacted FITC by column chromatography (gel filtration). It is then used to identify rhizobia in nodules by the direct FA technique. A modification of this method, referred to as the indirect FA technique is described in Appendix 13.

KEY STEPS/OBJECTIVES

1. Precipitate the immunoglobulins.
2. Precipitate the immunoglobulins for a second and third time.
3. Dialyze the serum globulins.
4. Determine the protein content of the dialysate.
5. Conjugate the immunoglobulins with FITC.
6. Purify the FA by column chromatography.
7. Test the quality of the FA.
8. Type nodules with the FA technique.

a. Precipitating Serum Globulins (Key Steps 1 and 2)

Place a 250-ml beaker filled with crushed ice onto a magnetic stirring plate. Immerse a 50-ml centrifuge tube containing 15 ml of antiserum into the ice and clamp the tube to a ring stand. Drop a 12-mm (0.5 in) stirring bar into the tube. To the same ring stand, attach a 30-ml burette filled with cold 3.9 M ammonium sulfate solution. The tip of the burette should be close to the surface of the antiserum. Add 15 ml of ammonium sulfate solution to the antiserum at the approximate rate of one drop per second while stirring continuously. Allow the resulting cloudy mixture to stand overnight (or for at least 2 h) at 4°C.

Separate the globulins by centrifugation in a refrigerated centrifuge at 10,000 × g for 30 min. Discard the supernatant and dissolve the precipitated globulins in enough saline to bring the solution back to the original serum volume (15 ml). Repeat the precipitation and centrifugation steps twice as previously described, but without the intermediate step of overnight refrigeration. Instead, allow the precipitates to settle for 5 min at 4°C before centrifugation. Three precipitations are usually sufficient to render the globulins completely white and free of hemoglobin.

b. Purifying the Serum Globulins by Dialysis (Key Step 3)

The next step is to remove the excess ammonium sulfate from the immunoglobulin solution by dialysis. Use a dialysis membrane-filter tubing of approximately 2 cm in diameter. The molecular cut-off rating should be at 16,000. This will keep the large immunoglobulin molecules inside the tubing while the small ammonium sulfate molecules can pass through the pores freely. Cut off a length of approximately 20 cm and soak it in distilled water for 2 h or overnight. Just before using, make a tight knot at one end of the tubing. Wear surgical gloves. Do not touch the tubing with bare hands.

Dissolve the final precipitate in approximately 7.5 ml of saline (half of original volume). Using a 10-ml pipette, transfer the immunoglobulin solution to the dialyzing bag. This is best accomplished by holding the dialyzing tubing in a beaker of water with one hand while pipetting with the other. The pipette must be inserted into the tubing until it is almost to the closed end. Slowly pull out the pipette while the solution is discharged. For this operation, the walls of the tubing must remain wet so that the pipette can slide in and out freely. Close the dialysis tubing with a knot or with a tubing clip. Trap approximately 1 ml of air inside the tubing. This will cause the dialysis bag to spin upright near the surface during dialysis.

Dialyze against 2 liters of dialyzing fluid (saline adjusted to pH 8 with 0.1 N sodium hydroxide) in a cold room with frequent changes of fluid until the ammonium sulfate is no longer detectable in the saline. Three changes of saline at intervals of 4, 10 (overnight), and 4 h again, with continued dialysis for another 4 h, is usually sufficient. Merthiolate may be added to the saline as a preservative at a concentration of 0.01% (w/v).

To determine the presence of sulfate, mix a few drops of the dialyzing fluid with an equal volume of a saturated barium chloride solution. If the mixture does not become cloudy, the dialysis can be considered complete.

If phosphate has been used as buffer for the dialyzing fluid, use Nessler's reagent to detect ammonium (Appendix 4) because phosphate will interfere with the sulfate precipitation. In a small test tube, mix a few drops of the dialyzing fluid with an equal amount of Nessler's reagent. A very fine brown precipitate will form in the presence of ammonium.

c. Determining the Protein Content of the Dialysate (Key Step 4)

After the globulin has been rendered free of ammonium sulfate, protein concentration is determined by the biuret test, which utilizes the following reaction:

$$\text{Protein} + CuSO_4 + NaOH \rightarrow \text{Purple color}$$

The amount of purple color formed is proportional to the amount of protein present (if alkaline $CuSO_4$ is in excess). By using several levels of protein and reading the purple color at the appropriate wavelength, a standard curve can be prepared showing protein concentration versus absorption.

Make a protein standard solution using bovine serum albumin (BSA) at a concentration of 20 mg ml^{-1}. Dissolve 200 mg of BSA in 10 ml of distilled water. Prepare a fresh biuret reagent (Appendix 4). In 15-ml test tubes set up standard and sample dilutions according to Table 13.1. Allow the tubes to stand for 30 min at room temperature. Use tube 6 to zero the spectrophotometer at 540 nm. Read and record the absorbance of the standards (tubes 1–5) and the unknowns (tubes 7–8).

Use the values obtained from tubes 1–6 (Table 13.1) to construct a standard curve plotting absorbance (y axis) against milligrams per milliliter protein per tube (x axis). Use this curve to read off the amount of protein milligrams per milliliter in the globulin test samples (tubes 7 and 8). Make a new curve for each protein determination. Usually at least one of the two unknowns will fall within the range of the curve. After determining the protein concentration, adjust the dialyzed immunoglobulin solution to 10 mg ml^{-1} by adding saline.

TABLE 13.1 Protocol for Total Protein Determination

Tube No.[1]	Biuret Reagent (ml)	Water (ml)	BSA Stock (ml)	BSA (mg ml^{-1})	Absorbance (at 540 nm)
1	8	1.0	1.0	20	
2	8	1.2	0.8	16	
3	8	1.4	0.6	12	
4	8	1.6	0.4	8	
5	8	1.8	0.2	4	
6	8	2.0	0.0	0	
7	8	1.2	0.8	—	
8	8	1.8	0.2	—	

[1]Tubes 1–6 contain BSA standards; tubes 7 and 8 contain globulin test samples.

d. Conjugating the Globulins with FITC (Key Step 5)

Place 10 ml of the immunoglobulin solution (10 mg ml^{-1}) (a total of 100 mg of protein) in a 50-ml beaker. Add 4 ml of 0.15 M sodium phosphate buffer (pH 9) (Appendix 4). In a separate 50-ml beaker dissolve 3.0 mg of FITC in 4 ml of a 0.1 M sodium phosphate buffer pH 8 (Appendix 4), continuously stirring with a magnetic stirrer.

Add this mixture to the buffered immunoglobulin solution. For the conjugation, the ratio of FITC to globulin is 0.03 mg of FITC per mg of protein. Adjust the pH of the FITC-immunoglobulin mixture to 9.2–9.5 with 0.1 N sodium hydroxide and increase the volume to a total of 20 ml with phosphate-buffered saline (PBS) pH 7.1 (Appendix 4).

Add Merthiolate solution. The Merthiolate should be present at a concentration of 1:10,000 to act as a preservative. Conjugate at room temperature for 8 h or overnight, using a magnetic stirrer for continuous mixing. Set the stirrer at the lowest possible speed to avoid frothing. To ensure that the sample is well insulated from heat generated by the stirrer, place a thin piece of a good insulation material, such as styrofoam, on the stirrer and clamp the sample container to a ring stand to elevate it slightly above the stirrer. At the end of the prescribed time, the reaction mixture will contain FITC-antibody conjugated (FAs) and unreacted FITC.

e. Purifying the FA (Key Step 6)

Separate the conjugated FAs from unreacted FITC by column chromatography or dialysis. For the column chromatography method, prepare a slurry of Sephadex G 25-150 or G 25-300 in PBS in a 1-liter Erlenmeyer flask. Use approximately 10 ml of PBS g^{-1} of dry Sephadex at this stage. The bed volume of G 25-150 Sephadex is 5 ml g^{-1} dry gel when swollen in PBS. Allow this to settle and remove fine particles by decanting. Repeat this procedure until the supernatant liquid is clear. Sephadex consists of tiny porous beads of cross-linked dextran (biopolymer) that swell on imbibing water. When contained in a chromatography column, the beads form a molecular sieve that will separate compounds according to molecular size. Large molecules of the conjugated immunoglobulins will meet little obstruction as they pass through the interstitial spaces between the beads and emerge with shorter elution times. The much smaller molecules of the free FITC will penetrate the lattice structure of the Sephadex, which increases the elution time.

Add Merthiolate (1:10,000) and leave at room temperature for 3 h to allow the Sephadex particles to swell. Alternatively, the slurry may be heat treated at 90°C for 1 h.

Plug a glass column approximately 2.5 × 30 cm with a small amount of glass wool and close the outflow. Add 2–3 ml of PBS. Premeasure the slurry to fill approximately 20 cm of the column when settled. Pour the slurry into the column in one continuous flow. The volume of a packed Sephadex column should be approximately three to five times the volume of the conjugate to be purified.

Equilibrate the column by passing at least three column volumes of PBS through it.

Control the outflow carefully so that the column bed remains covered with liquid. The column must be replaced, should it run dry. Measure the pH of the outflowing eluent. Repeated rinsing with buffer or distilled water is necessary if the pH is higher than neutral.

Allow the buffer to settle almost to the top of the bed without drying the bed, then add the conjugate with a Pasteur pipette. Permit the conjugate to penetrate the Sephadex until the conjugate level is slightly above the column bed. Gently wash the conjugate into the column with several 2-ml increments of PBS, added with a Pasteur pipette. After all the conjugate has penetrated the Sephadex to at least 3 cm into the column, a reservoir filled with PBS containing Merthiolate (0.01%) may be connected to the top of the column to maintain a PBS-filled column until the purified FA have been collected.

Collect the first yellow-banded fraction (FA) in a small (50 ml) beaker, taking care to stop the collection when no color is seen in the eluted buffer. The unconjugated FITC fraction is seen as a slow-moving, diffused yellow band. If the collected material is dilute, it may be concentrated using Carbowax (polyethylene glycol). The conjugate is placed in a beaker. Then a dialysis bag containing approximately 5 g of Carbowax is immersed into the conjugate and left in the cold for 4 to 8 h, or until the FA solution has reached a volume of 15 to 20 ml.

An alternate way to purify FA is through dialysis. Dialyze against PBS pH 7.1 until no color is detected in the dialysate. This may take more than 36 h. Distribute the purified FA in 1-ml volumes in labeled 2-ml screw-cap vials and store in the freezer. Lyophilization is also possible at this point if facilities are available. Often, some particulate matter accumulates at the bottom of the containers. This should be eliminated by centrifugation or by filtration through a 0.45-Mm membrane filter prior to use.

The Sephadex may be used repeatedly after thorough washing. The unconjugated FITC should be washed off the column by passing distilled water through it until no yellow color can be detected. The Sephadex may then be washed again batchwise and stored in a refrigerator.

f. Testing the Quality of the FA (Key Step 7)

Prepare twofold dilutions of the FA in saline (or PBS) for the titer determination. Dilute the FA in the range of 1:1, 1:2, 1:4, and so forth, up to 1:32. Using a small transfer loop, make thin smears on clean microscope slides from: (1) a young liquid culture of the homologous rhizobial strain for which the FA was prepared and (2) a young liquid nonhomologous rhizobial culture. Use a separate slide for each dilution of FA.

Air dry and heat fix the smears by passing them rapidly over the flame of a Bunsen burner. Cover each smear completely with one drop of each dilution of the FA. More FA material may be needed if the smears are too large to be covered by one drop. Incubate in a moisture-saturated chamber for 20 min at room temperature. A moisture-saturated chamber may be made from a large Petri dish into which a wet piece of filter paper is placed. Two glass rods are placed on the filter paper and spaced to provide a rail to support the slides. Larger incubation chambers can easily be improvised, but care has to be taken that the slides are resting level and that they are well separated from each other.

Wash off the excess FA with a gentle stream of PBS from a wash bottle or Pasteur pipette, taking care to avoid dislodging the cells in the smears. Then wash the smears by submerging the slides in saline or PBS for 20 min. Similarly, wash the smears in water for 15 min and air dry. Add a drop of mounting fluid (Appendix 4) and mount with a coverslip. Observe the smear under a UV microscope equipped with a mercury vapor light source and a suitable filter pack for FITC excitation. To ensure the validity of the results, compare the reactions with homologous and nonhomologous strains of rhizobia.

The intensity of the fluorescence decreases with the higher dilutions of the applied FA. Grade each smear for the intensity of the fluorescence using the following scale.

Grade	Fluorescence
4+	Brilliant yellow-green
3+	Bright yellow-green
2+	Yellow-green
1+	Dull green
0	No fluorescence

Ideally, FAs should show a 4+ reaction, even after they have been diluted by several twofold steps. Occasionally, FITC conjugations yield only FA of 3+ rating. The FA are diluted before use. The highest dilution that still results in an intensity of fluorescence comparable with the undiluted FA is used for strain identification. The nonhomologous reaction should show no more than background fluorescence. Strains that cross-react may show from 4+ down to 1+ reactions.

g. Typing Nodules Using the FA Technique (Key Step 8)

Stored soybean (*Glycine max*) nodules (previously oven dried at 60°C) from all treatments in Protocol 12 will be analyzed for nodule occupancy by the FA technique. At least 24 oven-dried nodules from each of the treatments in replications III and IV will be sampled.

Before smears of the bacteroids can be made, the oven-dried nodules must imbibe water. With a forceps, place one nodule in each well of a microliter U plate. The checkerboard numbering system on the U plate provides an identification method for each nodule being typed. With a Pasteur pipette, add three drops of sterile water to each well containing a nodule. Seal the wells with strips of cellophane tape and allow them to imbibe for 2 h at room temperature or overnight at 4°C.

At the end of the imbibition, pierce the nodule with a flat toothpick and squeeze it against the side of the well. Sufficient amounts of bacteroids will be loaded onto the end of the toothpick to make smears on microscope slides following the scheme in Table 13.1. (Such a drawing may be used as a template onto which the microscope slide is placed during sample application.) Label the slides. Use a fresh toothpick for each new nodule. Note that duplicate smears are made for each nodule. The size of the dots in Figure 13.1 indicates the size of the smear to be made on the microscope slides. Air dry and heat fix the smear.

With a Pasteur pipette, place a drop of gelatin-rhodamine isothiocyanate (RhITC) (Appendix 4) on the smears. This eliminates much of the background fluorescence normally caused by nodule debris. Before the rhodamine gel dries, add one drop of FA solution and allow it to react in a moisture-saturated chamber at room temperature. Incubate, rinse, wash, and dry the slides following the same procedures as described for determining the FA titer. After the smears have dried, circle the smears with a fine permanent marker or diamond pen on the reverse side of the slide. This will be helpful in locating them under the microscope.

Add sufficient mounting fluid, approximately two drops per slide for 12 or more smears. Place a long (4 cm) coverslip over the smears, taking care to exclude air bubbles. Observe the preparations with a UV microscope under a 40× or 60× objective. Also, observe under a 90× or 100× objective with oil immersion. If the microscope is equipped with a phase-contrast condenser, first focus on the smear using incandescent light before observing under UV light. This will greatly reduce the fading of the smear through prolonged exposure to UV light.

A strong, positive reaction is indicated by brilliant yellow-green fluorescence of the

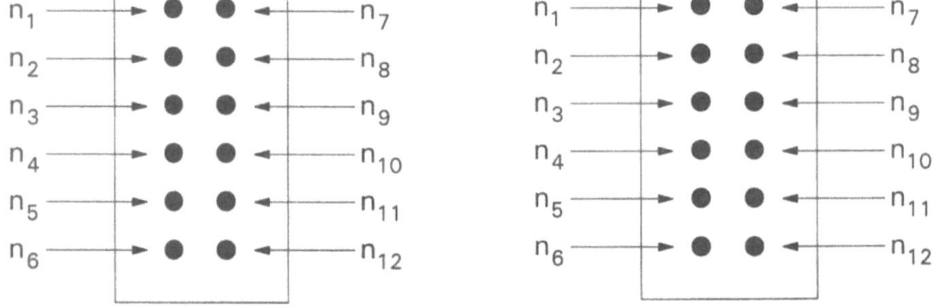

FIGURE 13.1 Scheme of nodule smears on microscope slides for identifying TAL 378 spc and TAL 379 str in mixed-strain inocula. [n_1–n_{12} nodules to be analyzed for strain occupancy; C_1 and C_2 cultured cell (controls) smears of TAL 378 and TAL 379, respectively.]

Table 13.2 FA Analysis of the Competition for Nodulation between Two Strains of *B. japonicum* Applied as Mixed-Strain Inocula

Mixed Inocula	No. of Nodules Examined	Nodule Occupancy (%)			Chi-Square Deviation (1 df)
		TAL 378	TAL 379	TAL 378 + TAL 379	
A					
C					
E					
G					

smear on a dark purple background. No cells will be visible (i.e., no fluorescence) if the specific strain is not present on the smear. A mixed infection [a nodule containing both TAL 379 streptomycin (str) and TAL 378 spectinomycin (spc)] is obvious when smears from a single nodule fluoresce with the FA stains of both strains. Record the results as indicated in Table 13.2. Compare results from this method with those of the other method used.

REQUIREMENTS

a. Precipitating Serum Globulin

Ring stand with three clamps
Magnetic stir plate with a 12-mm (0.5 in) stirring bar
Centrifuge
Refrigerator
Balance (for centrifuge tubes)
Burette
Graduated pipette, 10 ml
Centrifuge tubes with caps, 5 ml (two)
Beaker with crushed ice, 250 ml
Cold 3.9 M ammonium sulfate solution
Saline (0.85% NaCl filtered through 20-μm filter)
Rabbit antiserum (15 ml)

b. Purifying the Serum Globulins by Dialysis

Precipitated immunoglobulins [final precipitate from (a)].
Cold room or large refrigerator

Magnetic stir plate, 5–8-cm (2–3 in) stirring bar
Flasks or beakers, 3 liter (three)
Pipettes, 1 ml (two) and 10 ml (one)
Test tube
Dialyzing tubing (20 cm)
Surgical gloves
Six liters of saline adjusted to pH 8 with NaOH
Filtered saline
Merthiolate solution (filtered, 1%) (Sigma Chemical Co. St. Louis, MO)
Saturated barium chloride solution
Nessler's reagent (optional)

c. Determining the Protein Content of the Dialysate

Spectrophotometer, two cuvettes
Test tubes, 15 ml (eight)
Test tube rack
Pipettes, 5 ml (one) and 1 ml (two)
Distilled water
Filtered saline
Vial for dialysate
BSA solution (20 mg ml^{-1})
Dialyzed globulin solution from (b)
Biuret reagents

d. Conjugating the Globulins with FITC

Analytical balance, spatula, weighing paper
Magnetic stir plate, 12-mm (0.5 in) stirring bar
Ring stand with two clamps
pH meter
Beakers, 50 ml (two); Parafilm or foil for covering
Pipettes, 10 ml (two) and 1 ml (one)
Merthiolate solution (1% and filtered)
Sodium phosphate buffer 0.15 M (pH 9) (Appendix 4)
Sodium phosphate buffer 0.1 M (pH 8) (Appendix 4)
FITC (Sigma Chemical Co.)
PBS (Appendix 4)
Sodium hydroxide solution, 0.1 N
Dialyzed immunoglobulin from (c)

e. Purifying the FA

Suction pump or aspirator
Refrigerator, freezer

Centrifuge
Balance for centrifuge tubes
Two centrifuge tubes with caps
Chromatography column (approximately 2.5 × 20 cm)
Glass wool
Pasteur pipette with rubber bulb
Erlenmeyer flask, 1 liter with screw cap (or large glass bottle)
Glass beaker, 50–100 ml
Two-liter reservoir for PBS with connecting tubing and plug for column
PBS, 2 liters containing 0.01% Merthiolate
Sephadex G 25-150 (or G 25-300) (Sigma Chemical Co.)
Distilled Water
Carbowax (polyethylene glycol) (Sigma Chemical Co.)
Dialyzing tubing
Membrane filter unit with filter of 0.45-μm pore size
Screw-cap vials for storage of FA
FITC conjugate from (d)

f. Testing the quality of the FA

UV microscope (instrument with epifluorescence condenser preferable)
Transfer loop, flame
Microscope slides, coverslips, mounting fluid
Incubation chambers
Rinsing tank containing PBS, rinsing tank containing distilled water
Wash bottle containing PBS, wash bottle containing distilled water
Supply of PBS, 2 liters
Test tubes, rack
Pasteur pipettes, rubber bulb
Young cultures of *B. japonicum* strains TAL 378 and TAL 379
FA from (e)

g. Typing Nodules Using the FA Technique

UV microscope (instrument with epifluorescence condenser preferable)
Microscope slides, coverslips (long), mounting fluid, immersion oil
Inoculation loop, flame, forceps, scalpel
Incubation chambers, rinsing tanks as in (f)
Nodules containing TAL 378 spc and TAL 379 str (see Chapter 12)
Pure cultures of TAL 378 and TAL 379
FA of TAL 379 and TAL 378 (diluted for use)
Rhodamine gel (optional)

KEY REFERENCES

Bohlool, B.B. 1987. Fluorescence methods for study of *Rhizobium* in culture and in situ. pp. 127–147. *In* G.H. Elkan (ed.) Symbiotic Nitrogen Fixation Technology. Marcel Dekker, Inc., New York.

Schmidt, E.L., R.O. Bankole, and B.B. Bohlool. 1968. Fluorescent antibody approach to the study of rhizobia in soil. J. Bacteriol. 95:1987–1997.

Somasegaran, P., R. Woolfenden, and J. Halliday. 1983. Suitability of oven-dried root nodules for *Rhizobium* strain identification by immunofluorescence and agglutination. J. Appl. Bacteriol. 55:253–261.

14

Identifying Rhizobia by the Indirect Enzyme-Linked Immunosorbent Assay

The enzyme-linked immunosorbent assay (ELISA) is a colorimetric enzyme immunoassay used to detect antigens. Several other variations of the assay have been developed. An indirect method is described here that does not require the conjugation of the primary (rabbit) antibody with an enzyme. In this procedure, the antigen is immobilized in the wells of polystyrene microtiter plates and is reacted with specific (primary) antibodies. After a washing step, the primary antibodies bound to the antigens in the wells are reacted with alkaline phosphatase conjugated to secondary antibodies that bind specifically to the primary antibodies. Reaction with an enzyme substrate causes a color change, the intensity of which corresponds to the amount of primary antibodies present.

Unlike most other serological methods, the ELISA can be used to analyze a large number of samples simultaneously, making it especially useful for identifying strains in root nodules. In this chapter, the use of the ELISA is demonstrated in a competition study using two strains of *Bradyrhizobium japonicum*.

KEY STEPS/OBJECTIVES

1. Obtain antibodies.
2. Prepare cultured cell antigens.
3. Process nodule samples.
4. Immobilize the antigens.
5. Apply blocking solution.
6. Add primary antibodies.
7. Apply secondary antibodies.
8. React samples with substrate; record results.

a. Obtaining the Antibodies (Key Step 1)

Obtain or produce antisera for *B. japonicum* strains TAL 378 and TAL 379 to be used as sources for primary antibodies. Special purification is not necessary. Store antisera undiluted at 4°C until use. The goat anti-rabbit immunoglobulin (IgG) alkaline phosphatase conjugate used for the secondary antibody reaction may be purchased from a chemical supply house.

b. Preparing the Test Antigens (Key Steps 2 and 3)

Use fresh or dried nodules from representative plants grown in Chapter 12. Take 12 nodules from each of the four treatments labeled A, C, E, and G (Chapter 12, Table 12.1). Use nodules from replications VII and VIII for analysis here. Place them in the wells of a microtiter plate with 96 U-shaped wells. The bacteroid antigen from each nodule will be reacted against the antisera of strains TAL 378 and TAL 379. (The coding system on the microtiter plate facilitates precise identification of each nodule in the well by a combination of letter and number.) Add 200 μl of saline to each nodule in the wells.

Prepare cultured cell suspensions of TAL 378 and TAL 379 as homologous and heterologous controls. Centrifuge and wash the cells three times in distilled water and adjust the suspension to the same turbidity ($A_{540} = 0.45$). Pipette 100 μl of TAL 378 into one row of 12 wells. In the same manner, place the antigen of TAL 379 into the next row of wells.

Steam the plate for 20 min in a steam chamber or on a rack approximately 5 cm above boiling water in a covered pot or waterbath. With a Pasteur pipette or a pipettor, remove the saline from all nodule samples. Using a multiple pipettor system (Figure 14.1), add 200 μl of coating buffer (Appendix 4) to all wells including the controls.

Crush the nodules with sterile toothpicks. Do not homogenize the nodules. Instead, gently press each nodule against the side of each well to squeeze out the bacteroids (antigens). Remove the nodule residue with the same toothpick. Use a different toothpick for each nodule.

c. Applying the Antigens (Key Step 4)

Using a multiple-tip pipettor, remove 100 μl of nodule antigen from each well from the microtiter plate in step (b), and transfer it to the wells of a new microtiter plate (plate A, Figure 14.2). Duplicate this operation using a second, new microtiter plate (plate B, Figure 14.2). In addition, add a zero-cell control consisting of saline only to each of the duplicate plates. The two new microtiter plates are now filled with samples and controls as shown in Figure 14.2. Seal each well row with transparent tape to prevent evaporation and incubate at 4°C overnight (12–16 h).

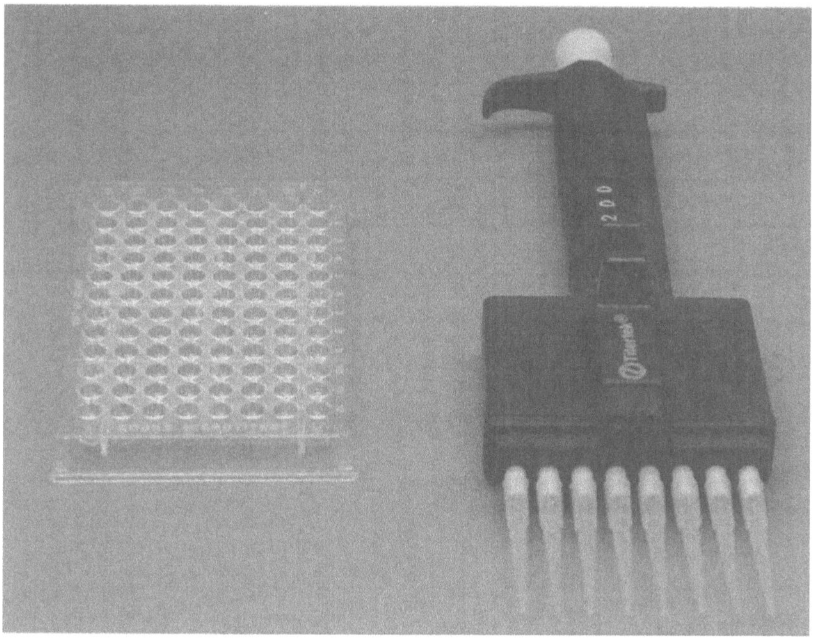

FIGURE 14.1 Multiple pipettor.

d. Blocking the Binding Sites (Key Step 5)

Carefully remove the transparent tape from the microtiter plates. Empty the wells by turning the plates upside down. Fill all wells with phosphate-buffered saline (PBS) contained in a squeeze bottle, allow to soak for 3 min and turn the plate over again. Repeat the washing step twice and shake out residual liquid.

The antigen is now bound (immobilized) to the well bottoms. To block nonspecific binding sites, add 200 μl of blocking buffer (Appendix 4) to each well. Incubate at 37°C in a moist chamber for 1 h. Empty plates and fill wells with PBS. Allow to soak for 3 min and wash twice. Shake out residual liquid.

e. Reacting the Antigens with Primary Antibodies (Key Step 6)

The antigens immobilized on the surface of the well bottoms will now be incubated with the antibodies for *B. japonicum* strains TAL 378 and TAL 379. Prepare the primary antibody solutions. The quality of the antisera determines the amount to be used. As in the immunoblot method, even low-quality antisera are usable. The higher the quality of the antiserum, the higher the dilution that can be used. Assuming the antisera have an agglutination titer of 1600, dilute 1:4000 by transferring 10 μl of antiserum to 40 ml of PBS with a micropipettor.

PLATE A: PRIMARY ANTISERUM TAL 379

	1	2	3	4	5	6	7	8	9	10	11	12
A												
B												
C												
D												
E												
F												
G												

PLATE B: PRIMARY ANTISERUM TAL 378

	1	2	3	4	5	6	7	8	9	10	11	12
A												
B												
C												
D												
E												
F												
G												

Well A1 - A12, bacteroid suspensions (Treatment A)
Well B1 - B12, bacteroid suspensions (Treatment C)
Well C1 - C12, bacteroid suspensions (Treatment E)
Well D1 - D12, bacteroid suspensions (Treatment G)
Well E1 - E12, cultured cell controls of TAL 379
Well F1 - F12, cultured cell controls of TAL 378
Well G1 and G2, saline controls

FIGURE 14.2 Application of samples, controls, and primary antisera for analysis of nodule occupancy by ELISA.

Dilute both antisera according to their quality and place them into shallow, wide containers (e.g., large Petri dishes) to allow access with the multi-tip micropipettor. Now pipette 100 μl of the diluted antiserum of TAL 379 into all wells of plate A. Similarly, pipette the same amount of the antiserum for TAL 378 into plate B. Incubate in a moist chamber at 37°C for 1 h. Repeat the wash procedure as described previously.

f. Reacting Globulin Antigens with the Secondary Antibody (Key Step 7)

The rabbit antibodies (IgG) are now attached to the rhizobial cells. They, in turn, are the antigens for the secondary antibody. The secondary antibody is goat anti-rabbit alkaline phosphatase (GAR-AP) conjugate. Dilute GAR-AP at the rate of 1:4000. Pipette 10 μl of GAR-AP to 40 ml of PBS. Add 100 μl of diluted GAR-AP conjugate to each well on both microtiter plates. Incubate in the moist chamber at 37°C for 3 h. Repeat the wash procedure as in previous steps.

g. Developing the Color (Key Step 8)

When the alkaline phosphatase reacts with the substrate solution, a yellow color results. Prepare the enzyme substrate solution (as described in Appendix 4) just before use. Transfer 100 μl to each well. Reincubate the samples in the moist chamber of 37°C for 20–60 min, or until color in the positive controls has developed to the desired intensity that can be measured, visually and/or spectrophotometrically. Add 50 μl of 3 M NaOH to stop the reaction. Gently swirl the microtiter plates to mix.

h. Reading and Recording the Results (Key Step 8)

For the most rapid quantitative measurements, use a microtiter reader. Alternatively, a spectrophotometer may be used. In this case, immediately transfer the samples from each well to tubes containing 3 ml of 0.2 M NaOH. Read the tubes at 405 nm.

In the absence of both a microtiter reader and a spectrophotometer, qualitative assessments can be made by visually comparing the samples with the controls. This should be sufficient for this competition study, where only a clear plus or minus is necessary to determine strain occupancy. To become familiar with ELISA reactions, inspect the simulated reactions in Figure 14.3.

If a color photo were used, all reactions would show up in shades of yellow. In this black and white representation, a dark well on plate A indicates the presence of TAL 379 and a dark well on plate B shows the presence of TAL 378. Note that the homologous controls are well defined (plate A, row E for TAL 379 and row F for TAL 378). The reactions from nodule samples show up as strongly as the controls. Note that positions A6, B2, and D5 show positive reactions on both plates, indicating simultaneous occupancy of both strains in one nodule.

Use the experience gained from the preceding example to interpret your results. If you use optical density (OD) readings (on a microtiter reader or a spectrophotometer), subtract the values obtained from heterologous controls from the sample readings. For

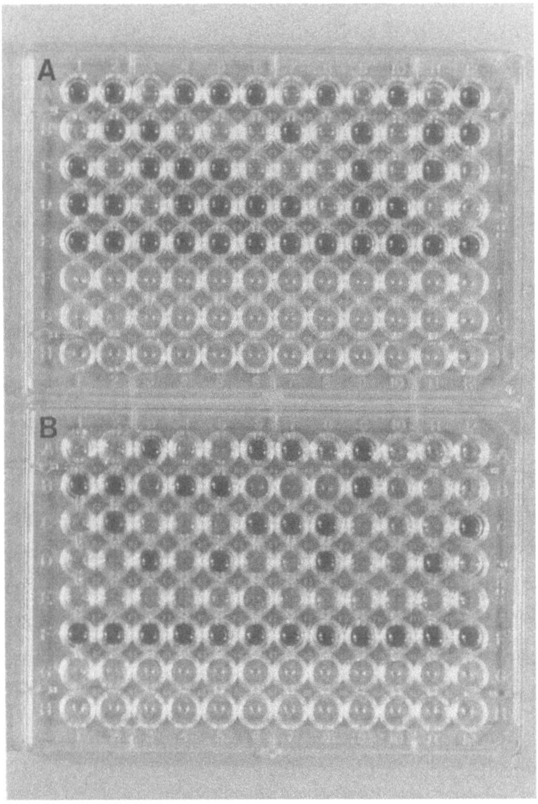

FIGURE 14.3 Example of ELISA reactions.

TABLE 14.1 ELISA Analysis of the Competition for Nodulation between Two Strains of *B. japonicum* Applied as Mixed-Strain Inocula

Mixed Inocula	No. of Nodules Examined	Nodule Occupancy (%)			Chi-square Deviation (1 df)
		TAL 378	TAL 379	TAL 378 + TAL 379	
A					
C					
E					
G					

visual assessment note that sample readings should show approximately as much or more color intensity as the homologous controls. There should be only little color development in the saline controls. Record results as shown in Table 14.1.

REQUIREMENTS

a. Obtaining the Antibodies

Antisera of TAL 378 and TAL 379
Goat anti-rabbit IgG alkaline phosphatase
Refrigerator

b. Preparing the Test Antigens

Nodules of each inoculant treatment (from Chapter 10)
Microtiter plates
Saline (85% NaCl)
Micropipettor, pipettor tips (Appendix 19)
Broth cultures of TAL 378 and TAL 379
Centrifuge tubes
Centrifuge
Steam chamber or pot with boiling water
Pasteur pipettes, 1-ml bulb
Coating buffer (Appendix 4)
Sterile toothpicks

c. Applying the Antigens

Multi-tip pipettor, pipetting tips
Polystyrene microtiter plates (Appendix 19)
Nodule samples from (b), cultured cell suspensions
Saline
Transparent tape
Refrigerator

d. Blocking the Binding Sites

Microtiter plates with samples from (c)
Squeeze bottle
PBS-Tween (PBST) (Appendix 4)
Micropipettor, tips
Incubator (37°C), moist chamber
PBS (Appendix 4)
Blocking solution (Appendix 4)

e. Reacting the Antigens with Primary Antibodies

Samples in microtiter plates from (d)
Antisera of TAL 378 and TAL 379
Micropipettor, tips
Shallow, wide container
Multi-tip micropipettor
Measuring cylinder, 50 ml
Antisera of TAL 378 and TAL 379
Squeeze bottle with PBST

f. Reacting Globulin Antigens with Secondary Antibody

Samples in microtiter plates from (e)
GAR-AP
Micropipettor, tips
Incubator (37°C), moist chamber
PBST in squeeze bottle
Measuring cylinder
Shallow container with large opening
Multi-tip micropipettor, tips

g. Developing the Color

Samples in microtiter plates from (f)
Enzyme substrate (Appendix 4)
Beaker, 1 liter
Micropipettor, tips
Measuring cylinders (i.e., 100 ml)
Distilled water
pH meter
Pipette, 1 ml
3 M NaOH solution
Incubator 37°C, moist chamber

h. Reading and Recording the Results

Samples in microtiter plates from (g)
Microtiter reader or spectrophotometer
Test tubes
Polyester sheets
Filter forceps

KEY REFERENCES

Kishinevsky, B., and M. Bar-Joseph. 1978. *Rhizobium* strain identification in *Arachis hypogaea* nodules by enzyme-linked immunosorbent assay (ELISA). Can. J. Microbiol. 24:1537–1543.

Kishinevsky, B., and D.G. Jones. 1987. Enzyme-linked immunosorbent assay (ELISA) for the detection and identification of *Rhizobium* strains. pp. 157–184. *In* G.H. Elkan (ed.) Symbiotic Nitrogen Fixation Technology. Marcel Dekker, Inc., New York.

Olsen, P.E., W.A. Rice, G.W. Stemke, and W.J. Page. 1983. Strain-specific serological techniques for the identification of *Rhizobium meliloti* in commercial alfalfa inoculants. Can. J. Microbiol. 29:225–230.

15

Identifying Rhizobia by Immunoblot

The immunoblot technique is an enzymatic immunoassay for the detection of antigens at picogram (10^{-12}) levels. Like the indirect enzyme-linked immunosorbent assay (ELISA), the method described here can be used with a commercially prepared secondary antibody-enzyme conjugate. In principle, the ELISA and the immunoblot assay are similar. In the immunoblot assay, the antigens are immobilized on a nitrocellulose membrane and reacted with strain-specific (primary) antibodies. After a washing step, the fixed primary antibodies bound to the membrane are reacted with alkaline phosphatase conjugate of antibodies specific to the primary antibodies. Reaction with an enzyme substrate causes appearance of purple dots on a white background, indicating the presence of the rhizobial strain assayed. Like the ELISA, but unlike most other serological methods, the immunoblot assay can be used to analyze a large number of samples simultaneously. This makes this method especially useful for establishing strain occupancy in root nodules. In this chapter, the use of the immunoblot assay is demonstrated in a strain competition experiment between two strains of *Bradyrhizobium japonicum*.

KEY STEPS/OBJECTIVES

1. Process nodule samples.
2. Prepare nitrocellulose sheets; apply samples.
3. Inactivate endogenous alkaline phosphatase; block non-specific binding sites.
4. React samples with the primary antibody.
5. React samples with the secondary antibody.
6. React samples with color developer; record results.

a. Obtaining Bacteroid Antigen from Nodules (Key Step 1)

Use the dried nodules of treatments A, C, E, and G of the strain competition experiment set up in Chapter 12. A total of 48 nodules will be analyzed. Take 12 nodules from each treatment and place into wells (one nodule per well) of a microtiter plate. Place the nodules of each treatment into one row. The bacteroid antigen of each nodule will be reacted with the antisera of TAL 379 and TAL 378. (The coding system on the microtiter plate facilitates precise identification of each nodule in the well by a combination of a letter and a number.)

Add three drops of distilled water to each nodule and cover the wells with transparent tape. Allow nodules to imbibe at 4°C for 2 h or overnight. Carefully remove the transparent tape. Inspect wells and add one drop of water to those wells in which nodules have absorbed all the water. Using flat toothpicks, squeeze out each nodule and remove the debris. Use a fresh toothpick for each nodule. With a Pasteur pipette, add more drops of water to each well as required. The liquid level of each well should be approximately 1 mm below the rim. The bacteroid antigen samples are now ready for application.

b. Preparing Nitrocellulose Membranes for Blotting (Key Step 2)

A nitrocellulose membrane (12 × 5 cm) is needed for blotting 48 test antigen samples and two rows of cultured cell control samples. Furthermore, the membrane should provide extra space for a safety margin and room for labeling. Do not touch the membranes with bare hands. Always use forceps.

Cut three pieces of nitrocellulose membrane measuring 12 × 5 cm. Prepare three membranes for this investigation. Label the membranes A, B, and C. In the later steps, membranes A and B will be treated with TAL 379 and TAL 378 antibodies, respectively. Membrane C serves as the control (secondary antibody).

Immerse the membranes in Tris-buffered saline (TBS). Hold each membrane at a 45° angle and slowly submerge to allow the membrane to take up water without developing dry spots. (This procedure should be followed in all immersion steps with membranes.) Soak the membranes for 5 min, then place them on filter paper to dry. Place the membranes on fresh, dry filter paper before sample application.

c. Applying the Antigen Samples (Key Step 2)

Inspect the nodule preparations from (a). Remove all residual nodule debris from the microtiter plates. The bacteroid antigens are in rows A, B, C, and D. As in Chapter 14, use row E for cultured cell controls of *B. japonicum* TAL 379 and row F for cultured cell controls of *B. japonicum* TAL 378 (Figure 15.1). Lay out the three prepared membranes on dry filter paper and apply the samples as described in the following paragraph.

Use a 96-prong replicator (Appendix 19). Dip the prongs into alcohol and flame to burn off any debris. Allow the prongs to cool for 1 min. Hold the applicator over the microtiter plate and match the prongs to the corresponding wells. Carefully insert the prongs until all are touching the bottom of the wells and are equally coated/wetted with the antigen samples. Allow the applicator to remain in the plate for 1 s. Carefully lift it and place it on the appropriate membrane for 1 s. Remove the applicator and allow

	1	2	3	4	5	6	7	8	9	10	11	12
A												
B												
C												
D												
E												
F												

Well A1 - A12, nodule bacteroid suspensions (from treatment A)
Well B1 - B12, nodule bacteroid suspensions (from treatment C)
Well C1 - C12, nodule bacteroid suspensions (from treatment E)
Well D1 - D12, nodule bacteroid suspensions (from treatment G)
Well E1 - E12, cultured cell controls of TAL 379
Well F1 - F12, cultured cell controls of TAL 378

FIGURE 15.1 Scheme of sample application for immunoblot.

the resulting wet blots to dry. Repeat this procedure for each one of the three membranes. When dry, label each membrane on the lower-right-hand corner using a ball point pen. A scheme of the sample antigen application is shown in Figure 15.1.

d. Planning the Assay

Perform all reactions at room temperature (25–30°C) unless otherwise indicated. In all reactions and washes, use sufficient solution to submerge the membranes completely. The minimum number of milliliters of solution required is approximately one-half of the area (square centimeters) of the membranes used. For the three membranes, (each 60 cm^2) a minimum of 90 ml of solution is required. It is, however, more practical to use 100 ml instead. Perform all incubation and washing steps under gentle and continuous agitation. For best results use an orbital or rocking platform shaker to maintain a uniform exposure of the membrane to the solution. Large Petri dishes, and small glass or plastic food containers are suitable containers for washing and incubation steps.

e. Inactivating Endogenous Alkaline Phosphatase (Key Step 3)

Root nodule bacteroid and cultural suspensions of rhizobia may contain endogenous alkaline phosphatase that could give false positives in the assay. The alkaline phosphatase is denature by acid treatment. Immerse the membranes in acidified TBS, pH 2.8 for 15 min to 1 h, depending on the amount of interference expected. Transfer membranes to Tris-buffered saline-Tween (TBST) (Appendix 4). Wash for 5 min, decant liquid and repeat washing step, and air dry.

f. Blocking of Nonspecific Binding Sites (Key Step 3)

In order to block nonspecific binding sites, immerse the membranes in the blocking solution. (Hold the membrane at a 45° angle when introducing the membrane into the blocking solution.) Incubate for 1 h under gentle agitation. Decant the blocking solution and add TBST to the membranes and wash for 10 min.

g. Reacting Antigens of Rhizobia with the Primary Antibody (Key Step 4)

The bacteroid and cultured cell antigens immobilized on the membrane will be reacted with the primary antibodies of *B. japonicum* strains TAL 379 and TAL 378. Obtain antisera of TAL 378 and TAL 379 previously set aside for this experiment (Chapter 14). As in the ELISA (Chapter 14), the amount of primary antiserum used depends on its quality. Assuming that both antisera have a titer of 1600, the recommended dilution is 1:4000.

Using a micropipette, transfer 25 μl of antiserum TAL 379 to 100 ml of TBST. Add membrane A to the diluted antiserum. Similarly, prepare diluted antiserum TAL 378 and add membrane B. Add the control membrane C to TBST, which does not contain antiserum. Incubate all three membranes for 1 h under gentle agitation. Remove the unbound primary antibody from membranes A and B by washing the two membranes for 5 min in TBST. Decant and repeat the wash with another portion of TBST for 5 min. Decant the TBST. Repeat both washing steps with the control membrane C separately.

h. Reacting Antigens with the Secondary Antibody (Key Step 5)

The secondary antibody solution is goat anti-rabbit alkaline phosphatase (GAR-AP) conjugate. Dilute GAR-AP in TTBS at the rate of 1:4000 by pipetting 25 μl to 100 ml of TBST. Submerge all three membranes in this solution. Incubate for 1 h under continuous agitation. Decant the conjugate solution and wash the membranes with TBST for 5 min with gentle agitation. Decant the TBST and repeat the wash step. Finally, wash the membranes in TBST for 5 min with gentle agitation to remove residual Tween 20 from the membrane surface.

i. Developing the Color (Key Step 6)

When the alkaline phosphatase reacts with 5-bromo-4-chloro-3-indolyl-phosphate (BCIP)/Nitro Blue Tetrazolium (NBT) substrate reagents, a purple color results. Positive reactions usually become visible as purple dots within several minutes. These dots become more intense with longer exposure to the developer.

Prepare the color-development solution just prior to use. Equilibrate 100 ml of carbonate buffer to room temperature. Using a glass pipette, transfer 1 ml of NBT stock and 1 ml of BCIP stock to the carbonate buffer and mix. Immerse the nitrocellulose membranes in this color-development solution. Remove all membranes as soon as the positive controls show dark purple dots. Stop the reaction by immersing the membranes in distilled water for 10 min under continuous agitation. Remove the residual color development solution B by decanting and wash in distilled water again for 10 min. A simulation of immunoblot reactions is shown in Figure 15.2.

j. Recording and Evaluating the Results (Key Step 6)

Place the membranes on filter paper and photograph them immediately. The purple color is enhanced while the membrane is still wet. Polaroid Type 108 and Polacolor 2

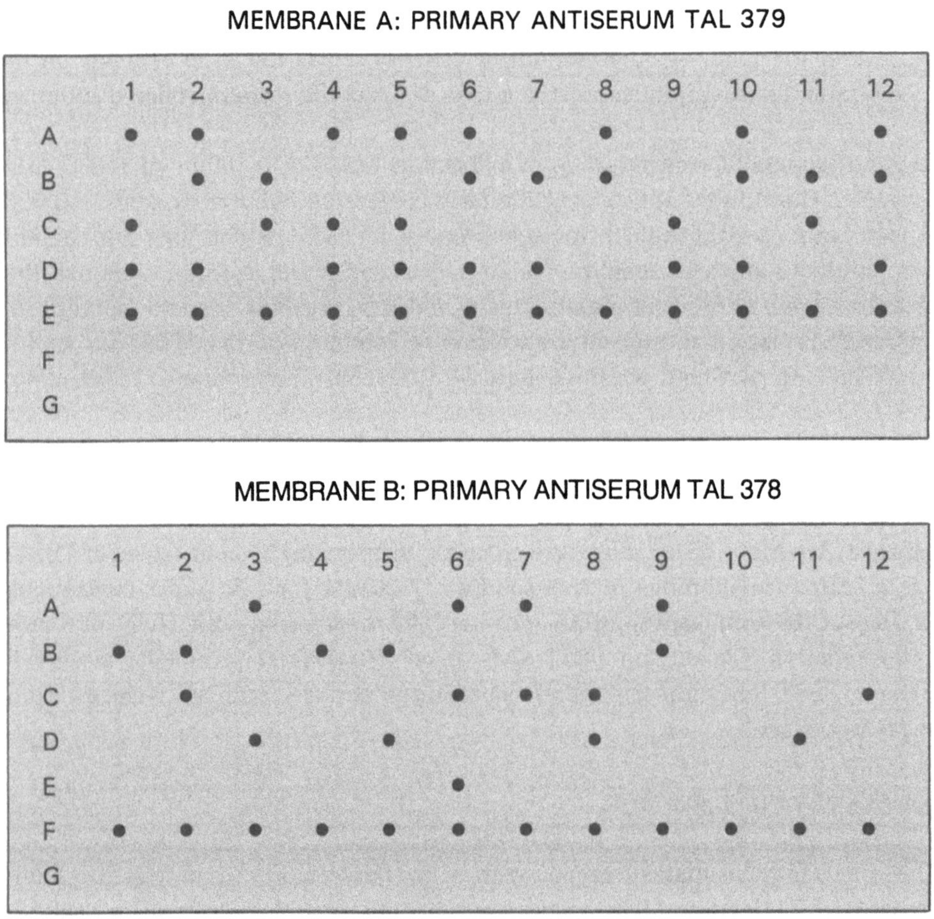

FIGURE 15.2 Example of immunoblot reactions.

Table 15.1 Immunoblot Analysis of the Competition for Nodulation between Two Strains of *B. japonicum* Applied as Mixed-Strain Inocula

Mixed Inocula	No. of Nodules Examined	Nodule Occupancy (%)			Chi-square Deviation (1 df)
		TAL 378	TAL 379	TAL 378 + TAL 379	
A					
C					
E					
G					

Land Film at f8 and 1-s exposure will yield acceptable photographs. Develop the film for 1 min. Allow the membranes to dry on the filter paper and store them between polyester sheets away from light to prevent fading. Alternatively, the membranes may be copied on a copy machine. Interpret the results and record them in Table 15.1.

It is important that the homologous controls show well-defined dots as they do in membrane A for TAL 379 (row E) and membrane B for TAL 378 (row F). Positives should show up as strongly as the controls. A dot on plate A means the presence of TAL 379 and a dot on plate B means the presence of TAL 378. Note that positions A6, B2, and D5 show positive reaction on membranes A and B. This indicates mixed infection or the occupancy of the nodule by TAL 379 and TAL 378. There should be no reaction on membrane C. However, if purple dots appear, it is assumed the indigenous alkaline phosphatase has not been completely inactivated by the acid treatment. Record the results in Table 15.1. Compare the results with those obtained by the other identification methods.

REQUIREMENTS

a. Obtaining Bacteroid Antigen from Nodules

Dried nodules of each inoculant treatment (Chapter 12)
Microtiter plates, U bottom (West Coast Scientific, Inc., Emeryville, CA)
Pasteur pipettes
Distilled water
Transparent tape
Toothpicks (flat type)
Filter forceps
Refrigerator
Forceps

b. Preparing Nitrocellulose Membranes for Blotting

Nitrocellulose membranes (BioRad Cat. no. 1620114; BioRad Laboratories, Richmond, CA)
Multiple inoculator, 96 prong (West Coast Scientific, Inc., Emeryville, CA)
Scissors
Forceps, flat-tipped (filter forceps)
Dish, small, rectangular, glass or plastic
TBS (Appendix 4)
Distilled water
Filter paper sheets, Whatman no. 1 or similar type
Filter forceps

c. Applying the Antigen Samples

Multiple inoculator
Large Petri dish filled with alcohol
Bunsen burner
Microtiter plate with samples from (b)
Filter forceps
Nitrocellulose membranes from (b)
Filter paper sheets, Whatman no. 1 or similar type

d. Planning the Assay

Incubation vessels of 10–15-cm length and width (sealable plastic food containers of the appropriate size)
Orbital or rocking platform shaker
Filter forceps

e. Inactivating Endogenous Alkaline Phosphatase

Nitrocellulose membranes with samples from (b)
TBST (Appendix 4)
Acidified pH 2.8, TBST (Appendix 4)
All requirements of (d)
Filter paper sheets
Filter forceps
Incubation vessels
Orbital or rocking platform shaker

f. Blocking the Nonspecific Binding Sites

Nitrocellulose membranes with samples from (e)
Blocking solution (Appendix 4)
Incubation vessels
Rotary or rocking platform shaker
TBST (Appendix 4)
Filter forceps

g. Reacting Antigens of Rhizobia with the Primary Antibody

Nitrocellulose membranes with samples from (f)
Primary antibodies of *B. japonicum* strain for TAL 378 and TAL 379
Micropipette and tips
TBST (Appendix 4)
Incubation vessels
Orbital or rocking platform shaker
Filter forceps

h. Reacting Antigens with the Secondary Antibody

Nitrocellulose membranes with samples from (h)
GAR-AP conjugate (Appendix 4)
Incubation vessels
Micropipette and tips
TBST and TBS (Appendix 4)
Filter forceps
Orbital or rocking platform shaker

i. Developing the Color

Nitrocellulose membranes with samples from (h)
NBT and BCIP stock solutions (Appendix 4)
Measuring cylinder
Carbonate buffer, pH 9.8 (Appendix 4)
Incubation vessels
Distilled water
Orbital or rocking platform shaker
Filter forceps

j. Recording and Evaluating the Results

Nitrocellulose membranes with samples from (i)
Filter paper sheets

Camera (Polaroid, if possible)
Polaroid Type 108 or Polacolor 2 Land Film (Polaroid Corp., Cambridge, MA)
Polyester sheets
Filter forceps

KEY REFERENCES

Ayanaba, A., K.D. Weiland, and R.M. Zablotowicz. 1986. Evaluation of diverse antisera, conjugates, and support media for detecting *Bradyrhizobium japonicum* by indirect enzyme-linked immunosorbent assay. Appl. Environ. Microbiol. 52:1132–1138.

Kang, U.G., P. Somasegaran, H.J. Hoben, and B.B. Bohlool. 1991. Symbiotic potential, competitiveness, and serological properties of *Bradyrhizobium japonicum* indigenous to Korean soils. Appl. Environ. Microbiol. 57:1038–1045.

Olsen, P.E., and W.A. Rice. 1989. *Rhizobium* strain identification and quantification in commercial inoculants by immunoblot analysis. Appl. Environ. Microbiol. 55:520–522.

16

Isolating Spontaneous Antibiotic-Resistant Mutants of Rhizobia

Rhizobia bearing genetic markers are obtained through a mass selection technique. Rhizobia, like other bacteria, contain small numbers of naturally occurring mutants that are resistant to high concentrations of certain antibiotics. This resistance may be used for recognizing rhizobial strains. In this chapter, two strains of *Bradyrhizobium japonicum* are used to produce spontaneous mutants resistant to streptomycin and spectinomycin. *B. japonicum* mutants, with simultaneous resistance to both antibiotics, are also produced.

KEY STEPS/OBJECTIVES

1. Culture rhizobia in yeast-mannitol broth (YMB).
2. Prepare yeast-mannitol agar (YMA) plates containing antibiotics.
3. Spread selected culture(s) onto appropriate antibiotic and nonantibiotic plates.
4. Check for natural resistance.
5. Transfer resistant colonies to YMA slants.
6. Culture resistant isolates in YMB.
7. Spread broth culture(s) resistant to streptomycin onto plates containing spectinomycin (and vice versa).
8. Transfer double-resistant mutants to YMA slants. Confirm resistance to streptomycin and spectinomycin; streak onto plates containing both antibiotics.
9. Confirm retention of symbiotic effectiveness of resistant strain.

a. Culturing Selected Strains (Key Step 1)

Select strains for the development of antibiotic-resistant mutants. Culture the strains in duplicate flasks containing 50 ml of YMB. Place on a shaker for 3–7 days, according to the growth rates of the strains chosen.

b. Preparing YMA Plates Containing Antibiotics (Key Step 2)

Prepare a stock solution of streptomycin (str) with a concentration of 4 mg ml^{-1} (Appendix 3). Filter sterilize 5 ml of the stock through a sterile Millipore filter of 0.20-μm pore size. Add the filtrate to 500 ml of YMA (in a 1-liter Erlenmeyer flask) kept molten in a water bath at 50°C. Mix well, but avoid vigorous shaking to minimize the formation of air bubbles. Return the flasks to the water bath for 10 min to reequilibrate the temperature and to allow the air bubbles to dissipate from the agar. Pour the plates. These plates will have str 40 μg ml^{-1} agar.

Similarly, prepare plates containing 250 μg ml^{-1} of spectinomycin (spc) from a stock solution containing 25 mg ml^{-1} (Appendix 3). Also prepare YMA plates containing a mixture of both the antibiotics in the previously mentioned concentrations for use in the selection of rhizobia with resistance to two antibiotics. Prepare an equal number of YMA plates without antibiotics. All the plates may be stored under refrigeration.

c. Selecting Spontaneous Mutants with Resistance to One Antibiotic (Key Step 3, 4, and 5)

Spread 0.1 ml of each broth culture on plates containing: (1) no antibiotics, (2) str (40 μg ml^{-1}), and (3) spc (250 μg ml^{-1}). The growth rates of rhizobial strains may be retarded in the presence of antibiotics. Prepare to incubate up to 12 days but check for emerging colonies every day after day 5. The plates of treatment 1, which contain no antibiotics, should have abundant rhizobial growth. The plates of treatment 2 and 3 should have very little growth compared with treatment 1. Not more than 30 resistant colonies are expected since the rate of mutation is 1 in 10^5 to 1 in 10^7 with most strains of rhizobia.

Pick four colonies each from treatments 2 and 3, and transfer to separate YMA slants (containing no antibiotics) in culture tubes. Incubate at 25–30°C for 5–9 days, then store at 4°C. These four isolates must be kept separate until the end of the selection process. Confirm the antibiotic resistance of str and spc isolates. Streak the mutants on YMA containing antibiotics and on a control plate containing plain YMA. Incubate at 25–30°C for 5–9 days.

d. Selecting Strains of Rhizobia with Resistance to Two Antibiotics (Key Steps 6, 7, and 8)

To develop strains resistant to both str and spc, spread 0.1 ml of broth culture of an spc mutant on a plate containing streptomycin (in a similar manner, an str mutant should be spread on YMA containing spectinomycin). Incubate at 25–30°C for 5–9 days. Check for growth of colonies on the plates containing antibiotics. Again, because of a similar mutation rate as with resistance to one antibiotic, no more than 30 colonies of spontaneous mutants with double resistance (str · spc) are expected.

Transfer four of these colonies to YMA slants (containing no antibiotics) in culture tubes, incubate, and store. Confirm resistance to both antibiotics by streaking on plates containing both streptomycin (40 μg ml^{-1}) and spectinomycin (250 μg ml^{-1}), and on control plates of plain YMA. Incubate at 25–30°C and compare growth on antibiotic and control plates.

e. Effectiveness Test on Resistant Strains (Key Step 9)

Str, spc, and str-spc-resistant strains usually retain their N_2-fixing capability. Mutant strains should be compared with their parent strain in a symbiotic effectiveness test as described in Chapter 20 prior to use in ecological experiments. To be useful, mutant strains should not show significant differences in N_2 fixation from the parent strain.

REQUIREMENTS

a. Culturing Selected Strains

Transfer hood, incubator, shaker
Bunsen burner
Inoculation loop
Two flasks, 150 ml, containing 30 ml of YMB each
Culture of rhizobia

b. Preparing YMA Plates Containing Antibiotics

Filled water bath adjusted to 50°C
Suction pump or aspirator with moisture trap
Two sterile filter-sterilizing units with sterile Millipore filter (0.20 μm) (Millipore Corp., Bedford, MA)
Pipettes, 10 ml (three)
Wash bottle with distilled water
Stock solution of str (4 mg ml^{-1})
Stock solution of spc (250 mg ml^{-1})
Sterile molten YMA, 3 liters, in three 2-liter or six 10-liter Erlenmeyer flasks
Petri dishes, sterile

c. Selecting Spontaneous Mutants with Resistance to One Antibiotic

Incubator, bunsen burner, transfer loop, small beaker of alcohol
YMA plates containing str (40 μg ml^{-1})
YMA plates containing spc (250 μg ml^{-1})
YMA plates
Spreading stick

Broth culture
YMA slants in culture tubes (six)
Graduated pipette, 1 ml

d. Selecting Strains of Rhizobia with Resistance to Two Antibiotics

Transfer or laminar flow hood, tools and incubator as in (c)
Antibiotic stock solutions and YMA plates as in (c)
YMA slants (six)
Mutant broth inoculum resistant to streptomycin
Mutant broth inoculum resistant to spectinomycin
Alcohol, spreading stick

e. Effectiveness Test on Resistant Strains

For requirements refer to Chapter 20

KEY REFERENCE

Schwinghamer, E.A., and W.F. Dudman. 1973. Evaluation of spectinomycin resistance as a marker for ecological studies with *Rhizobium* species. J. Appl. Bacteriol. 36:263–272.

17

Analyzing Nodule Occupancy Using Antibiotic-Resistant Markers

Antibiotic-resistant marked strains of rhizobia may be identified by their ability to grow on media containing antibiotics. The antibiotic marker technique is applied in ecological studies where strain identification is not possible by serology due to cross-reactions of the strains, or because of unavailability of antisera. Antibiotic markers also provide useful confirmatory data.

In this chapter, the method is demonstrated by typing nodules from a strain competition experiment. Competing strains are resistant to streptomycin (str) and spectinomycin (spc), respectively.

KEY STEPS/OBJECTIVES

1. Set aside inoculated soybean (*Glycine max*) plants (from Chapter 12).
2. Prepare antibiotic plates for nodule typing.
3. Harvest soybean plants; clean, trim, and sterilize roots; type nodules.
4. Read results.
5. Compare results to those obtained by the serological methods (Chapter 12, 13, and 16).

a. Obtaining Plants Inoculated with Antibiotic-Resistant Rhizobial Strains (Key Step 1)

Soybean plants that were set up in Chapter 12 will also be used in this experiment. They were inoculated separately with *Bradyrhizobium japonicum* TAL 379 str and *B. japonicum* TAL 378 spc, and a mixture of TAL 379 str and TAL 378 spc. Set aside the plants from replication IV for the work performed in this Chapter.

b. Preparing YMA-Containing Antibiotics for Nodule Typing (Key Step 2)

Prepare plates containing: (1) str (40 $\mu g\ ml^{-1}$ YMA), (2) spc (250 $\mu g\ ml^{-1}$ YMA), and (3) plain yeast-mannitol agar (YMA) as in Chapter 16. Draw a grid pattern on the bottom of each plate. Draw approximately 20 squares, each of which can be individually identified by a letter and a number (Figure 17.1). Squares with identical number and letter combinations on each of the three plates are meant to correspond to the same nodule.

c. Typing Nodules Using Antibiotic-Resistant Markers (Key Step 3)

Harvest one of each inoculation treatment from Leonard jars, saved from Chapter 12. Detach and surface sterilize nodules as in Chapter 1. Pick up one nodule with a pair of sterile, blunt-tipped forceps. While holding the nodule between the tips of the forceps, apply just enough pressure until the milky nodule content emerges. Spread this nodule material within its allotted square of the grid pattern on each plate.

Inoculate the plain YMA plate last to check for sufficiency of nodule inoculum. Process at least 20 nodules from each replication in this way. Flame forceps thoroughly between fresh nodules. Alternatively, sterile toothpicks or pins may be used to transfer bacteroids from the nodules to the plates. This method is especially useful for smaller nodules.

In some experiments in which it is necessary to identify not only which nodules on the plants were formed by the introduced, marked strain, but also the specific location and distribution of those nodules in the root system, use the following procedure. Trim off all nonnodulated roots with a pair of scissors and discard. Sterilize the trimmed nodulated part of the root and place onto sterile filter paper in a sterile Petri dish. Make a sketch of the nodulated root system on a record sheet and assign reference numbers

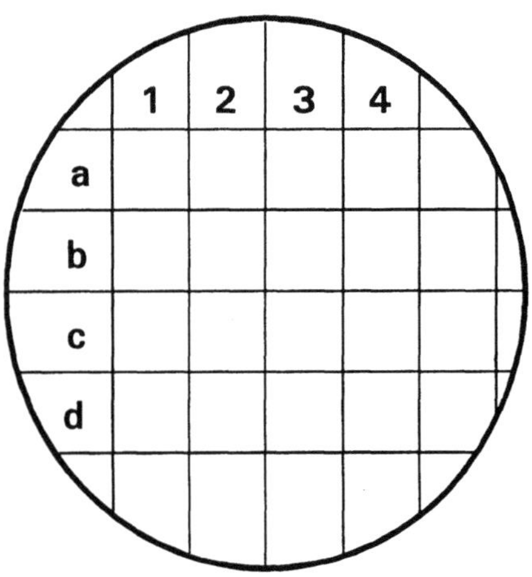

FIGURE 17.1 Plate with grid pattern for nodule identification.

to the nodules. Detach the nodules with sterile forceps, one at a time starting with the first nodule on the upper part of the root.

Incubate the plates at 25–30°C and make daily observations. Some contaminants (bacteria and fungi) may be resistant to the levels of spc and str used. Therefore, if the nodules have not been properly surface sterilized, these contaminants may appear on the plates earlier than the rhizobia.

d. Interpreting the Growth Patterns
(Key Steps 4 and 5)

Five to 8 days after inoculation, inspect the plates for signs of growth. Since corresponding squares on the three different plates have been inoculated with bacteroids from the same nodule, it should now be possible to determine which strain or strains occupied the nodule by the presence and absence of growth (Figure 17.2). Note in Figure 17.2 that since there was positive growth on 4c positions in YMA + str and YMA + spc plates, the nodule contained both TAL 378 spc and TAL 379 str. This is an example of a mixed infection. Tabulate results as shown in Table 17.1.

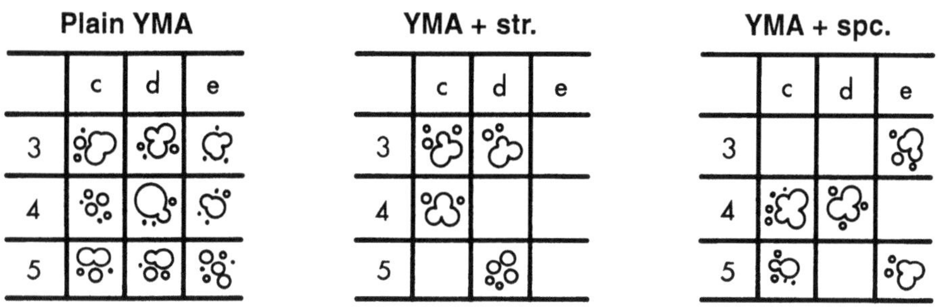

Treatment/Observation	Interpretation
3c, 3d, 5d	Streptomycin resistant strain
3e, 4d, 5c, 5e	Spectinomycin resistant strain
4c	Double or mixed infection
4e	Anomalous: re-streaked on antibiotic plates from plain medium

Figure 17.2 Interpreting growth patterns of antibiotic-resistant marked strains.

TABLE 17.1 Antibiotic-Resistant Marker Analysis of the Competition for Nodulation between Two Strains of *B. japonicum* Applied as Mixed-Strain Inocula[1]

Mixed Inocula	No. of Nodules Examined	Nodule Occupancy (%)			Chi-square Deviation (1 df)
		TAL 378 spc	TAL 379 str	TAL 378 spc + TAL 379 str	
A					
C					
E					
G					

[1]Antibiotic-resistant markers on TAL 379 and TAL 378 are str and spc, respectively.

REQUIREMENTS

a. Obtaining Plants Inoculated with Antibiotic-Resistant Rhizobial Strains

Nodules from Chapter 12 set aside for typing by antibiotic marker method.

b. Preparing YMA-Containing Antibiotics for Nodule Typing (Chapter 16)

Plates containing YMA + str (40 μg ml^{-1})
Plates containing YMA + spc (250 μg ml^{-1})
YMA plates
Felt pen with permanent ink, ruler

c. Typing Nodules Using Antibiotic-Resistant Markers

Incubator (25–30°C)
Requirements for sterilizing nodules (Appendix 10)
Soybean plants listed in (a)
Running water
Scissors, forceps (two)
Plates prepared in (b)
Sterile toothpicks or pins (optional)

d. Interpreting the Growth Patterns

Inoculated plates from (c)

KEY REFERENCES

Kuykendall, L.D., and D.F. Weber. 1978. Genetically marked *Rhizobium* identifiable as inoculum strain in nodules of soybean plants grown in fields populated with *Rhizobium japonicum*. Appl. Environ. Microbiol. 36:915–919.

Materon, L.A., and J.M. Vincent. 1980. Host specificity and interstrain competition with soybean rhizobia. Field Crops Res. 3:215–224.

18

Distinguishing between Strains of Rhizobia by Rhizobiophage Susceptibility

Most groups of microorganisms can serve as hosts for viruses. The soil environment abounds in viruses that prey on rhizobia (rhizobiophages). Because the interactions between rhizobiophages and their hosts are highly specific, they can be valuable tools in ecological research with rhizobia. Strains of rhizobia vary in their resistance to rhizobiophages. The different patterns of susceptibility that result from exposure of a rhizobial strain to a range of rhizobiophages can be used to differentiate between rhizobial populations. The use of phagetyping for strain identification may be possible but is not well documented. However, phagetyping has proven to be a valuable tool to distinguish between rhizobial strains of the same serogroup. In this chapter, rhizobiophages are isolated from soil and used to differentiate between several strains of *Bradyrhizobium japonicum*.

KEY STEPS/OBJECTIVES

1. Collect soil samples from various soybean (*Glycine max*) fields.
2. Inoculate yeast-mannitol broth (YMB) with *B. japonicum* strain TAL 379.
3. Inoculate broth cultures of TAL 379 with soil from soybean fields.
4. Filter broth cultures.
5. Enrich phage suspensions by filtration.
6. Assay filtrates for phage concentrations.
7. Inoculate YMB with rhizobial strains to be typed.
8. Spread plate rhizobial suspensions and spot phages.
9. Inspect plates for phage-forming units (PFUs) and tabulate results.
10. Phage type rhizobial strains.

a. Isolating Bacteriophages (Key Steps 1–5)

Collect soil samples from sites where soybeans are growing or have been grown. Obtain the soil from the rhizosphere of individual plants. Include some root material and, if possible, nodules. Collect samples from eight locations. Mix each soil thoroughly and store the samples at 4°C until use.

Four 150-ml flasks, each containing 50 ml of sterile mannitol nitrate broth (MNB) (Appendix 3), are required for each soil sample. Inoculate the flasks with *B. japonicum* strain of choice (e.g., TAL 379) in batches of eight, with a lag period of 1 day between each batch. This will provide cultures in the exponential growth phase when needed for subsequent phage inoculation from the eight soils. Incubate cultures on a rotary shaker at 25–30°C.

When the first batch of cultures has reached its exponential phase of growth (2–4 days after inoculation), add 1 g of soil to each flask of batch 1. Make sure that each flask is inoculated with soil from a different location. Incubate for 18–20 h at 25–30°C.

Remove cells and soil by centrifugation (10,000 × g for 15 min) and filter the supernatant through a sterilized membrane filter (0.20 μm). This filtrate contains the rhizobiophages that are small enough to pass through the filter. Add 10 ml of each filtrate to a fresh culture (second batch) of the same strain of rhizobia, incubate on a shaker for 18–20 h, and again centrifuge and filter. Repeat this procedure two more times, making certain that the filtrate is matched to the corresponding culture.

The turbidity in the bacterial cultures should diminish noticeably 8–10 h after the addition of the phage filtrate. The last filtrate is the phage suspension and should contain 10^6–10^9 phage particles. Dispense the filtrate into 20-ml tubes, add four drops of chloroform, and store in the refrigerator at 4°C.

b. Assaying for Phage Using the Overlay Method (Key Step 6)

Make tenfold serial dilutions of the phage filtrates in phosphate-buffered saline (PBS at pH 7.1, Appendix 4). Add 0.1 ml of each dilution to tubes containing 2.5 ml of molten mannitol nitrate agar (MNA) (Appendix 3) kept at 50°C in a water bath. Stir in 0.5 ml of a fresh culture of TAL 379 and immediately pour the yeast-mannitol agar (YMA) mixture over plates of MNA and distribute evenly. Prepare controls with only PBS and TAL 379 (no phage) added to agar and poured over YMA plates. Allow plates to stand for 10–15 min. Incubate inverted plates for 24–72 h and look for plaques (small clear zones). Count the plaques per plate. To calculate the number of Plaque Forming Units (PFUs) in 1 ml of the original filtrate, multiply the number of plaques per plate by 10 and by the dilution factor. If 20 plaques were counted on a plate containing a 10^{-5} dilution, the number of PFUs is $20 \times 10 \times 10^5 = 2 \times 10^7$ PFU/ml.

c. Characterizing Rhizobia Using Phages (Key Steps 7, 8, and 9)

Because of the specificity of bacteriophages for their bacterial hosts, each strain of bacterium exhibits a unique pattern of susceptibility against a large number of different bacteriophages. This unique pattern can be used to identify (phage type) the organisms of interest. However, it is possible that strains may share phage susceptibility patterns. When this occurs amongst strains of the same serogroup, other strain differentiation methods must be employed (Chapter 17).

Select several strains of *B. japonicum* and distinguish between them by their susceptibility to a range of phages. Incubate each strain in duplicate flasks of 50 ml of MNB at 25–30°C. Include strains of *B. japonicum* TAL 379 and TAL 378.

After 5–9 days incubation, spread 0.1 ml of each of the cultures to be typed over a separate MNA plate with a sterile glass spreader. Spot the surface of the bacterial lawn with a small loopful of each of the collected phage suspensions. The location of the spots should be marked on the back of the plates. Allow the plates to stand for 10–15 min, invert and incubate for 24–48 h.

Inspect the plates for a clear zone where each of the phages was spotted. Record presence (+) or absence (−) of plaques in a table similar to the example in Table 18.1. The susceptibility of a rhizobial strain to a range of phages can be regarded as its fingerprint, enabling it to be recognized in ecological investigations.

d. Phage-Typing Strains of Rhizobia (Key Step 10)

In order to establish a phage-typing scheme, a collection of bacteriophages with different degrees of specificity is needed. You can obtain well-characterized phage cultures from other investigators, or isolate and characterize a wide range of phage isolates from different soils and nodules as demonstrated in Table 18.1. To determine strain occupancy in a competition experiment as described in Chapter 12, test the susceptibility of a broth culture made from fresh nodule isolates and then compare the resulting pattern with the fingerprints obtained in the characterization scheme for phage typing (Table 18.1).

TABLE 18.1 An Example of Bacteriophage Susceptibility Patterns amongst Strains of *B. japonicum*

Phage	Strains of *B. japonicum*[1]						
Isolate	A	B	C	D	E	F	TAL 379
1	+	−	−	+	+	−	+
2	−	+	+	+	−	−	+
3	−	−	+	+	−	−	+
4	+	−	−	−	+	−	+
5	+	−	+	−	+	−	+
6	+	+	+	−	−	+	+
7	−	−	+	−	+	+	+
8	−	−	−	+	+	+	−

[1]A, B, C, D, E, and F are *B. japonicum* isolates from soybean nodules.

REQUIREMENTS

a. Isolating Bacteriophages

Refrigerator, rotary shaker, centrifuge balance
Desiccator, centrifuge tubes, 50 ml; rack, pipettes, 10 ml
Membrane filter units, sterile with Millipore filters of 0.20-μm pore size (Millipore Corp., Bedford, MA)
Soil samples (from four locations where inoculated soybeans are/were grown)
Digging tools, plastic bags
Erlenmeyer flasks, 150 ml, containing 50 ml of MNB each
Transfer loop, flame
Chloroform solution (1%)
Slant culture of *B. japonicum* (TAL 379)

b. Assaying for Phage Using the Overlay Method

Incubator, water bath, rotary shaker
PBS (pH 7.1)
Pipettes, 1 ml
Tubes containing 2.5 ml of liquid YMA (50°C)
Plates of MNA
Phage filtrates from (a)
Broth cultures of TAL 379

c. Characterizing Rhizobia Using Phages

Inoculation loop, flame
Pipettes, 1 ml
Spreaders
Erlenmeyer flasks, 125 ml, containing 50 ml of MNA
MNA plates
Slant cultures of a range of *B. japonicum* strains (including TAL 379)
Phage isolates from (a) as well as others, if available

d. Phage Typing Strains of Rhizobia

Materials as in (c)
Root nodules from Chapter 12

KEY REFERENCES

Billings, E. 1969. Isolation, growth and preservation of bacteriophages. pp. 315–329. *In* J.R. Norris and D.W. Ribbons (eds.) Methods in Microbiology, vol. 3B. Academic Press, New York.

Evans, J., Y.M. Barnet, and J.M. Vincent. 1979. Effect of bacteriophage on the colonization and nodulation of clover roots by paired strains of *Rhizobium trifolii*. Can. J. Microbiol. 27:974–978.

Staniewski, R. 1970. Typing of *Rhizobium* by phages. Can. J. Microbiol. 16:1003–1009.

Additional References and Recommended Reading

Berger, J.A., S.N. May, L.R. Berger, and B.B. Bohlool. 1979. Colorimetric enzyme-linked immunosorbent assay for the identification of strains of *Rhizobium* in culture and in the nodules of lentils. Appl. Environ. Microbiol. 37:742–746.

Brockwell, J., and W.F. Dudman. 1968. Ecological studies of root-nodule bacteria introduced into field environments. II. Initial competition between seed inocula in the nodulation of *Trifolium subterraneum* L. seedlings. Aust. J. Agric. Res. 19:749–757.

Brockwell, J., E.A. Schwinghamer, and R.A. Gault. 1977. Ecological studies of root-nodule bacteria introduced into field environments. V. A critical examination of antigenic and streptomycin-resistance markers for identification of strains of *Rhizobium trifolii*. Soil. Biol. Biochem. 9:19–24.

Bushby, H.V.A. 1982. Direct quantitative recovery of *Rhizobium* from soil and rhizospheres. pp 59–67. *In* J.M. Vincent (ed.) Nitrogen Fixation in Legumes. Academic Press, Sydney, Australia.

Franco, A.A., and J.M. Vincent. 1976. Competition amongst rhizobial strains for the colonization and nodulation of two tropical legumes. Plant Soil 45:27–48.

Goldman, M. 1968. Fluorescent antibody methods. Academic Press, New York.

Gollobin, G.S., and R.A. Levin. 1974. Streptomycin resistance in *Rhizobium japonicum*. Arch. Microbiol. 101:83–90.

Gornall, A.G., D.J. Bardawell, and H.M. David. 1949. Determination of serum protein by means of the biuret reaction. J. Biol. Chem. 177:751–766.

Graham, P.H. 1963. Antigenic affinities of the root nodule bacteria of legumes. Antonie van Leeuwenhoek J. Microbiol. Serol. 29:281–291.

Hagedorn, C. 1979. Relationship of antibiotic resistance to effectiveness in *Rhizobium trifolii* populations. Soil. Sci. Soc. Am. J. 43:921–925.

Hubbell, D.H. 1970. Studies on the root hair 'curling factor' of *Rhizobium*. Bot. Gaz. (Chicago) 113:337–343.

Johnston, A.W.B., and J.E. Beringer. 1976. Mixed inoculations with effective and ineffective strains of *Rhizobium leguminosarum*. J. Appl. Bacteriol 40:375–380.

Jones, D.G., and P.E. Russell. 1972. The application of immunofluorescence techniques to host plant/nodule bacteria selectivity experiments using *Trifolium repens*. Soil Biol. Biochem. 4:277–282.

Josey, D.P., J.L. Beynon, A.W.B. Johnston, and J.E. Beringer. 1979. Strain identification in *Rhizobium* using intrinsic antibiotic resistance. J. Appl. Bacteriol. 46:343–350.

Kawamura, A. 1969. Fluorescent Antibody Techniques and Their Applications. University of Tokyo Press, Tokyo; University Park Press, Baltimore, MD.

Lennette, E.H., E.H. Spaulding, and J.P. Truant. 1974. Manual of Clinical Microbiology. American Society for Microbiology, Washington, DC.

Materon, L.A., and C. Hagedorn. 1982. Nodulation of crimson clover by introduced rhizobia in Mississippi Soils. Soil Sci. Soc. Am. J. 46:553–556.

Obaton, M. 1973. The use of spontaneous antibiotic resistant mutants for the ecological study of *Rhizobium*. Bull. Ecol. Res. Comm. (Stockholm) 17:170–171.

Parker, C.A., and P.L. Grove. 1970. The rapid serological identification of rhizobia in small nodules. J. Appl. Bacteriol. 33:248–252.

Russell, P.E., and D. Gareth-Jones. 1975. Immunofluorescence studies of selection of strains of *R. trifolii* by S184 white clover (*T. repens* L.) Plant Soil 42:119–129.

Sadowsky, M.J., B.B. Bohlool, and H.H. Keyser. 1987. Serological relatedness of *Rhizobium fredii* to other rhizobia and the bradyrhizobia. Appl. Environ. Microbiol. 5:1785–1789.

Schwinghamer, E.A. 1964. Association between antibiotic resistance and ineffectiveness in mutant strains of *Rhizobium* spp. Can. J. Microbiol. 10:221–233.

Schwinghamer, E.A. 1967. Effectiveness of *Rhizobium* as modified by mutation for resistance to antibiotics. Antonie van Leeuwenhoek J. Microbiol. Serol. 33:121–136.

Skrdleta, V. 1969. Serological analysis of eleven strains of *Rhizobium japonicum*. Antonie van Leeuwenhoek 35:77–83.

Trinick, M.J. 1969. Identification of legume nodule bacteroids by the fluorescent antibody reaction. J. Appl. Bacteriol. 32:181–186.

Vincent, J.M. 1982. The basic serology of rhizobia. pp. 13–26. *In* J.M. Vincent (ed.) Nitrogen Fixation in Legumes. Academic Press, Sydney, Australia.

Wilson, M.H.M., B.A. Humphrey, and J.M. Vincent. 1975. Loss of agglutinating specificity in stock cultures of *Rhizobium meliloti*. Arch. Microbiol. 103:151–154.

Yao, P.Y., and J.M. Vincent. 1969. Host specificity in the root hair 'curling factor' of *Rhizobium* spp. Aust. J. Biol. Sci. 22:413–423.

Yao, P.Y., and J.M. Vincent. 1976. Factors responsible for the curling and branching of clover root hairs by *Rhizobium*. Plant Soil 45:1–16.

SECTION III
Evaluating Symbiotic Potential of Rhizobia

SIGNIFICANCE OF SYMBIOTIC NITROGEN FIXATION TO AGRICULTURE

The value of legumes in improving and sustaining soil fertility was well known to agriculturalists, but it was the work of Lawes and Gilbert in 1891 which showed that legumes had the inherent ability to add nitrogen to the soil. In 1888, Hellriegel and Wilfarth demonstrated that nitrogen gains in peas (*Pisum sativum*) took place only in the presence of soil microorganisms and that the root nodules of legumes were necessary to the process. Finally, in 1888, Beijernick isolated the N_2-fixing bacteria in the root nodules. The names *Rhizobium* and *Bradyrhizobium* are now given to these organisms.

The symbiotic association between legumes and rhizobia is by far the most important contributor to the world's supply of biologically fixed N_2 to agriculture. Effective symbiosis can only be achieved when the nodules are formed by efficient and effective rhizobia. Researchers now accept the term *symbiotic effectiveness* to describe the ability of a nodulated legume to fix N_2, and this can be expressed qualitatively (high, moderate or intermediate, ineffective) or quantitatively (e.g., total plant N, shoot or nodule dry weight). Quantitative symbiotic effectiveness is measured by comparison with the performance of standard rhizobial strains, with reference to the legume receiving adequate mineral N, or with noninoculated legumes. Efficiency is expressed as a qualified rate, such as milligrams of N_2 fixed per gram of nodule weight.

Effectively nodulated legumes can fix substantial amounts of N_2. Estimates of N_2 fixation have been made for various legume-rhizobial symbioses. Examples of the amounts of N_2 fixed (kilograms of N_2 fixed per hectare) by some legume-rhizobial symbioses are as follows: alfalfa (*Medicago sativa*), 125–335; red clover (*Trifolium pratense*), 85–190; pea (*Pisum sativum*), 80–150; soybean (*Glycine max*), 65–115; cowpea (*Vigna unguiculata*), 85; sweet clover (*Melilotus* sp.), 100–150; *faba bean* (*Vicia faba*), 240–325; peanuts (*Arachis hypogaea*) 50; mung bean (*Vigna radiata*), 55; and lentils (*Lens culinaris*), 100. The symbiotic N_2 fixation process and its potential for increasing protein production by legume inoculation is one of the means of providing improved nutrition in developing countries.

Pigeon pea or red gram (*Cajanus cajan*) is a major food (pulse) legume in India, while chickpea (*Cicer arietinum*) is the third most widely grown grain legume in the world and is very important in the semiarid tropics. The common bean (*Phaseolus vulgaris*) is an important source of dietary protein in many of the Latin American countries. Soybean has become a widely accepted crop in many parts of the world because of its food value, both for humans and animals. Besides their importance as a source of fodder and food, tree legumes play an important part in agroforestry and cropping systems. Species that are significant in agroforestry include *Leucaena leucocephala*, *L. diversifolia*, *Gliricidia sepium*, various species of *Acacia*, *Sesbania rostrata*, *S. grandiflora*, and numerous others.

THE NEED TO INOCULATE LEGUMES

The benefits of symbiotic N_2 fixation by legumes can only be appreciated when the major crop legumes in a country show responses to inoculation, as shown by yield increases in experimental field trials and in farmers' fields. Demonstrating the inoculation response is the necessary first step in the adopting symbiotic N_2 fixation technology for legume production, leading to the development of inoculant production capability. An economic analysis comparing costs and benefits of N fertilizer applications versus inoculation is also needed.

The inoculation practice is warranted whenever a new legume is being introduced, especially in areas where there are no indigenous species belonging to the same cross-inoculation group as the introduced legume. Inoculation is also recommended if a field has not been cropped to a legume in the past 3–4 years, or if temperature extremes or other soil conditions are likely to decrease rhizobial populations in the soil.

SOME FACTORS AFFECTING INOCULATION RESPONSE

Since N_2 fixation by legumes is a symbiotic process, environmental factors that affect the host legume and the rhizobia must be optimal for establishing an effective N_2-fixing symbiosis.

Soil pH is an important environmental factor. Many legumes respond to liming when grown in acid soils. Many legumes will grow and nodulate well at soil pH 5.6–6.8. For the rhizobia, optimal pH levels for growth in culture are variable (pH 5.8–7.2), depending on the rhizobial species. Generally, tropical rhizobia survive well in soils with pH 5.8–6.8.

Aluminum and Mn toxicities are likely to be encountered in many tropical soils with soil pH 4.5 or less. Both plant roots and nodulation are adversely affected. Shortage of P will severely limit the formation of nodules and N_2 fixation. Therefore, researchers selecting effective rhizobia in field soils must consider adequate P fertilization.

Soil N_2 (NO_3^-) has an inhibitory effect on the nodulation and N_2 fixation of the legume-rhizobial symbiosis. The size, effectiveness, and competitiveness of the native or indigenous rhizobial population are also important factors that influence the ability to achieve increased crop yield through inoculation.

Molybdenum is an essential micronutrient to all plants and is required for the formation and function of the nitrogenase enzyme complex. Soils deficient in Mo produce poor and ineffectively nodulated legumes.

SPECIFICITY IN THE SYMBIOSIS

The legume-rhizobial symbiosis exhibits widely differing degrees of specificity. In some instances, the symbiosis is highly specific in that a particular species or strain of *Rhizobium* or *Bradyrhizobium* can form an effective symbiotic association with only one particular legume species or variety. This category includes the temperate legumes *Trifolium, Cicer, Phaseolus,* and *Medicago,* and tropical species like *Glycine max, Leucaena,* and *Lotononis.* There are also intermediate cases that exhibit varying degrees of cross-inoculation capability, as in *Centrosema, Phaseolus acutifolius, P. lunatus,* some *Desmodium* spp., and *Acacia* spp. At the opposite extreme are the promiscuous associations, in which diverse legumes may be infected by one or more of several rhizobia. This condition is more prevalent in the tropical legumes than in the temperate species. Because the earlier studies of symbiotic N_2 fixation were initiated in temperate regions, the taxonomy of the genera *Rhizobium* and *Bradyrhizobium* was based on a host-dependent classification system that emphasizes temperate associations (see Section I). Several tropical rhizobia that form symbiotic associations with *Vigna, Macroptilium, Arachis, Cajanus, Lablab,* and other genera of legumes are simply labeled as the "cowpea miscellany" or *Bradyrhizobium* spp.

In some cases, it is desirable to select a strain for a wide range of hosts. An example would be the *Bradyrhizobium* sp. (CB756; TAL 309) isolated from the nodule of *Macrotyloma africanum.* This strain effectively nodulates approximately 40 of the promiscuous tropical legumes. This broad-spectrum-strain characteristic would be advantageous if this superior strain of *Bradyrhizobium* sp. were to be introduced to locations where those diverse legumes are to be grown. In a different situation, it might be advisable to work with a very specific symbiosis to ensure infection by a particular inoculant strain that competes with native soil rhizobia. Due to these and other considerations, characterizing rhizobial associations is of utmost importance when a legume cultivar is being developed through breeding or when a legume is being introduced into a new environment.

STRAIN SELECTION

After rhizobial strains have been isolated from nodules, they must be evaluated for their ability to form nodules and fix N_2 with targeted legumes. The source of rhizobial strains for a strain selection program can range from local isolates, to strains already tested in other parts of the region or country, to cultures from various overseas collections. Preliminary screening is performed in the greenhouse, where numerous strains can be tested on several host varieties. If the inoculated plants form nodules and produce healthy green leaves when grown in N-free media, it can be assumed that an effective symbiosis has been established. Rhizobia selected in greenhouse trials, where conditions are usually optimal, must then be evaluated in the field. Rhizobia that adapt to the agronomic conditions under which the host legumes will be cultivated and that enhance crop production through N_2 fixation can then be selected for inoculant production.

Field evaluation of effective rhizobia is critical because the symbiosis may be affected by many environmental factors discussed earlier in this section. The ability of an inoculant strain to persist in a particular environment, while in some cases competing against a resident soil population of rhizobia, is of critical importance. A combination of the these factors should be anticipated in the selection process to ensure good performance at different geographical locations. The task of introducing superior strains into soils that are already inhabited by effective rhizobia is difficult, and evaluation methods are an important key to success.

The standard approach for isolating rhizobia from nodules is to seek out legumes that appear successful in native pastures and/or stable natural ecosystems. Nodules collected and desiccated in glass vials are later used in isolating rhizobia in the laboratory.

The standard process of selecting a strain involves evaluating authenticated rhizobia in step-by-step screening experiments in controlled environments, greenhouses, and under field conditions. Considerable time is involved before a rhizobial strain finally gets an approval for use in inoculant production. It is not uncommon to encounter inoculation failure attributable to the selected strain in spite of all the careful methods of strain evaluation. Such failures may be avoided if the rhizobial strains made available for inoculant production are genetically compatible with the intended host legume. This is often not possible if the standard collection, isolation, and strain testing procedures are followed.

Recent approaches to isolating rhizobia and selecting a strain have resorted to screening site soil containing indigenous rhizobia for symbiotic potential as a preliminary step before a decision is made to isolate rhizobia from nodules. This approach has been tentatively described as the *whole-soil inoculation* technique, and has recently been successfully applied in demonstrating the need to inoculate *Trifolium subterraneum* and *Medicago sativa* under field conditions.

The whole-soil inoculation technique has also been adapted for strain selection work for soybean using soils from the center of origin of the soybean. The results of the research demonstrated clearly that soils containing indigenous populations of *B. japonicum* vary in their symbiotic potential. Further, it was shown that highly effective and competitive strains of *B. japonicum* (compared with widely used inoculant strains) can be found in the center of origin of the soybean. The conclusion from this work is that the symbiotic potential of indigenous rhizobia needs to be compared alongside recommended or imported exotic strains in screening experiments before embarking on inoculant production.

This approach (i.e., whole-soil inoculation) presents excellent potential for strain selection work for new legumes (e.g., tree legumes) for which effective strains are not yet available from various rhizobial germplasm resources. Another application of the technique is its ability to map out the symbiotic potentials of indigenous rhizobial populations of the sites under legume cultivation or projected for future planting. Such information can yield useful information and aid in isolating highly effective rhizobia that are available locally for inoculant production.

PARAMETERS FOR MEASURING SYMBIOTIC NITROGEN FIXATION

Final evaluation of the symbiosis will be based on several measurable parameters. Short-term trials with Leonard jars or sterile sand culture pots can provide an adequate basis for gross comparison of strains. The shoot dry weight of plants harvested at floral initiation or after significant plant biomass accumulation is the generally accepted criterion for N_2-fixing effectiveness, but nodule dry weight may also be employed. Nodule number is a less reliable indicator of strain effectiveness. The measurement of activity in the nodules by the N_2-fixing enzyme, nitrogenase, may also be done. This is accomplished by means of the acetylene reduction assay, which is a measure of ethylene production and indicates nitrogenase activity. However, the results of this assay should not be used to conclude on the actual amounts of N_2 fixed. This assay requires the availability of a gas chromatograph and other rather sophisticated equipment and materials. Total N accumulation in the shoot can be measured by the Kjeldahl method. Since total N content and nodule dry weight frequently correlate well with shoot dry weight, the latter parameter provides an acceptable basis for strain comparison.

In recent years, the ureide technique has been developed for measuring N_2 fixation. Ureides are a group of nitrogenous compounds including allantoin and allantoic acid. Some legumes produce large quantities of ureides when N is fixed symbiotically, but not when assimilated from soil mineral sources. Isotopic techniques using N^{15} are also used for measuring N_2 fixation, but the analysis is costly. The final proof of inoculation response must come from the field, when the seed and N yields at harvest are determined for grain legumes or from the dry matter production for forage legumes.

19

Testing For Genetic Compatibility between Rhizobia and Legumes

The occurrence of nodulation, and whether it is effective or ineffective, depends on the genetic compatibility between the rhizobia and the legume. In this chapter, symbiotic promiscuity and specificity for nodulation are demonstrated using inoculated legumes grown in Leonard jars and seedling-agar tubes. The disadvantages of the cross-inoculation concept in classifying rhizobia are demonstrated. However, the practical significance of the cross-inoculation concept in the correct practice of seed inoculation is highlighted.

KEY STEPS/OBJECTIVES

1. Culture rhizobia.
2. Prepare seedling-agar tubes and Leonard jars.
3. Prepare water-agar plates.
4. Select, surface sterilize, and germinate seeds.
5. Plant pregerminated seeds in seedling-agar tubes and Leonard jars.
6. Thin seedlings in Leonard jars.
7. Inoculate seedlings in Leonard jars and tubes.
8. Make periodic observations of nodulation.
9. Harvest after 5 weeks.
10. Evaluate results.

a. Culturing Rhizobia (Key Step 1)

Culture each of the *Rhizobium* spp. and *Bradyrhizobium* spp. listed in Table 19.1 in 100 ml of yeast-mannitol broth (YMB) in 250-ml Erlenmeyer flasks.

TABLE 19.1 Genetically Compatible Rhizobia-Legume Combinations that Result in Effective Nodulation

TAL No.	Rhizobial Species	Host Legume
169	*Bradyrhizobium* sp.	*Macroptilium atropurpureum* (siratro)
169	*B.* sp.	*Vigna unguiculata* (cowpea)
182	*R. leguminosarum* bv. phaseoli	*Phaseolus vulgaris* (bean)
379	*B. japonicum*	*Glycine max* (soybean)
380	*R. meliloti*	*Medicago* spp. (alfalfa and sweet clover)
382	*R. leguminosarum* bv. trifolii	*Trifolium* spp. (clover)
1145	*R.* sp. (*Leucaena*)	*Leucaena* sp. (leucaena)
620	*R.* sp. (*Cicer arietinum*)	*Cicer arietinum* (chickpea)
634	*R. leguminosarum* bv. viceae	*Lens culinaris* (lentil)

b. Preparing Seedling-Agar Tubes and Leonard Jars (Key Step 2)

Prepare 54 seedling-agar slants or NifTAL tubes (Appendix 7) using 30 × 250-mm tubes. The composition of the seedling-agar is detailed in Appendix 3 and its preparation in Appendix 7. A simple set-up for dispensing the melted agar into the tubes is illustrated in Appendix 7 (Figure A7.1). Set up 108 Leonard jars as explained in Appendix 11. Nitrogen-free nutrient solution for use in Leonard jars is of similar composition as that used for making seedling-agar.

Each treatment (rhizobial species–legume host combination and controls) in this exercise will be done in duplicate. Refer to Figure 19.1 for the treatments and the various combinations to test genetic compatibility between rhizobia and legumes.

c. Preparing Media for Germination (Key Step 3)

Make 450–500 ml of 0.75% (w/v) water-agar in a 1-liter flask and sterilize. Pour 25 ml of melted water-agar into 18 or more Petri dishes and allow to cool. Surface sterilized seeds will be pregerminated in these plates.

d. Surface Sterilizing Seeds (Key Step 4)

Check percentage germination of each legume species in advance of the experiment. Batches of seeds with more than 70% viability will be suitable. Select undamaged seeds for uniformity in size and color. Surface sterilize enough seeds (at least 200 of each species) to give at least 100 germinated seeds.

Surface sterilize the seeds (Appendix 10) by immersion in a 3% sodium hypochlorite solution for 3–5 min. [To prepare 3% sodium hypochlorite solution, add 10 parts of commercial bleach (5.25% sodium hypochlorite) to 7.5 parts of water.] Hard seed-coated

Legume / Rhizobia	Soybean	Cowpea	Bean	Lentil	Leucaena	Chickpea	Alfalfa	Clover	Siratro
B. japonicum TAL 379									
Bradyrhizobium sp. TAL 169									
R. leguminosarum bv. phaseoli TAL 182									
R.leguminosarum bv. viceae TAL 634									
Rhizobium sp. (leucaena) TAL 1145									
Rhizobium sp. (chickpea) TAL 620									
R. meliloti TAL 380									
R. leguminosarum bv. trifolii TAL 382									
Uninoculated									

FIGURE 19.1 Scheme for recording observations on symbiotic association between legumes and rhizobia.

species (e.g., leucaena and siratro) are scarified and sterilized simultaneously by immersion for 10 min in concentrated sulfuric acid. Drain off all excess acid prior to rinsing with sterile water. (If acid is used, the first rinse should be done quickly to prevent loss of viability of the seeds caused by the heat generated when water is added to the acid.)

Rinse seeds with six to eight changes of sterile water after surface sterilization. Allow the seeds to imbibe water by soaking for 1 h and then rinse twice. Transfer the seeds aseptically to agar plates with a spoon-shaped spatula.

Each batch of 100 seeds should be dispensed evenly in two or more water-agar plates (depending on the seed size) and incubated at 25–30°C. (The large-seeded species, e.g., *Phaseolus* and *Cicer* may need more water-agar plates.) Invert the plates containing the small-seeded species to provide straight radicles that are much easier to handle in later steps of the chapter.

e. Planting and Inoculating (Key Steps 5, 6, and 7)

Soybean, cowpea, bean, chickpea, lentil, and leucaena seeds will be planted in Leonard jars. Make three well-spaced holes in the rooting medium to a depth that will accommodate the pregerminated seeds 1 cm below the surface. Pick up well-germinated seeds with sterile forceps and place one seed in each hole with the radicle entering first. (Proper orientation of the radicle during planting is important to ensure proper emergence of the shoot and establishment of the seedling.) After placement of the seed, inoculate (1 ml per seed) with the rhizobial culture and cover the hole with the rooting medium. If vermiculite is used as the rooting medium, autoclaving will cause swelling and loosening of the vermiculite. This leads to poor anchorage of the root. Therefore, gentle compacting of the vermiculite will be required before planting/sowing of the seeds. Firmness of the rooting medium can be restored by pressing it down with the bottom (sterilized by flaming) of a 125-ml Erlenmeyer flask. Place the inoculated jars on benches in the green house.

After planting and inoculating are completed, add sterile gravel over the surface of the rooting medium. Set up 18 jars for each species. Siratro, clover, and alfalfa will be cultured on agar slants in tubes. Select and plant one seedling on the agar surface. Observe the usual aseptic precautions, taking care to sterilize the hands with 70% alcohol, flame sterilizing the inoculating loop and mouth of the tube when transferring the seedling. Using an inoculating loop, pick up the pregerminated seedlings and transfer them into the tubes. The seedling radicles should be 0.5–1.0 cm long and straight. After planting, tubes should be kept in a slant position for the radicles to adhere to the agar surface for at least 2 h. Dispense 1 ml of culture over the roots of the seedlings in the agar slants. Use a fresh pipette for each new rhizobial species or strain. Aluminum foil wrapped around the lower part of the tubes will shield the roots from light and heat. Seedling-agar tubes need to be placed in suitable wooden racks and kept in a growth chamber (environmental growth chamber or in a temperature-controlled greenhouse) for proper seedling development. Thin plants in the Leonard jars to two uniform plants per jar after 5 days. Excise the shoot of the unwanted plant aseptically using scissors.

f. Observing Periodically and Harvesting (Key Steps 8 and 9)

Examine the plants over a period of 5 weeks. Note color and growth. Replenish tubes and Leonard jars with sterile water as required. At the end of the fifth week, excise the tops and determine their dry weight (dry for 48 h at 70°C). Remove roots from the jars and tubes, and wash them free of rooting medium. Where nodules are present, describe nodule shape, size, pigmentation, and distribution.

g. Evaluating the Experiment (Key Step 10)

Note cross-inoculation groups as recorded in Figure 19.1 and the ineffectiveness and effectiveness of each rhizobial species-legume combination. Effectiveness will be apparent from the green coloration of the plant and the abundant nodules that are red/pink when sliced open. Record the results in Figure 19.1. Enter the observations as follows:

E = Plants with healthy green leaves, nodules pink/red when sliced open
I = Plants chlorotic and nodules with white interior
e = Plants chlorotic and not nodulated

REQUIREMENTS

a. Culturing Rhizobia

Transfer chamber
Slant cultures of rhizobia
YMB in flasks

b. Preparing Seedling-Agar Tubes and Leonard Jars

Seedling-agar slants (Appendix 3 and Appendix 7)
Leonard jars

c. Preparing Media for Germination

Agar powder; Petri dishes; flasks, 500 ml
Balance

d. Surface Sterilizing Seeds

Seeds of cowpea, bean, soybean, alfalfa, clover, leucaena, siratro, chickpea, and lentil

3% sodium hypochlorite solution or other sterilants (Appendix 10)
Concentrated sulfuric acid
Sterile water
Sterile flasks or beakers
Incubator

e. Planting and Inoculating

Pregerminated seeds of the various species
Leonard jars
Seedling-agar slants, wooden racks, growth chamber
Aluminum foil
Alcohol, forceps
Sterile pipettes, 1 ml; cultures of rhizobia
Alcohol burner

f. Observing Periodically and Harvesting

Sterile water
Scissors, paper bags or envelopes
Drying oven (70°C)
Scalpel or razor blades

g. Evaluating the Experiment

Scalpel or razor blades

KEY REFERENCES

Burton, J.C. 1952. Host specificity among certain plants in the cowpea cross-inoculation group. Soil Sci. Soc. Am. Proc. 16:356–358.

Fred, E.B., I.L. Baldwin, and E. McCoy. 1932. Root-nodule bacteria and leguminous plants. University of Wisconsin, Madison.

Turk, D., and H.K. Keyser. 1991. Rhizobia that nodulate tree legumes: Specificity of the host for nodulation and effectiveness. Can. J. Microbiol. 38:451–460.

20

Screening Rhizobia for Nitrogen-Fixation Potential

The N_2-fixation potential of several strains of pure cultures of *Bradyrhizobium japonicum* in symbiotic association with soybean (*Glycine max*) is compared. The linear relationship between shoot and nodule dry weights is determined by simple correlation analysis. The most effective strains from this experiment will be compared (in Chapter 21) in potted field soil.

KEY STEPS/OBJECTIVES

1. Prepare Leonard jars.
2. Culture rhizobia.
3. Prepare water-agar plates.
4. Sterilize and plate seeds for germination.
5. Plant and inoculate seedlings in Leonard jars.
6. Observe the progress of the experiment.
7. Harvest the plants in the experiment.
8. Analyze data.

a. Experimental Design and Treatments

The experiment is set up as a randomized complete block design with three blocks or replications (Figure 20.1). There are 14 inoculation treatments, a plus-N control with no inoculation, and a noninoculated control with no N. The plus-N control will contain 70 mg liter^{-1} of N applied as a 0.05% KNO_3 (w/v) solution. The N is added to the nutrient solution in the reservoir of the Leonard jar assembly.

BLOCK I				BLOCK II				BLOCK III			
1	14	7	6	13	5	14	2	14	8	3	9
13	2	8	15	15	16	4	12	7	2	4	10
5	9	3	12	3	10	1	11	15	16	1	13
10	4	11	16	9	8	7	6	11	12	5	6

FIGURE 20.1 An example of a randomized complete block design experiment with three blocks and 16 treatments. Each treatment is replicated once within each block. (A table of random numbers should be consulted for every experiment.)

b. Preparing Leonard Jars (Key Step 1)

A total of 48 Leonard jar assemblies will be required. Prepare the jars as explained in Appendix 11.

c. Culturing the Rhizobia (Key Step 2)

Each of the 14 strains of *B. japonicum* to be evaluated is cultured for 5–7 days prior to planting. Grow the rhizobia in 100-ml Erlenmeyer flasks containing 20 ml of yeast-mannitol broth (YMB). Incubate these at room temperature (25–30°C) on a rotary shaker for 5–7 days.

d. Surface Sterilizing the Seeds (Key Steps 3 and 4)

Check the germination (percentage viability) of the soybean seeds and surface sterilize a sufficient number of uniform, undamaged seeds to give about 200 germinated seeds. Sterilize by immersing seeds in 3% sodium hypochlorite solution for 3–5 min as described in Appendix 10. Germinate the seeds by plating on sterile water-agar [0.75% (w/v)] and incubate at room temperature (25–30°C) until the radicles are 0.5–1.0 cm long. Avoid overcrowding agar plates with the seeds. [Contact between seeds in an overcrowded plate increases the risk of cross-contamination from a partially sterilized seed to neighboring seeds. Uncrowded plates (approximately 25–30 seeds) produce more uniform and better germination due to better availability of moisture.]

e. Planting and Inoculating Seeds (Key Steps 5 and 6)

Follow the method for planting and inoculating the seeds described in Chapter 19. Plant three well-germinated seeds in each jar. Plant three jars per treatment. Label the jars and indicate block (replicate) assignment. Group the treatments according to block assignment and keep them separated.

Remove all Leonard jars of block I to the growth room (or glasshouse) bench. Randomize the placement of the jars within block I. Similarly, randomize the placement of the Leonard jars of block II and block III.

Keep in mind that growing conditions such as temperature and light intensity during this experiment must be in the range to which the species are adapted. Excessive temperatures are particularly damaging and can severely impair the infection process, nodule development, and nodule function.

Make daily observation of the experiment. Five to 10 days after planting, thin to two uniform plants per jar. Begin by thinning down the controls first. Excise the shoot of the unwanted plant with sterilized scissors. Plants may green-up gradually at the time that nodules begin to function, delivering fixed N_2 for plant metabolism. Plants inoculated with ineffective strains of rhizobia, and also the uninoculated controls, will remain yellow (chlorotic) and stunted.

f. Harvesting the Plants (Key Steps 7 and 8)

Harvest the plants after 30 days. To minimize errors during harvest, the stem should be cut at the point of cotyledon attachment. This point is marked by a scar on the stem. These scars are not visible in some species. The stem should then be cut at the level of the growth medium. Place the plant shoots in labeled paper bags. Dry to a constant weight at 70°C for 2 days. Each bag should contain the plant shoots from only one jar. (Paper envelopes may be substituted for smaller plants, e.g., *Centrosema, Trifolium, Desmodium*, etc.)

Roots and adhering rooting medium are dislodged into a coarse sieve. Wash the rooting medium from the roots using a gentle stream of water. Describe the nodule distribution mentioned in Appendix 1 (e.g., prolific taproot nodulation, occasional nodules on lateral roots and distant from the taproot, large numbers of small nodules, or small numbers of large nodules). Detach the nodules, count them, determine their total fresh weight, and place them in vials or aluminum foil weighing boats for drying. Dry the nodules to constant weight at 70°C for 2 days. (Nodule harvest from each Leonard jar must be treated individually as in the case of the shoots.) Do not pool nodules of the three replicates of any one treatment into a single vial.

Determine the dry weight of shoots and nodules for all treatments. Perform an analysis of variance on the dry weight data (shoots and nodules) using the method described in Appendix 17. Plot the mean shoot weight (y axis) against the mean nodule dry weight (x axis). Determine the correlation coefficient (r) of the plot and test the significance of r at the 5% and 1% levels of confidence. Draw the best regression line on your plot after determining the regression equation for the regression line.

Shoot weight and nodule weight are usually highly correlated; thus, shoot weight is used routinely as an indicator of relative strain effectiveness. Other parameters that are highly correlated with shoot weight are total N of shoot and nodule dry weight.

Nitrogenase activity (acetylene reduction) may not easily correlate unless done under very controlled conditions.

REQUIREMENTS

a. Experimental Design and Treatments

No special requirements

b. Preparing Leonard Jars

Leonard jars (48) (Appendix 11)

c. Culturing the Rhizobia

Transfer chamber
Agar-slant cultures of *B. japonicum*
YMB
Shaker

d. Surface Sterilizing the Seeds

Soybean seeds
Sodium hypochlorite solution (3%) or commercial bleach (Chlorox)
Water-agar plates
Incubator

e. Planting and Inoculating Seeds

Broth cultures of *B. japonicum* from (c)
Pregerminated seeds
Sterile pipettes, 1 ml, or Pasteur pipettes
Alcohol lamp and matches
Forceps, glass rods, alcohol spray bottle
0.05% KNO_3 (w/v) solution
Bench space in greenhouse

f. Harvesting the Plants

Scissors, paper envelopes or bags
Coarse sieve, vials or aluminum foil weighing boats

Drying oven (70°C)
Weighing balance

KEY REFERENCES

Date, R.A. 1975. Principles of *Rhizobium* strain selection. pp. 137–150. *In* P.S. Nutman (ed.) International Biological Programme, Vol. 7. Cambridge University Press, Cambridge, UK.

Gibson, A.H. 1987. Evaluation of nitrogen fixation by legumes in the greenhouse and growth chamber. pp. 321-363. *In* G.H. Elkan (ed.) Symbiotic Nitrogen Fixation Technology, Marcel Dekker, Inc., New York.

21

Screening Effective Strains of Rhizobia in Potted Field Soil

In the Leonard jar method of selecting rhizobia for effectiveness, the rhizobia interact with the legume roots growing in sterile sand or vermiculite irrigated with an N-free nutrient solution. In nature, the legume roots interact with inoculant rhizobia in the soil environment in the presence of other microorganisms, native/indigenous rhizobia, and soil N. Therefore, selected strains must be tested for effectiveness under soil conditions to weed out less effective ones. In this experiment, rhizobia previously screened in Leonard jars are evaluated further in potted field soil. The effectiveness of mixed- and single-strain inocula are compared. Infective, native rhizobial populations in field soil are determined.

KEY STEPS/OBJECTIVES

1. Culture rhizobia.
2. Collect the soil from the test field.
3. Prepare the soil; determine the pH and total N content.
4. Pot the soil.
5. Determine the water-holding ability (field capacity) of the soil.
6. Apply fertilizer.
7. Plant and inoculate surface sterilized seeds.
8. Thin seedlings to desired number.
9. Inspect for nodulation and perform most-probable-number (MPN) counts.
10. Water and observe the plants.
11. Harvest the plants; examine nodulation.
12. Analyze the data.

a. Designing the Experiment and Treatments

The experimental design is a randomized complete block design with three blocks, as in Chapter 20. There are 18 treatments: 15 inoculated (14 single-strain inoculations and one treatment receiving a mixed-broth inoculum comprising the three best strains selected in Leonard jars from Chapter 16); a plus-N control without inoculation; and two sets of noninoculated controls. At 2 weeks the extra set of the noninoculated controls is removed for inspection for nodulation by native rhizobia. If nodulation is observed in the noninoculated controls, initiate MPN counts (Chapter 6) of the native population using the soil set aside for this purpose. Pots are sown with eight seeds and four plants are maintained for the experiment upon thinning.

b. Preparing the Inoculum (Key Step 1)

All 14 cultures of *Bradyrhizobium japonicum* used in Chapter 20 are evaluated in the soil. Inoculate each strain into 70 ml of yeast-mannitol broth (YMB) contained in 125-ml Erlenmeyer flasks. Allow strains to grow for 5–7 days to reach maximum turbidity (approximately 1×10^9 cells ml^{-1}). To prepare the mixed inoculum, pipette 10 ml of the fully grown broth culture of each of the three best strains into a clean 125-ml Erlenmeyer flask. Use a fresh pipette for each strain. Mix the contents thoroughly by swirling.

c. Choosing the Site for Collecting Soil

The ideal site for soil collection is the one where the field experiment (which follows the pot experiment) is to be conducted. The site soil should be low in N. The native rhizobial population should be less than 10^3 rhizobia per gram of soil, have no previous history of inoculation and cultivation with the intended legume, and have no water-logging or salinity problems. In practice, these prerequisites may not be met in the chosen site. This, however, should not deter experimentation with a particular soil.

d. Collecting, Preparing, and Potting Field Soil (Key Steps 2, 3, and 4)

With a steel spade or other suitable implement, obtain field soil from a depth of 10–15 cm. Soil samples should be taken randomly within a soil type. Collect and transport the soil (approximately 150 kg) in strong plastic bags to a clean room. Spread large pieces of clean cardboard on the floor and cover with thick, clean, plastic sheets or tarpaulins. Empty the bags of soil onto the plastic to pool all the collected soil. Spread the soil and allow it to air dry. Mix the soil thoroughly and remove debris (e.g., stones, roots, leaves, etc.). Break lumps with a wooden mallet. Sift the soil using a 5-mm mesh screen. Take a sample to determine the soil pH using a pH meter. If the soil is acidic, add lime to bring the pH to 6.0–6.5. Mix the soil and lime thoroughly and allow to equilibrate for at least 7 days. During the equilibration period, cover the soil with a plastic sheet. Use one of the methods shown in Appendix 16 to calculate the amount of lime needed to adjust the pH level of the soil.

As an additional measure, soil N can be immobilized by incorporating finely milled sugarcane bagasse at the rate of 10 g kg^{-1} of soil. Other sources of C may also be suitable. The C source should be added before equilibrating the soil.

Obtain strong PVC (polyvinyl chloride) pots. Pots of 15–16-cm diameter, and 18-cm height with a capacity of just over 3 liters and with at least one hole on the bottom, are suitable for potting. Pots should be clean. Plastic bags of suitable size and thickness will be used as inner liners for the pots. Punch holes (1-cm diameter) in the bottom of the bags to allow for drainage. Position the bags in the pots and fold the open end of the bag over the rim of the pot.

Pots of the recommended size will hold approximately 2.4–2.7 kg of a soil high in organic matter. Tropical soils with less organic matter, but occupying a similar volume will be heavier. Weigh 2.4 kg of soil in each plastic bag and place in the pot. (Any coarse balance is suitable for weighing the soil, as high precision is not required.) Gently tamp the pots on the floor to compact the soil. Soil in all pots must be tamped down to occupy nearly the same volume to achieve similar bulk density. Set aside 250 g of soil in a refrigerator (4°C) for an MPN count of the native rhizobial population following the method described in Chapter 6.

e. Adjusting Moist Field Soil to Field Capacity (Key Step 5)

A soil moisture content at field capacity is suitable for most plants. Because the field capacity varies with different soils, determine the field capacity for the soil under investigation. At sowing and during initial phase of seed germination and seedling establishment, the soil moisture should be maintained at field capacity for better plant performance. Determine the field capacity of the moist field soil using the simple method described in Appendix 21. Record the volume of water needed to adjust the soil to field capacity. Use this data to bring the soil to field capacity after the seeds are sown.

f. Applying Fertilizer (Key Step 6)

The fertility of the soil must be adjusted to optimal levels to obtain good plant growth. The following fertilizer treatments are recommended. Rates per pot have been calculated on the basis of 2.4 kg of soil per pot.

Phosphorus, (P)

100 kg P ha^{-1}; applied as 500 kg ha^{-1} triple superphosphate (TSP[1]); 529 mg pot^{-1} (or 468 mg KH_2PO_4 pot^{-1}). Higher P levels may be needed for many tropical soils.

---

[1]TSP is approximately 11% Ca. If KH_2PO_4 is added, there is no Ca addition. If the soil is limed, it will remove Ca, otherwise it should be added as $CaSO_4 \cdot 2H_2O$ at the same rate as Mg.

Potassium (K)

200 kg K ha^{-1}; applied as 382 kg ha^{-1} KCl; 404.2 mg pot^{-1} (K_2SO_4 may also be used).

Magnesium (Mg)

5 kg Mg ha^{-1}; applied as 50 kg ha^{-1} $MgSO_4 \cdot 7H_2O$; 53.3 mg pot^{-1}.

Zinc (Zn)

10 kg Zn ha^{-1}; applied as 46.8 kg ha^{-1} $ZnSO_4 \cdot 7H_2O$; 49.5 mg $pot.^{-1}$

Molybdenum, (Mo)

1.0 kg Mo ha^{-1}; applied as 1.76 kg $(NH_4)_6 Mo_7O_{24} \cdot H_2O$ ha^{-1}; 1.95 mg $pot.^{-1}$

Nitrogen, (N) (for N-Control Pots)

100 kg N ha^{-1}; applied as 222 kg ha^{-1} urea, $CO(NH_2)_2$; 219 mg pot^{-1}. 25% of N is applied at planting and the remaining 75% at 3 weeks. (Ammonium nitrate may be substituted for urea for the N-controls. Also, the level of N may be increased considerably to realize the full yield potential of the legume. Consult with an agronomist for the N recommendations.)

Prepare the fertilizers (except the insoluble TSP) in the form of solutions and pipette them onto the soil surface and allow to dry. Add the TSP. Mix the soil in each pot thoroughly to ensure uniform distribution of the nutrients (mixing is easily achieved by removing the bag of soil from the pot and massaging it).

g. Planting and Inoculating the Seeds (Key Steps 7 and 8)

At the planting rate of eight seeds per pot, a total of 24 seeds are needed for each treatment in triplicate. A grand total of 408 seeds are needed for all 17 treatments. From a batch of seeds with good germination, select 500 seeds and surface sterilize, as in Appendix 10. Allow the sterilized seeds to imbibe water for 1 h. Give the seeds a final rinse and plant the seeds at a 2-cm depth. Inoculate each seed with 1 ml of the culture following the method described in Chapter 19. Label the treatments and assign block numbers.

Water the soil in the pots to field capacity using the data from (e). Add sterilized, coarse sand mulch to control contamination. Randomize the pots on the greenhouse bench. When plants are 5 days old, thin to four uniform plants per pot as described in Chapter 19.

h. Inspecting Noninoculated Control Plants for Nodulation by Native Rhizobia (Key Step 9)

When plants are 3 weeks old, remove the extra set of noninoculated controls to inspect for nodulation by native rhizobia. Carefully remove the plastic bag containing the plants from the pot and place it in a shallow basin. Slit the bag open and wash the roots with a gentle stream of water. Examine the roots for nodulation. Similarly, observe the remaining two pots set up for inspection. If nodules are present, make preparations for performing the MPN count of rhizobia in the soil set aside for this purpose in (d). The count may be done in growth pouches or Leonard jars following the method described in Chapter 6.

Weigh 100 g of the soil, dilute it in 900 ml of sterile water and prepare a fourfold dilution series ranging from 4^1–4^{10} dilution. Inoculate each dilution in quadruplicate. A fourfold series gives more precision than a tenfold series, especially for soils when populations are less than 1×10^3 rhizobia per gram of soil. Note that the starting sample has been diluted 1:10. More details on the method and calculations are given in Chapter 6.

i. Watering the Pots and Observing Periodically (Key Step 10)

During active growth and fixation, legumes will use a considerable amount of water each day. During this period, the pots need to be watered regularly. Water the pots more than once each day, if needed. Weigh sample pots showing vigorously growing plants to determine the volume of water needed to replace the water lost. If there are large differences in plant growth, pots should be watered to weight on a pot-by-pot basis. Measure out the required volume of water in a measuring cylinder and pour into the pot without excessively disturbing the soil. Keep plants well watered and make growth observations periodically as in Chapter 19.

j. Harvesting the Experimental Plants (Key Steps 11 and 12)

Harvest the plants at 35 days. Determine dry weight of shoots and nodules for all treatments. Analyze yield data as in Chapter 19.

REQUIREMENTS

a. Designing the Experiment and Treatments

No special requirements

b. Preparing the Inoculum

Transfer chamber
Agar slant cultures of rhizobia
YMB
Shaker
Erlenmeyer flasks
Pipettes

c. Choosing the Site for Collecting Soil

Soil analysis data

d. Collecting, Preparing, and Potting Field Soil

Steel spade
Strong plastic bags, plastic sheets, cardboard
Wooden mallet
Mesh screen, 1 cm
pH meter
Lime ($CaCO_3$)
PVC pots and plastic bags (inner liners for pots)
Balance for weighing potted soil

e. Adjusting Moist Field Soil to Field Capacity

Determine field capacity (Appendix 21)
Balance
Oven (110°C)
Soil from (d)
Hand shovel
Measuring cylinder, 250 ml, with hole in bottom (Appendix 21)
Water
Metal spatula
Metal weighing boots

f. Applying Fertilizer

Weighing balance
Spatulas
TSP, potassium chloride, zinc sulfate, ammonium molybdate, urea, lime, magnesium sulfate, potassium phosphate
Pipettes, 1 ml and 10 ml

g. Planting and Inoculating the Seeds

Seeds, sodium hypochlorite solution (3%), sterile water
Sterile empty beakers
Sterile pipettes, forceps, marker pens
Balance, water
Scissors, alcohol lamp, matches

h. Inspecting Noninoculated Control Plants for Nodulation by Native Rhizobia

Tap water, scissors, shallow basin

KEY REFERENCES

Gibson, A.H. 1987. Evaluation of nitrogen fixation by legumes in the greenhouse and growth chamber. pp. 321-363. *In* A.H. Elkan (ed.) Symbiotic Nitrogen Fixation Technology, Marcel Dekker, Inc., New York.

Singleton, P.W., H.M. Abdel-Magid, and J.W. Tavares. 1985. The effect of phosphorus on the effectiveness of strains of *Rhizobium japonicum*. Soil Sci. Soc. Am. J. 49:613–616.

Somasegaran, P., and B.B. Bohlool. 1990. Single-strain versus multistrain inoculation: Effect of soil mineral N availability on rhizobial strain effectiveness and competition for nodulation on chickpea, soybean, and dry bean. Appl. Environ. Microbiol. 56:3298–3303.

22

Verifying the Nitrogen-Fixing Potential of Glasshouse-Selected Soybean Rhizobia in the Field Environment

Soybean (*Glycine max*) rhizobia, previously selected in potted field soil, are evaluated in the field environment so as to further identify the most effective strains for inoculant production. The effectiveness of a multi-strain inoculant is compared with single-strain inoculants.

KEY STEPS/OBJECTIVES

1. Select rhizobial strains and prepare the inoculants.
2. Prepare the field and apply fertilizers.
3. Inoculate the seeds and plant.
4. Determine the number of rhizobia on the inoculated seeds.
5. Inspect the field and weed as necessary.
6. Harvest at 50% flowering (early harvest).
7. Harvest for grain yield (final harvest).
8. Analyze the data.

a. Setting Up the Experiment

Set up the experiment as a randomized complete block with four replications (Figure 22.1). Set up eight treatments: six inoculated (five single-strain and one multi-strain), a plus-N, and a noninoculated control without N.

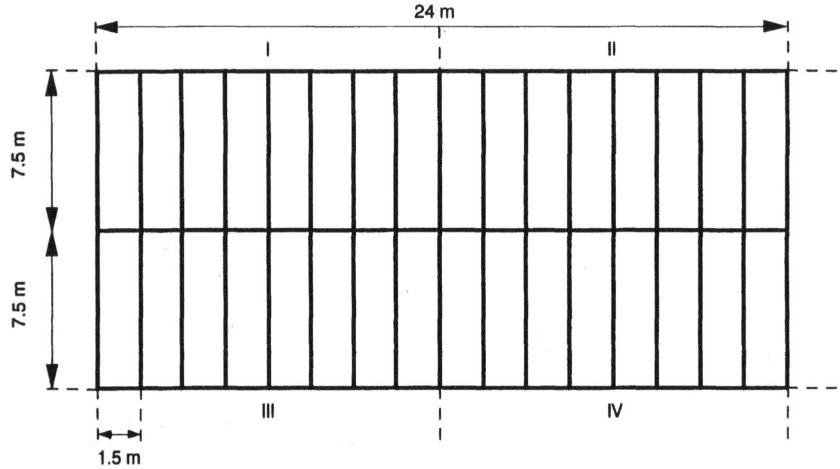

FIGURE 22.1 Field layout and dimensions. I, II, III, and IV denote replications. Each replication has eight treatment plots.

Field Dimensions

A field area of 360 m^2 (24 × 15 m) is required. Make rows 7.5 m in length and 0.5 m apart. Each treatment plot is flanked by an uninoculated guard (border) row along each side, with two center harvest rows (Figure 22.2). The area of each plot is 11.25 m^2 (0.001125 ha). The area harvested for grain yield is 3.75 m^2.

Choice of rhizobial strains

Use five of the best strains, according to their order of ranking in Chapter 21. From this group, select three serologically distinct strains for the preparation of the multi-strain inoculant for use in this experiment and later in Chapter 24.

b. Selecting Rhizobia for the Experiment (Key Step 1)

Serologically distinct and/or antibiotic resistant labeled rhizobia may be selected using methods described in Section II. If selection of serologically distinct strains is not possible (because of cross-reactions amongst the strains chosen), a multi-strain inoculant can still be prepared, but may not be suitable for studying aspects of strain ecology (competition and persistence) in the soil by serological methods. However, cross-reacting strains of rhizobia may still be used if their antisera are made strain specific by cross-absorption (Appendix 25). Antibiotic labeling offers an alternative to the use of serologically distinct strains. However, the antibiotic labeling method is suggested as more reliable only when each of the component strains in the multi-strain inoculant has double antibiotic resistance labels. Single-label strains may also be used, but with caution.

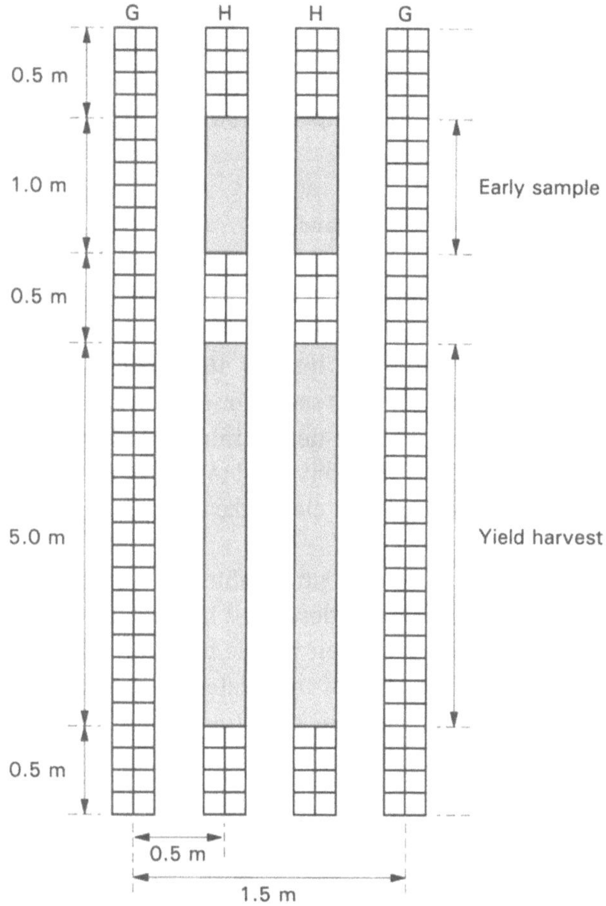

FIGURE 22.2 Diagram of a plot showing guard (G) and harvest (H) rows. Shaded boxes indicate harvest areas.

Since three strains are used in the multi-strain inoculant, the process of labeling (multi-labeling) and identifying the strains may become too involved, especially when more antibiotics are needed in developing the resistant strains. Moreover, the labeled strains need to be confirmed for retention of symbiotic effectiveness (Leonard jar screening, Chapter 20) when compared with the parent strains prior to their use in the inoculants.

c. Preparing Inoculants (Key Step 1)

In this experiment, inoculate the seeds (except controls) with peat cultures. Prepare the peat inoculants of the five strains following procedures described in Chapter 27 using gamma-irradiated or autoclaved peat.

Prepare the inoculants in advance of the experiment and allow them to mature for at least 2 weeks at 25–30°C. Determine and record the quality of each inoculant (number of viable rhizobia per gram of peat) by plate counts (Chapter 5 and 6). After the 2 weeks of curing, the inoculants may be stored up to 4 weeks in a refrigerator (4°C). Based on the quality check, mix appropriate weights of inoculants of the selected strains to obtain a mixed-strain inoculant containing a 1:1:1 ratio of the component strains. Prepare this mixture just before use.

d. Preparing the Seeds for Inoculating and Planting (Key Steps 3 and 4)

A planting distance of 3.7 cm between seeds is optimal for good soybean yields. Based on this planting distance, approximately 203 seeds are needed per 7.5-m row. Since there are two inoculated rows per plot and four replications, a total of 1624 seeds will be needed for each inoculated treatment. Count or weigh 2000 seeds for each treatment to make allowances for losses and for samples to be taken for determining the number of rhizobia per seed at planting. The seed numbers should be converted to weight measures for convenience. Weigh out the seeds for each treatment in clean plastic bags and label accordingly.

For soybean, 10 g of peat-based inoculant and 3.0 ml of gum arabic for 100 g of seed are recommended for experiments. Inoculate the seeds as described in Chapter 29. Inoculate the seeds just before planting. Keep the seeds in their plastic bags and in a cool place away from direct sunlight. Set aside 20 seeds of each inoculated treatment and, with minimum delay, determine the number of rhizobia per seed (inoculation rate) as described in Chapter 29.

e. Preparing the Field (Key Step 2)

Conduct the experiment in the field site from where soil was previously collected for Chapter 21. Drive posts into the soil at the four corners of the field to indicate the boundary of the experimental site. Clear and remove all surface vegetation and treat the field with an herbicide. Plow the field after sufficient time has been given for the herbicide to take effect in killing the weeds. Remove large rocks, plant roots, and other forms of debris. Till the soil to break up lumps and prepare a smooth, firm seedbed. Alternatively, the sowing may be done without plowing. This will minimize disturbance to the soil and release of soil N. Mark the plots and designate treatments for the different plots. (Treatment should be randomized in advance of planting and recorded.)

f. Controlling Cross-Contamination by Modifying Irrigation Methods (Key Step 2)

Rhizobia are soil bacteria and are easily spread when soilborne or in soil suspension. Surface overflow resulting from heavy rains and the flood-irrigation method may cause serious cross-contamination. In this particular exercise, where several different strains of rhizobia are tested, the methods of irrigating the field may be modified to control heavy cross-contamination.

Cross-contamination from rainwash may be controlled by the preparation of elevated seed beds (bunds). This method will result in the creation of shallow ditches between the seedbeds (rows). Alternately, an elevated plot with a surrounding ditch would be suitable for areas of heavy rains. Elevated plots may be preferred over elevated rows, as the latter are more susceptible to erosion. Rainwater can be efficiently drained away during heavy rains if the rows are prepared so as to follow the general inclination of the slope if the slope is not too great.

In locations of very low rainfall, where irrigation water is obtained from canals or rivers, flood irrigation is frequently practiced (Egypt and Sudan). In this situation, ditches between rows are preferred because they deliver water more efficiently to the roots of plants growing on elevated rows than to rows of plants on an elevated plot. However, if plots are not elevated, irrigation by flooding the entire surface of the plot may be done. This would require the construction of an elevated bund around each treatment plot to prevent water flow from one plot to neighboring plots. Water must be controlled to flow only from the mainstream into the plot. Backflow into the mainstream must be prevented. Successive irrigation by channeling water from one plot to neighboring plots must be prevented. Sprinkler and drip-irrigation methods may be used if these are available. If inoculated seed are handled carefully, and free peat is not windblown at planting, cross-contamination will be quite minimal unless there is massive soil erosion.

g. Applying the Fertilizer (Key Step 2)

Fertilize the field soil to optimize growth conditions. Follow levels of fertility as recommended for the potted soils (Chapter 21). Lime the soil to pH 6.0–6.5. The quantity of lime may vary from 500–10,000 kg ha^{-1} (depending on the soil and its initial pH) to bring about appreciable changes in the soil pH. Apply the lime 2 weeks prior to the application of the other fertilizers. Use the lime requirement data from Chapter 21.

To facilitate applying of the fertilizers, each of the four blocks is fertilized individually by broadcasting. The rates per block (90 m^2) are as follows: triple superphosphate (TSP), 4.5 kg; potassium chloride, 3.44 kg; zinc sulphate, 0.42 kg; ammonium molybdate, 0.016 kg; and magnesium sulfate, 0.45 kg.

Weigh out the fertilizer quantities in containers (plastic bags or buckets) of adequate size and apply by broadcasting. The smaller quantities, for example, zinc sulphate, ammonium molybdate, and magnesium sulfate, may be mixed with an inert carrier (e.g., sand) and broadcasted or sprayed on. Do not apply the urea with the other fertilizers because this is applied at planting only to the plus-N controls. Till in the fertilizers soon after application. The field is ready for planting 1 day after applying the fertilizers.

h. Planting the Experiment (Key Step 3)

Make furrows 7.5 m long, 0.5 m apart, and four per plot and 3.0–3.5 cm in depth. Make furrows for only a few plots at a time so that open furrows are not subjected to drying out from prolonged exposure in the sun. A straight 2-m-long wooden stick with 3.7-cm graduations, placed alongside the furrow, is a useful guide for even placement of seeds. Plant the controls and guard rows first and cover the seeds on completing each row.

Seeds inoculated with one inoculant should be sown in all blocks at the one time. However, if weather is extreme (very warm or impending rain) at planting, planting by block will reduce variation and bias compared with planting by treatment.

Prevent contamination of the seeds by sterilizing your hands when handling each batch of seeds inoculated with a different strain. Hands are easily sterilized by thorough washing with soap and water, followed by swabbing with alcohol after the hands are dry.

Apply urea, only to the plus-N controls, at the rate of 0.23 kg urea per plot with 25% (58 g per plot) at planting and the remaining 75% (174 g per plot) at 4 weeks. Weigh out 58 g each of urea in four bags, one for each of the four replicates. A plus-N control with higher level(s) of N may be instituted to demonstrate the yield potential of the legume. Consult an agronomist for the N recommendation.

Make a furrow 4–5 cm deep, parallel to and 4–5 cm away from the planted row. Evenly distribute the urea with your hands. Cover the furrows immediately after application. Exposing of the urea will result in hydrolysis and loss of N (as ammonia) to the atmosphere.

i. Monitoring the Trial and Harvest (Key Steps 5, 6, and 7)

Inspect the field frequently for plant damage by disease and insect pests. Take appropriate measures to control these pests. Weed the plots whenever necessary. Appropriate measures should be taken to prevent cross-contamination during weeding and other operations. Make frequent observations of plant growth and color. Note the treatments with early signs of N fixation. Record the time taken for 50% of plant population to initiate flowering. Make an early harvest at this time.

The area of the plots for early harvest and harvest for grain yield are indicated in Figure 22.2. Harvest plants for dry matter yield. Observe nodule size, color, and distribution on the root. Obtain the fresh and dry weight of nodules. Check on any cross-contamination by comparison with controls and, if possible, by testing nodule occupancy.

If facilities are available, perform the acetylene reduction assay to determine nitrogenase activity as described in Appendix 15. Also, analyze the dry matter of the plant tops (shoot) for total N by the Kjeldahl procedure. Record the time for the plants to reach maturity. Process the plants for determining grain yield (dried to 5–6% storage moisture). Express grain yield on a kilogram-per-hectare basis.

j. Analyzing the Data (Key Step 8)

Analyze the data from the early harvest for correlation (Appendix 18): tops versus nodule weight, tops versus nodule numbers, tops versus nitrogenase activity (if available), and nodule weight versus nitrogenase activity. In addition, perform a correlation analysis to correlate total N with all the parameters measured.

1. Rank the strains according to N_2-fixing potential, and compare your data with that from the Leonard jars and potted soil experiments.
2. Compare the performance of the multi-strain inoculant with single-strain inoculants. What could be the advantage of a multi-strain inoculant?
3. Rank the data obtained for grain yield. Does ranking of strains according to dry matter production at early harvest and at grain yield agree?

REQUIREMENTS

a. Setting Up the Experiment

Measuring tape, 50 m
Field site, 24 × 15 m
Five best rhizobial strains from Chapter 21

b. Selecting Rhizobia for the Experiment

Serologically distinct or antibiotic-labeled strains of rhizobia

c. Preparing Inoculants

Agar slant cultures from (a)
Erlenmeyer flasks, 250 ml (five), each containing 100 ml of yeast-mannitol-broth
Sterile plastic syringes, 50 ml (six); sterile needles, 3/4 in, 18 gauge (six)
Bags of peat, 50 g per bag (six), autoclaved or irradiated
Sterile pipettes, 10 ml
Incubator
Quality check of inoculants (materials as in Chapter 26)

d. Preparing the Seeds for Inoculating and Planting

Soybean seeds, balance, plastic bags
Peat inoculants; gum arabic solution; pipette with wide-bore tip (for pipetting gum arabic solution), 10 ml
Samples of inoculated seed

e. Preparing the Field

Field area
Wooden posts for marking field perimeter (four)

Herbicide(s) and spraying equipment
Plowing and tilling machinery
Other field preparation accessories

f. Controlling Cross-Contamination by Modifying Irrigation Methods

Suitable field design to control cross-contamination

g. Applying the Fertilizer

Magnesium sulfate: 0.45 kg × 4 blocks = 1.8 kg
TSP: 4.5 kg × 4 blocks = 18 kg
Potassium chloride: 3.44 kg × 4 blocks = 13.76 kg
Zinc sulfate: 0.42 kg × 4 blocks = 1.68 kg
Ammonium molybdate: 0.016 kg × 4 blocks = 0.064 kg
Balance, plastic bags or plastic buckets
Tiller or hoes

h. Planting the Experiment

Inoculated and noninoculated soybean seeds from (d)
Irrigation water
Metric tape, hoes or suitable equipment for making furrows
Planting guide for even placement of seeds
Soap, water, clean rags, alcohol in a spray bottle
Urea for N controls
Covered container to keep seeds

i. Monitoring the Trial and Harvest

Insecticides and spraying equipment
Weeding tools, hoes
Scissors/snips, paper bags, aluminum weighing boats
Coarse sieve
Drying oven (70°C)
Balance

j. Analyzing the Data

Calculators and statistical tables
Statistical assistance

KEY REFERENCES

Abaidoo, R.C., T. George, B.B. Bohlool, and P.W. Singleton. 1989. Influence of elevation and applied nitrogen on rhizosphere colonization and competition for nodule occupancy by different rhizobial strains on field-grown soybean and common-bean. Can. J. Microbiol. 36:92–96.

Brockwell, J., A. Diatloff, R.J. Roughley, and R.A. Date. 1982. Selection of rhizobia for inoculants. pp. 173–191. *In* J.M. Vincent (ed.) Nitrogen Fixation in Legumes. Academic Press, Sydney, Australia.

23

Evaluating the Symbiotic Potential of Indigenous Rhizobial Populations of Soils Using the Whole-Soil Inocula Technique

To estimate the N_2-fixing capacity of a population of rhizobia in a specified soil, the usual approach is to make several rhizobial isolates from nodules, test them for effectiveness on a given legume, and then integrate the various measurements to obtain a single estimate for the whole population. An alternative estimate of the symbiotic potential of a rhizobial population can be made by using the soils as inocula. This experiment is designed to quantify the effectiveness of the indigenous soil rhizobial populations infective on a specific legume species or on several species in a cross-inoculation group.

In this chapter, four legumes peanut (*Arachis hypogaea*), siratro (*Macroptilium atropurpureum*), lima bean (*Phaseolus lunatus*), and cowpea (*Vigna unguiculata*) of the cowpea cross-inoculation group are used to evaluate the symbiotic potential of bradyrhizobial populations indigenous to three soils. Even though these four species of legumes are classified in the same cross-inoculation group, they vary significantly in their rhizobial requirements. Seeds of the test legume sown in Leonard jars are inoculated with diluted soil suspensions (whole-soil inocula) containing the population of rhizobia. The level of effectiveness of the indigenous soil population is compared with that of known or recommended strains of bradyrhizobia. Three soils are selected for the assay. The sizes of the indigenous populations are determined by the plant infection method and isolates are made from nodules to characterize the indigenous soil populations for their diversity.

KEY STEPS/OBJECTIVES

1. Collect soil samples.
2. Determine moisture content of soil samples.
3. Pregerminate seeds for plant infection counts.

4. Set up plastic pouches for plant infection counts.
5. Isolate rhizobia from nodules.
6. Prepare Leonard jars for the assay.
7. Pregerminate seeds for Leonard jar planting.
8. Prepare whole-soil inocula.
9. Plant seeds in Leonard jars and inoculate.
10. Harvest.
11. Analyze data.
12. Screen isolates.
13. Harvest and analyze screening experiment.

a. Collecting Soil Samples for the Assay (Key Steps 1 and 2)

Select three soils based on land usage and management (e.g., pasture soil, farm soil, forest soil, etc.) or belonging to different soil orders (e.g., Mollisol, Oxisol, Ultisol, etc.). The presence of nodulated legumes in these soils would show the presence of rhizobia, but provides no information as to specificity. Soils chosen for the bioassay should not have been previously used for inoculation studies.

Obtain soil samples with a 25-mm soil core sampler from the top 25 cm of the soil profile. At least 25–30 soil core subsamples from each site should be taken and pooled in a plastic bag. Break up large clumps of soil, and remove stones and other debris by sieving. Thoroughly wash the soil core sampler, then dry. Sterilize by spraying with alcohol, then burn off the alcohol. The soil core sampler needs to be sterilized before samples are taken from a different soil. To determine the moisture contents in the pooled samples, remove 20 g from each soil and oven dry at 110°C for 24 h.

b. Using the Plant Infection Method to Determine the Size of the Rhizobial Population (Key Steps 3 and 4)

A promiscuous legume such as *M. atropurpureum* may be acceptable for estimating the total number of infective indigenous bradyrhizobia. However, plant infection determinations should be performed using each of the intended host legumes that will be tested in the bioassay rather than employing a promiscuous substitute such as *M. atropurpureum*.

Determine the size of the bradyrhizobial population by the plant infection technique (Chapter 6) using all four legume species (*A. hypogaea*, *P. lunatus*, *V. unguiculata*, and *M. atropurpureum*) for each soil. An indication of the effectiveness of rhizobia with these hosts may also be secured. Store the rest of the soil in a refrigerator (not more than 3 weeks). The stored soils will be used later in the experiment to prepare whole-soil inocula. At the end of 3 weeks terminate the test and calculate the most probable number (MPN) of the indigenous bradyrhizobia.

c. Isolating Rhizobia from Nodules of Plants Used in the Plant Infection Counts (Key Step 5)

Isolate rhizobia (see Chapter 1) from the nodules of plants in the treatments inoculated with 10^{-1} diluted soil suspension and also at the highest dilution at which nodulation occurs. (The highest dilution at which nodulation occurs may differ with each of the four species investigated and also with the soils in question.) Isolate from at least 12 nodules of 10^{-1} dilution and 12 nodules of the highest dilution. The number of nodules sampled for isolation of the rhizobia will depend on the number of replications used in the plant infection counts. For each legume species a total of 72 nodules (2 dilutions $\times$ 12 nodules $\times$ 3 soils) will be sampled for isolation. All isolates need to be checked for purity by Gram staining and authenticated (Chapter 1) before isolates are used in characterization studies.

d. Preparing Leonard Jars (Key Step 6)

Prepare 128 Leonard jars (Appendix 11).

e. Pregerminating Seeds for Leonard Jar Planting (Key Step 7)

Obtain high viability seeds of *A. hypogaea*, *P. lunatus*, *V. unguiculata*, and *M. atropurpureum*. Surface sterilize and pregerminate 50–60 seeds of each species. Since the seeds of the different species will germinate at different times, it is essential that all the seeds are ready for planting at the same time. Therefore, before pregerminating large batches of seeds for the experiment, pretest at least 10 seeds of each species to determine germination times. Use the observed times for germination to stagger the pregermination so that seeds of all species are ready for planting at the same time.

f. Preparing Whole-Soil Inocula (Key Step 8)

Whole-soil inocula are the soil-water suspensions of the intended test soils containing the indigenous rhizobia. Prepare whole-soil inocula of 10^{-1} dilution by suspending 10 g of soil sample in 90 ml of sterile water in a 160-ml dilution bottle. Shake the mixture for 15–20 min with a wrist-action shaker. Transfer 10 ml of the 10^{-1} diluted soil suspensions into 90 ml of sterile water to obtain whole-soil inocula at 10^{-2} dilution. Mix thoroughly. Finally, obtain 10^{-3} diluted whole-soil inocula by transferring 10 ml of 10^{-2} diluted whole-soil inocula into 90 ml of sterile water. Mix thoroughly. Refrigerate the whole-soil inocula until needed.

g. Setting Up and Monitoring the Assay (Key Steps 9 and 10)

The effectiveness of indigenous soil rhizobia on each legume species should be compared against rhizobial strains or inocula of known effectiveness. The following bradyrhizobia of known effectiveness are suitable controls: TAL 1371 (Nitragin 8A11) for *A. hypogaea*, TAL 22 for *P. lunatus*, and TAL 169 (Nitragin 176A22) for *V. unguiculata* and *M. atropurpureum*. Other strains of known effectiveness may be substituted.

Prepare single-strain peat inoculants of TAL 22, TAL 169, and TAL 1371 as described in Chapter 27). Perform viable counts on these inoculants as described in Chapter 5. In place of peat inoculants, yeast-mannitol broth (YMB) cultures of the strains are also suitable. The advantage of using peat inoculants is that peat inoculants of known strains can be prepared ahead of time and stored in a refrigerator without loss of viability for at least 6 months or more.

The known strains are inoculated as water suspensions of the peat inoculant. Prepare a 10^{-3} dilution of peat inoculant suspension of each known strain by suspending 10 g of inoculant in 90 ml of sterile water and then diluting further to obtain 10^{-3} dilution. Peat inoculants will generally have 10^{9} rhizobia g^{-1}. Based on this assumption, at 10^{-3} dilution the peat inoculant-water suspension will contain approximately 10^{6} rhizobia ml^{-1}. The treatments needed to determine the symbiotic potential of the indigenous rhizobial population in three soils are shown in Table 23.1. Whole-soil inocula are applied at 10^{-1} and 10^{-3} dilutions. Compute the number of rhizobia in the whole-soil inocula at 10^{-1} and 10^{-3} dilutions using the MPN information from (b).

Sow three pregerminated seeds in each Leonard jar. Inoculate each seed at planting with 1 ml of the appropriate inocula. At 7 days, thin to two plants per jar. Make regular observations of the experiment to note treatment differences. If actively fixing and non-fixing chlorotic plants are observed in the treatments given the whole-soil inocula, there is evidence that the inoculation treatment has a significant effect.

Harvest the experiment at 30–35 days of growth. Process the plant material for shoot and nodule dry weights, and nodule numbers described in Chapter 20. Also, determine the shoot total N. The C_2H_2 reduction assay (Appendix 15) may be done as a useful measurement to detect possible low levels of nitrogenase activity in some treatments.

h. Data Analysis of the Assay and Tabulation (Key Step 11)

The variables in this experiment are soils, legume species, and inoculation. Therefore, the data should be analyzed by factorial analysis. Consult a statistician. The analyzed data may be presented as shown in Table 23.2. Construct similar tables for *P. lunatus*, *V. unguiculata*, and *M. atropurpureum*, and strains TAL 22 and TAL 169.

i. Characteristics and Effectiveness of Isolates of Indigenous Rhizobia (Key Steps 12 and 13)

Isolates made in (c) may be characterized for diversity in serological properties (Section II), intrinsic antibiotic resistance patterns (Chapter 4), effectiveness on homologous host (Chapter 20), or cross-inoculation properties (Chapter 19).

TABLE 23.1 Experimental Treatments Needed for Determining the Symbiotic Potential of Indigenous Bradyrhizobia in Three Soils by the Whole-Soil Inocula Assay.

		Dilution of Whole-Soil Inocula		Controls	
Soil Sample	Legume Species	10^{-1}	10^{-3}	Non-inoculated	Known Strain
1	*A. hypogaea*	X-1	X-1	X	X-TAL 1371
	P. lunatus	X-1	X-1	X	X-TAL 22
	V. unguiculata	X-1	X-1	X	X-TAL 169
	M. atropurpureum	X-1	X-1	X	X-TAL 169
2	*A. hypogaea*	X-2	X-2	ND[2]	ND
	P. lunatus	X-2	X-2	ND	ND
	V. unguiculata	X-2	X-2	ND	ND
	M. atropurpureum	X-2	X-2	ND	ND
3	*A. hypogaea*	X-3	X-3	ND	ND
	P. lunatus	X-3	X-3	ND	ND
	V. unguiculata	X-3	X-3	ND	ND
	M. atropurpureum	X-3	X-3	ND	ND

[1]The prefix X indicates experimental treatments done in triplicate or quadruplicate and the numerical suffix identifies the test soil sample or control rhizobial strain.
ND[2] Indicates treatments not done.

TABLE 23.2 Symbiotic Potential of Indigenous Bradyrhizobia in Three Soils Analyzed by the Whole-Soil Inocula Bioassay for *A. hypogaea*

	Controls		Whole-Soil Inocula					
			Soil 1		Soil 2		Soil 3	
Parameters	Non-inoculated	TAL 1371	10^{-1}	10^{-3}	10^{-1}	10^{-3}	10^{-1}	10^{-3}
Shoot dry weight								
Shoot total N								
Nodule dry weight								
Nodule no.								
C_2H_2 reduction								

In this chapter it is important to screen individual isolates for effectiveness on the homologous legume host (Chapter 20). Evaluating the results of the bioassay can determine the need to screen isolates for effectiveness. For example, with *A. hypogaea*, if 10^{-1} whole-soil inocula treatment of soil 2 fixed N_2 significantly better than the control strain TAL 1371, then it is reasonable to expect isolates with superior effectiveness in the screening. Following the procedure described in Chapter 20, screen the 24 isolates of soil 2 for effectiveness on *A. hypogaea* using TAL 1371 as a control of known effectiveness.

REQUIREMENTS

a. Collecting Soil Samples for the Assay

Soil sites
Soil core sampler, 25 mm
Plastic bags
Sieves
Alcohol and sprayer
Oven (110°C)
Water, balance, spatula
Metal weighing boats

b. Using the Plant Infection Method to Determine the Size of the Rhizobial Population

Plastic growth pouches and racks
Seeds of *A. hypogaea*, *P. lunatus*, *V. unguiculata*, and *M. atropurpureum* soil samples from (a)
90 ml of sterile water in dilution bottles (160-ml capacity)
Weighing balance
Wrist-action shaker
Pipettes, 1 ml
Growth room
Refrigerator

c. Isolating Rhizobia from Nodules of Plants Used in the Plant Infection Counts

See Chapter 1—Requirements for isolating rhizobia

d. Preparing Leonard Jars

See Appendix 11

e. Pregerminating Seeds for Leonard Jar Planting

Erlenmeyer flasks, sterile, 250 ml
Seeds of *A. hypogaea*, *P. lunatus*, *V. unguiculata*, and *M. atropurpureum*
Sterile water
Commercial bleach or sodium hypochlorite solution
Concentrated sulfuric acid
Sterile and moist vermiculite (autoclaved)
Measuring cylinders

f. Preparing Whole-Soil Inocula

Weighing balance, weighing paper, spatula
90 ml of sterile water in milk dilution bottles
Wrist-action shaker
Pipettes, 10 ml
Refrigerator

g. Setting Up and Monitoring the Assay

Pure cultures of TAL 1371 (*A. hypogaea*), TAL 22 (*P. lunatus*), and TAL 169 (*V. unguiculata*)
Gamma-irradiated peat
YMBroth, 100 ml, in 250-ml Erlenmeyer flasks
Incubator (28°C)
Transfer chamber
90 ml of sterile water in milk dilution bottles
Whole-soil inocula
Pregerminated seeds of test legumes
Leonard jars
Alcohol and sprayer
Tweezers/forceps and alcohol lamps
Pipettes, 1 ml
Greenhouse bench space

h. Data Analysis of the Assay and Tabulation

No special requirements

i. Characteristics and Effectiveness of Isolates of Indigenous Rhizobia

Isolates from (c)
Requirements as in Chapter 20

KEY REFERENCES

Bonish, P.M. 1979. Clover rhizobia in soils: Assessment of effectiveness using plant infection count method. N.Z. J. Agric. Res. 22:89–93.

Brockwell, J., R.A. Holliday, and A. Pilka. 1988. Evaluation of the symbiotic nitrogen-fixing potential of soils by direct microbial means. Plant Soil 108:163–170.

Kang, U.G., P. Somasegaran, H.J. Hoben, and B.B. Bohlool. 1991. Symbiotic potential, competitiveness, and serological properties of *Bradyrhizobium japonicum* indigenous to Korean soils. Appl. Environ. Microbiol. 57:1038–1045.

24

Investigating the Importance of Optimal Soil Fertility in the Response of a Legume to Inoculation with Rhizobia[1]

The nutrition of a nodulated legume is different from other plants because of the additional nutritional demand placed on the legume by the N_2-fixing nodules. The efficiency of the symbiosis is affected by the levels of the various nutrients in the soil and variability in the response to inoculation. This experiment is designed to compare inoculation response in unamended soil (except for liming) and in soil fertilized to optimal levels. Inoculation response is evaluated using the three basic treatments: (1) inoculated, (2) plus-N without inoculation, and (3) no nitrogen without inoculation. Each of these treatments is set up at two different fertility levels. A multi-strain peat-based inoculant is used to study the effect of soil fertility on the competition of rhizobia for nodulation.

KEY STEPS/OBJECTIVES

1. Prepare the mixed inoculant.
2. Prepare the field and apply fertilizers.
3. Inoculate the seeds and sow.
4. Determine the number of rhizobia on the inoculated seed.
5. Inspect the field and weed.
6. Harvest at 50% flowering.

---

[1]The experiment described in this chapter is based on the design used in the International Network of Legume Inoculation Trials (INLIT) promoted by the NifTAL Project, Department of Agronomy and Soil Science, College of Tropical Agriculture and Human Resources, University of Hawaii, Honolulu.

7. Harvest for grain yield.
8. Analyze yield and nodule identification data.

a. Setting Up the Experiment

Experimental Design and Treatments

The experiment is designed as a randomized complete block with four replicates. There are three basic treatments: (1) inoculated, (2) plus-N without inoculation, and (3) no nitrogen without inoculation. Each of these basic treatments is set up at two fertility levels, giving a total of six treatments. Peat inoculant containing a mixture of three strains of rhizobia are used to inoculate seeds. The randomized treatments and field layout are indicated in Figure 24.1.

Field Dimensions

Plants are raised in plots of 7.5 × 2.4 m (0.0018 ha). The rows are 60 cm apart with four rows per plot. A total field area of 0.0432 ha (28.8 × 15 m) is needed for the experiment. Details of a plot showing areas reserved for early sampling and grain yield determinations are presented in Chapter 22 (Figure 22.2).

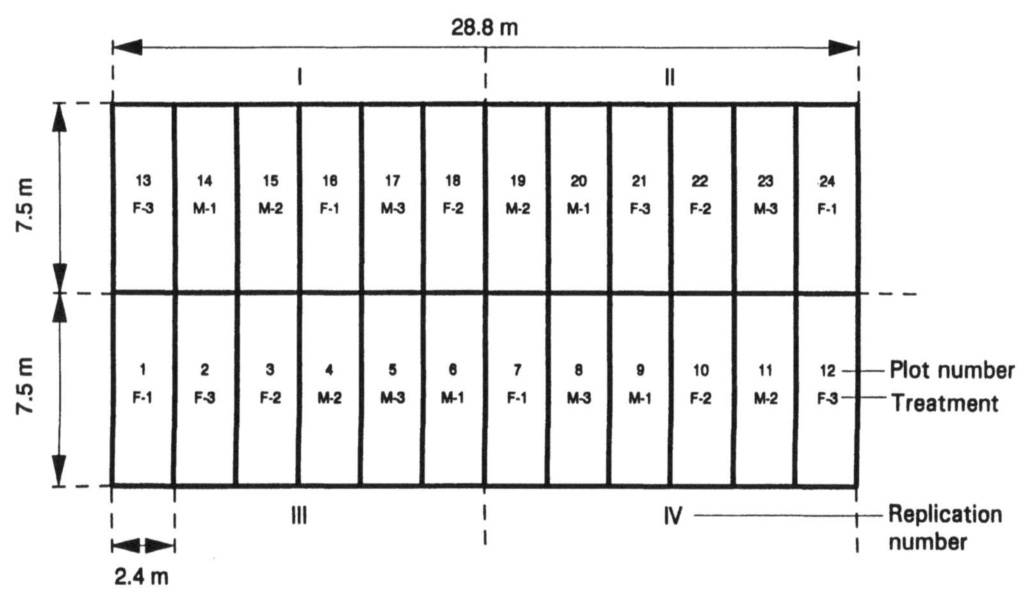

FIGURE 24.1 Field layout and dimension. The various treatments are randomized. Farm fertility and maximal fertility plots are indicated by F and M, respectively. Treatment details for each plot are given in Table 24.1.

b. Preparing the Mixed Inoculant and Inoculating the Seeds (Key Step 1)

The inoculants are prepared in advance of the experiment. The three antigenically distinct strains selected in Chapter 22 are used here.

Culture each of the three strains separately in 150 ml of yeast-mannitol broth (YMB) in Erlenmeyer flasks. Inoculate fully grown broth cultures into 50 g of gamma-irradiated or autoclaved peat as described in Chapter 27. Each package of the peat should be inoculated with only one strain.

Incubate the bags for 2 weeks at 25–30°C. At the end of the maturity period, determine the quality of the peat inoculants of the different strains. Aseptically remove 1-g samples in duplicate from each bag and plate serially diluted samples as described in Chapter 27. Refrigerate the inoculant bags immediately after sampling. Immediate refrigeration for 2–3 weeks or longer will maintain the original population at sampling without significant changes.

From the quality check (plate counts) of the inoculants, determine the number of rhizobia per gram inoculant for the three strains. From this information, determine the weights of the inoculants to be mixed to give a 1:1:1 ratio as shown in the following example.

In a quality check of three strains, the number of viable rhizobia per gram of each inoculant were as follows:

Strain A:	1.8×10^9
Strain B:	2.6×10^9
Strain C:	3.4×10^9

To establish a 1:1:1 ratio of A, B, and C in a mixture, determine factors that will convert strains A and B to 3.4×10^9 rhizobia per gram of peat. Start with the strain that has the higher count of rhizobia per gram.

$$\text{For strain A, conversion factor} = \frac{3.4}{1.8} = 1.9$$

This means that 1.9 g of peat inoculant of strain A will contain 3.4×10^9 rhizobia.

$$\text{For strain B, conversion factor} = \frac{3.4}{2.6} = 1.3$$

This means that 1.3 g of peat inoculant of strain B will contain 3.4×10^9 rhizobia. Therefore, a peat inoculant mixture containing 1.9 g of strain A, 1.3 g of strain B, and 1 g of strain C will result in a 1:1:1 ratio of the three component strains.

Alternatively, if equal weights of the inoculants of the three strains were mixed, a ratio of 1:1:1 will not be established. The ratio would then be 1.8:2.6:3.4, or approximately 2:3:3. Competition of the three strains may still be studied using this approximated ratio, but it is not recommended.

Calculate the weight of seed (Chapter 22) for the four rows of each inoculated plot and for all the inoculated treatments in the experiment using a planting distance of 3.7-cm between seeds. With this information and the recommended inoculation rate of 10 g of mixed peat inoculant per 100 g of soybean (*Glycine max*) seed, calculate the total weight of inoculant needed for all the inoculated treatments. Alternatively, 1 g of mixed inoculant may be used to inoculate 100 g of seed to achieve a lower inoculation rate.

Mix the inoculants 1 day ahead of planting. Remove the bags of inoculants from the refrigerator. Weigh out calculated amounts into a sterile 1-liter beaker. Mix the inoculants thoroughly with a spatula. (Observe aseptic techniques throughout the preparation.) After mixing, cover the beaker with aluminum foil and refrigerate immediately. Inoculate the seeds (Chapter 29) just before planting. Save seed samples and determine the number of viable rhizobia on the seeds at sowing (Chapter 29).

c. Choosing a Site and Preparing the Field (Key Step 2)

A field site having soil conforming to the description outlined in Chapter 21 will be suitable. A farmer's field is preferred, if available. The fertility status of the site soil has to be determined. If facilities are available, analyze soil samples for: free nitrate; extractable P; exchangeable K, Ca, and Mg; exchangeable Al and Mn, if soil pH is below 5.2 and 5.6, respectively; soil pH; and organic matter. Collect the soil samples and determine the population of native soybean rhizobia using the most-probable-number (MPN) method (Chapter 6). Prepare the field as outlined in Chapter 22. The field preparation may need modifications to control cross-contamination resulting from heavy rainwash or from flood-irrigation method.

d. Applying Fertilizers (Key Step 2)

Since two fertility levels are used, namely the farmer's fertility without amendments (F) and maximal fertility (M), each treatment plot has to be marked to facilitate recognition during fertilizer applications. This is especially important because each plot is fertilized individually.

Mark boundaries by driving in short stakes at the four corners of each plot. About 6–9 in (approximately 15–23 cm) of the stake should remain exposed to allow for good visibility and for ready recognition of plot boundaries. Erect a suitable-sized signboard in front of each treatment plot indicating its treatment according to the field layout presented in Figure 24.1.

Lime the soil in the F plots only if the pH is below 5.4. Lime all M plots to pH 6.0–6.5. Allow the soil to equilibrate for at least 2 weeks after the lime application. Determine the lime requirement of the soil using one of the methods described in Appendix 16.

The fertilizer recommendations for maximal fertility are similar to rates used in Chapters 21 and 22. The amount of fertilizer applied per plot is as follows: triple superphosphate (TSP), 0.9 kg; potassium chloride, 0.69 kg; zinc sulfate, 0.08 kg; ammonium molybdate, 0.0033 kg; magnesium sulfate, 0.09 kg; and urea, 0.373 kg (in the plus-N plots). (Consult with an agronomist if a higher N level needs to be applied to demonstrate the yield potential of the legume.)

Weigh out the quantities of the fertilizers in plastic bags and assign the treatment labels. Mix smaller quantities of the fertilizers (zinc sulphate, magnesium sulfate, and ammonium molybdate) with an inert carrier (e.g., sand) to facilitate application by broadcasting or by spraying.

Apply the fertilizers individually to all plots of the various treatments (Table 24.1). Till the fertilizers into the soil after broadcasting. The N (urea) is applied as side-dressing in furrows at sowing (0.09 kg $plot^{-1}$) and after 4 weeks from emergence (0.28 kg $plot^{-1}$).

e. Planting the Experiment (Key Steps 3 and 4)

Inoculate the soybean seeds as described in Chapter 29. Set aside about 100 inoculated seeds for determining the number of rhizobia per seed. (Use only 20 randomly selected seeds from this sample for the determination.) Make furrows 7.5 m long, 0.6 m apart, and 3.0–3.5 cm deep. Plant the soybean seeds at approximately 35 seeds per meter of row, planting the uninoculated treatments first. Thin rows evenly to 27 plants per meter at 2 weeks. Irrigate the field. Take precautions against cross-contamination during sowing (see Chapter 22). Note that in the experiment in Chapter 22, a single, uninoculated guard row was common to two adjacent plots. In this experiment, each plot does not share common guard rows with adjacent plots. Each plot in this experiment has four rows (two guard and two harvest rows) and the guard rows are of the same treatment as the harvest rows. After sowing, side-dress the recommended amount of urea in furrows of the plus-N controls and irrigate the field if necessary.

f. Monitoring the Trial and Harvest (Key Step 5)

Carry out regular field inspections, weeding and harvesting activities as described in Chapter 22.

TABLE 24.1 Summary of Treatments

Fertility Levels	Fertilizers Applied[1]						Inoculation
	P	K	Zn	Mo	Ca	N	
M-1	+	+	+	+	+	−	−
M-2	+	+	+	+	+	+	−
M-3	+	+	+	+	+	−	+
F-1[2]	−	−	−	−	−	−	−
F-2	−	−	−	−	−	+	−
F-3	−	−	−	−	−	−	+

[1]+ and − indicate application and no application, respectively.

[2]Ca is added as calcium carbonate if pH is below 5.3–5.4

g. Harvesting Nodules for Strain Identification (Key Steps 6 and 7)

Because the seeds of the inoculated treatments were coated with a mixed-peat inoculant containing equal proportions of three antigenically distinct strains, the nodules may be harvested and processed for establishing nodule identity. Data from nodule identification can then be analyzed for strain competition for nodulation at the two fertility levels.

Nodules for strain identification are obtained by harvesting at 50% flowering. Obtain plants from each inoculated plot (F-3 and M-3). Excavate plants only from the sample rows of each plot. Select the plants randomly. Wash the roots to clean off soil and adhering debris. Pick the nodules from the roots. Count and pool all the nodules of the two plants obtained from the same plot to obtain the plot sample.

To test serological cross-reaction of indigenous rhizobia, obtain samples of nodules from all uninoculated plots (F-1 and M-1). Select plants from sample rows. Identify at least 30 nodules at random from each inoculated plot sample using one of the serological methods described in Section II. Store the rest of the nodules by: (1) desiccation over silica gel; (2) freezing; or (3) storing in small plastic bags or vials after oven drying at 70°C. Large batches of nodules require much more time for identification. If nodulation is poor (less than 10 nodules per plant), identify all nodules from each plot sample. Small nodules (1-mm diameter or less) may not contain sufficient antigen for identification against three antisera using the agglutination or gel-diffusion techniques. However, small nodules are better identified using the fluorescent-antibody technique because this method requires only antigen smears from the nodule.

Apply the chi-square (X^2) method to determine from the data whether or not the frequencies observed in the nodule identification depart significantly from the expected frequencies of 1:1:1 for the three strains at each fertility level. If there is a significant departure from the expected ratio of 1:1:1, the data indicate competition. Is the competition pattern the same at both fertility levels?

h. Analyzing the Yield Data (Key Step 8)

The harvest data at 50% flowering and from grain yield will reveal valuable information on the importance of fertility to ensure a good response to inoculation with effective strains of rhizobia. Using the harvest data, carry out statistical analysis; determine the treatment giving the highest yield. The Duncan's new multiple range test is suggested for preliminary analysis of the data. Other statistical approaches may treat the data more vigorously to detect significant interactions in the treatments. Such analysis may require expert statistical assistance.

REQUIREMENTS

a. Setting Up the Experiment

Measuring tape, 50 m
Field site, 28.2 × 15 m

b. Preparing the Mixed Inoculant and Inoculating the Seeds

Transfer chamber
Agar slant cultures of antigenically distinct strains used in Chapter 22
Erlenmeyer flasks, 250 ml, (three), each containing 150 ml of YMB
Sterile plastic syringes, 50 ml (three); sterile needles, 3/4 in, 18 gauge (three)
Bags of peat, 50 g per bag (three) autoclaved or irradiated
Incubator, refrigerator
Quality check of inoculants (materials as in Chapter 27)
Sensitive balance (to weigh peat)
Coarse balance (to weigh seeds)
Beaker, 1 liter; spatula; aluminum foil
Gum arabic solution, pipettes (wide-bore tip), plastic bags

c. Choosing a Site and Preparing the Field

Farmer's field or alternative site
Soil analysis data
Soil sample for MPN counts
Wooden posts to mark field perimeter
Herbicide(s) and spraying equipment
Plowing and tilling machinery
Suitable field drainage to control cross-contamination
Other field preparation accessories
Lime

d. Applying Fertilizers

Wooden stakes (boundary markers)
Signboards for all M and F plots
TSP: 0.9 kg × 12 plots = 10.8 kg
Potassium chloride: 0.69 kg × 12 plots = 8.3 kg
Zinc sulphate: 0.96 kg × 12 plots = 11.5 kg
Ammonium molybdate: 0.0033 kg × 12 plots = 0.04 kg
Magnesium sulfate: 0.09 kg × 12 plots = 1.1 kg
Urea: 0.0373 kg × 8 plots = 0.3 kg

e. Planting the Experiment

Irrigation
Inoculated and noninoculated soybean seeds from (b)
Metric tape, hoes or other suitable equipment for making furrows

Planting guide for even placement of seeds
Urea for N controls

f. Monitoring the Trial and Harvest

Insecticides and spraying equipment
Weeding tools, hoes
Scissors/snips, paper bags, aluminum weighing boats
Coarse sieve
Drying oven (70°C)
Balance

g. Harvesting Nodules for Strain Identification

Water source
Hoe for digging, coarse sieve
Glass or plastic vials (for nodule storage, Appendix 2), marker pens
Plastic bags
Freezer space (or silica gel in vials)

h. Analyzing the Yield Data

Calculator and statistical tables
Statistical assistance

KEY REFERENCES

Thies, J.E., P.W. Singleton, and B.B. Bohlool. 1991a. Influence of the size of indigenous rhizobial populations on establishment and symbiotic performance of introduced rhizobia on field-grown legumes. Appl. Environ. Microbiol. 57:19–28.

Thies, J.E., P.W. Singleton, and B.B. Bohlool. 1991b. Modeling symbiotic performance of introduced rhizobia in the field by use of indices of indigenous population size and nitrogen status of the soil. Appl. Environ. Microbiol. 57:29–37.

Additional References and Recommended Reading

Broughton, W.J., and M.J. Dilworth. 1971. Control of leghaemoglobin synthesis in snake beans. Biochem. J. 125:1075–080.

Burton, J.C. 1979a. *Rhizobium* species. pp. 29–58. *In* H.J. Peppler (ed.) Microbial Technology, 2nd ed. Academic Press, New York.

Burton, J.C. 1979b. *Rhizobium* inoculation and soybean production. pp. 89–100. *In* F.T. Corbin (ed.) World Soybean Research Conference II: Proceedings. Raleigh, NC. March 26–29, 1979. Westview Press, Boulder, CO.

Burton, J.C., O.N. Allen, and K.C. Berger. 1954. Response of beans (*Phaseolus vulgaris* L.) to inoculation with mixtures of effective and ineffective rhizobia. Soil Sci. Soc. Am. Proc. 18:156–159.

Cassman, K.G., A.S. Whitney, and R.L. Fox. 1981. Phosphorus requirements of soybean and cowpea as affected by mode of nutrition. Agron. J. 73:17–22.

Chapman, H.D., and P.F. Pratt. 1961. Methods of Analysis for Soils, Plants and Waters. Division of Agricultural Sciences, University of California, Riverside.

Date R.A. 1970. Microbiological problems in the inoculation and nodulation of legumes. Plant Soil 32:703–725.

Date, R.A., and J. Halliday. 1979. Selecting *Rhizobium* for acid infertile soil of the tropics. Nature (London) 277:62–64.

Dreyfus B.L., and Y.R. Dommergues. 1981. Nodulation of *Acacia* species by fast- and slow-growing tropical strains of *Rhizobium*. Appl. Environ. Microbiol. 41:97–99.

Erdman, L.W., and U.M. Means. 1952. Use of total yield for predicting nitrogen content of inoculated legumes grown in sand culture. Soil Sci. 73:231–235.

Graham, P.H., and J.C. Rosas. 1979. Phosphorus fertilization and symbiotic nitrogen fixation in common bean. Agron. J. 71:925–926.

Halliday, J. 1978. Field responses by tropical forage legumes to inoculation with *Rhizobium*. pp. 123–137. In P.A. Sanchez and L.E. Tergas (eds.) Pasture Production in Acid Soils of the Tropics, Centro Internacional de Agricultura Tropical, Cali, Colombia.

Hardy, R.W.F., R.D. Holsten, E.K. Jackson, and R.C. Burns. 1968. The acetylene-ethylene assay for N_2 fixation: Laboratory and field evaluation. Plant Physiol. 43:1185–1207.

Haydock, K.P., and D.O. Norris. 1967. Opposed curves for nitrogen per cent on dry weight given by *Rhizobium* dependent and nitrate dependent legumes. Aust. J. Sci. 29:426–427.

Hohenberg, J.S., D.N. Munns, and C.L. Tucker. 1982. *Rhizobium* host specificities in *Phaseolus coccineus* L. and *Phaseolus vulgaris* L. Crop Sci. 22:455–459.

Iruthayathas, E.E., and K. Vlassak. 1982. Symbiotic specificity in nodulation and nitrogen fixation between winged bean and *Rhizobium*. Hortic. Sci. (Stuttgart) 16:313–322.

Lange, R.T. 1961. Nodule bacteria associated with the indigenous Leguminosae of South-Western Australia. J. Gen. Microbiol. 61:351–359.

Lawrie, A.C. 1983. Relationships among rhizobia from native Australian legumes. Appl. Environ. Microbiol. 45:1822–1828.

Little, T.M., and F.J. Hills. 1978. Agricultural Experiment. Design and Analysis. John Wiley & Sons, New York.

May, S.N., and B.B. Bohlool. 1983. Competition among *Rhizobium leguminosarum* strains for nodulation of lentils (*Lens culinaris*). Appl. Environ. Microbiol. 45:960–965.

Mengel, D.B., and E.J. Kamprath. 1978. Effect of soil pH and liming on growth and nodulation of soybeans in Histosols. Agron. J. 70:959–963.

Norris, D.O., and R.A. Date. 1976. Legume bacteriology. pp. 134–174. *In* N.H. Shaw and W.W. Bryan (eds.) Tropical Pasture Research: Principles and Methods, Bull. 51. Commonwealth Bureau of Pastures and Field Crops, Hurley, England.

Peters, D.B. 1965. Water availability. pp. 279–281. *In* C.A. Black et al. (ed.) Methods of Soil Analysis, Part 2, Agronomy Monograph 9. American Society of Agronomy, Madison, WI.

Postgate, J. 1971. The acetylene reduction technique for nitrogen fixation. pp. 343–356. *In* J.R. Norris and D.W. Ribbons (eds.) Methods in Microbiology, Academic Press, London.

Schwinghamer, E.A., H.J. Evans, and M.D. Dawson. 1970. Evaluation of effectiveness of mutant strains of *Rhizobium* by acetylene reduction to other criteria of nitrogen fixation. Plant Soil. 33:192–212.

Skrdleta, F., and J. Karimova. 1969. Competition between two serotypes of *Rhizobium japonicum* used as a double strain inocula in varying proportions. Arch. Microbiol. 66:26–29.

Sloger, C. 1969. Symbiotic effectiveness and N_2 fixation in nodulated soybean. Plant Physiol. 44:1666–1668.

Somasegaran, P., H.J. Hoben, and L. Lewinson. 1990. Symbiotic interactions of *Phaseolus acutifolius* × *P. vulgaris* hybrid progeny in symbiosis with *Bradyrhizobium* spp. and *Rhizobium leguminosarum* bv. phaseoli. Can. J. Microbiol. 37:497–503.

Thies, J.E., B.B. Bohlool, and P.W. Singleton. 1991. Subgroups of the cowpea miscellany: Symbiotic specificity within *Bradyrhizobium* spp. for *Vigna unguiculata*, *Phaseolus lunatus*, *Arachis hypogaea*, and *Macroptilium atropurpureum*. Appl. Environ. Microbiol. 57:1540–1545.

Trinick, M.J. 1965. *Medicago sativa* nodulation with *Leucaena leucocephala* root-nodule bacteria. Aust. J. Sci. 27:263–264.

Trinick, M.J. 1968. Nodulation of tropical legumes. I. Specificity in the *Rhizobium* symbiosis of *Leucaena leucocephala*. Exp. Agric. 4:243–253.

Trinick, M.J. 1973. Symbiosis between *Rhizobium* and the non-legume Trema aspera. Nature (London) 244:459–460.

Trinick, M.J. 1980. Relationships among the fast-growing rhizobia of *Lablab purpureus*, *Leucaena leucocephala*, *Mimosa* spp., *Acacia farnesiana* and *Sesbania grandiflora* and their affinities with other rhizobial groups. J. Appl. Bacteriol. 49:39–53.

Turk, D., H.H. Keyser, and P.W. Singleton. 1993. Response of tree legumes to rhizobial inoculation in relation to the population density of indigenous rhizobia. Soil Biol. Biochem. 25:75–81.

Vincent, J.M. 1970. A Manual for the Practical Study of Root-Nodule Bacteria. IBP Handbook no. 15, Blackwell Scientific Publications, Oxford.

Wagner, G.H., G.M. Kassim, and S. Martyniuk. 1978. Nodulation of annual Medicago by strains of *R. meliloti* in a commercial inoculant as influenced by soil phosphorus and pH. Plant Soil 50:81–89.

Weaver, R.W. 1975. Growing plants for *Rhizobium* effectiveness tests. Soil Biol. Biochem. 7:77–78.

Wilson, J.K. 1944. Over five hundred reasons for abandoning the cross-inoculation groups of legumes. Soil Sci. 58:61–69.

Zablotowicz, R.M., and D.D. Focht. 1981. Physiological characteristics of cowpea rhizobia: Evaluation of symbiotic efficiency in *Vigna unguiculata*. Appl. Environ. Microbiol. 41:679–685.

SECTION IV
Inoculant Technology

The agricultural benefits possible from use of selected, high-N_2 fixing strains of rhizobia can be realized only when farmers obtain and properly use high-quality inoculants on their legume seeds or soil before planting. Technology on growing rhizobia, preparing inoculants with suitable carrier materials, and distributing viable inoculants to farmers is essential. This section is concerned with inoculant production and use.

CULTURING RHIZOBIA

Rhizobia are easy to grow in the laboratory. These bacteria are aerobic and also microaerophilic. They require aeration, which may be provided by using a mechanical shaker or by bubbling sterile air through the medium. Rhizobia grow best at 25–30°C. The medium must supply energy, a source of N, certain mineral salts, and growth factors. Most commonly used is a yeast extract mannitol mineral salts medium, but if cost or availability is a concern, sucrose or glycerol may be substituted.

Commercial-scale inoculant production requires the culturing of rhizobia in large volumes of liquid culture media. For cost-effective production, the ingredients for these media must be inexpensive and readily available. Various industrial by-products have been used with acceptable results. Molasses, corn steep liquor, and proteolyzed peahusk have been used as sources of C and growth factors for propagating various rhizobial species.

Yeast extract is most frequently used as the N source. Successful replacements for yeast extract have been soybean (*Glycine max*), chickpea (*Cicer arietinum*), and malt sprout extract. Whey, a by-product of the cheese industry, has also been successfully used as an N source. Mass culturing rhizobia is done in fermentors in which the growth media are heat sterilized prior to use. Vessels for fermentors vary in size from a few liters to several thousand liters.

Inoculant production starts with a pure slant culture. This culture is used to inoculate yeast-mannitol broth (YMB) in a small flask. The resulting broth culture will serve as inoculum (starter culture) for a greater volume of broth or medium contained in a large flask or a 2–4-liter glass fermentor. The volume of a starter culture should be a minimum of 1% of the broth volume in the fermentor. Thus, a 1-liter starter culture would be required to inoculate 100 liters of medium in a steel fermentor. Often, a larger inoculum is used to reduce the incubation time needed to obtain 1×10^9 rhizobial cells ml^{-1}. This population level is considered necessary, particularly when using nonsterile carrier material. Aseptic conditions are maintained throughout the production period. The broth culture is checked frequently for purity.

INOCULANT CARRIERS

Most inoculants are a mixture of the broth culture and a finely milled, neutralized carrier material. The properties of a good carrier material are: (1) nontoxic to rhizobia, (2) good moisture absorption capacity, (3) easy to process and free of lump-forming materials, (4) easy to sterilize by autoclaving or gamma-irradiation, (5) available in adequate amounts, (6) inexpensive, (7) good adhesion to seeds, and (8) good pH buffering capacity.

The best researched and most frequently used carrier material for inoculant production is peat. A large number of studies have shown that peat provides better protection for the rhizobia in the package and on inoculated seed than other carriers. The physical and chemical analysis of well-known peats are shown in Tables IV.1 and IV.2. However, physical and chemical analysis of a peat are only partial assessments of its

TABLE IV.1 Characteristics of Sedge Peat Used for Commercial Inoculant Production in the United States[1]

Sedge Peat Contents	Amounts
Total N (%)	1.62
Organic matter (%)	86.80
Ash (500°) (%)	13.20
Exchangeable K (ppm)	62
N as NH_4 and NO_3 (ppm)	94
Available P (ppm)	12
pH	4.5–5.0
Moisture (%)	7–8
Analysis of Ash (%)	
K	1.12
P	0.33
Ca	5.21
Mg	1.14
Fe	2.10
Si	28.00
Al	6.32
Na	0.52

[1]From Burton, 1979.

TABLE IV.2 Characteristics of Badenoch Peat Used for Commercial Inoculant Production in Australia[1]

Characteristics	Range	Mean
Organic matter (%)	28.8–75.4	64.3
Organic C (%)	16.4–42.1	36.1
Mineral matter (%)	10.0–16.0	12.1
Total soluble salts (%)	0.09–1.5	0.87
Cl (%)	0.01–0.31	0.11
N (%)	0.89–2.30	1.83
K_2O (%)	0.12–0.17	
P_2O_5 (%)	0.09–0.22	
$CaCO_3$ (%)	0	0
Water-holding capacity (%)	216–522	320
CN ratio	15.0–17.5	16.7
pH	5.8–7.8	6.8

[1]Adapted from Roughley, 1970.

suitability as a carrier. Only a test related to growth and survival of rhizobia can confirm its acceptability.

Peat is not available in some countries. A wide range of substitutes, e.g., coal, charcoal, bagasse, filter mud, vermiculite, polyacrylamide, mineral soils, vegetable oils, and ground plant residues have been tested as alternative carriers. Regardless of which type of carrier is chosen, a rhizobia-carrier interaction cannot be avoided. A particular rhizobial strain may survive well in one carrier but not another. Therefore, it is important that the producer knows the strain-carrier interactions for all of his strains of rhizobia.

Carrier processing, e.g., mining, drying, and milling, are the most capital-intensive aspects of inoculant production. Generally, the wet peat is mined, drained, and screened to remove stones and roots, then shredded and dried. The peat is then ground in a high-speed hammer mill and passed through a sifting machine. Material with a particle size of 10–40 μm (0.001–0.004 mm) is collected for seed coating. Peat with a particle size of 500–1500 μm (0.5–1.5 mm) is used for the production of soil implant (granular) inoculant. The carriers are neutralized with precipitated calcium carbonate (pH 6.5–7.0). Both sterilized and nonsterilized peat are used in commercial production systems.

Peat carrier designated for sterilization is prepackaged in thin-walled polyethylene bags. The sealed bags are then gamma-irradiated at 5.0 Mrads. Alternatively, the carrier may be autoclaved in partially opened, thin-walled, polypropylene bags for 60 min at 15 lb/in^2 pressure and 121°C. However, heat sterilization of some peats has been found to produce undesirable changes and to release toxins. Gamma-irradiation sterilization is preferred.

Shredded peat designated for the nonsterile carrier-based production is often flash dried by passage through a revolving drum with an inlet air temperature of 650°C and an outlet temperature of 121°C. They are then stored in bulk until incorporating the broth culture.

INCORPORATING BROTH CULTURE INTO CARRIERS

Under commercial conditions in the United States, quality-tested broth cultures are incorporated into peat at the rate of 1 liter per kilogram of peat. After a curing period, the mixture is packaged in thin-gauge (0.05 mm) polyethylene bags. Bags of this specification permit gas exchange while minimizing moisture loss from the inoculant. The expiration date for inoculants based on nonsterile carriers is usually 6 months.

Inoculant producers in some countries, such as South Africa, Australia, and New Zealand, produce inoculants with sterilized carriers. In this case, the carrier is first packaged and then sterilized by gamma-irradiation or autoclaving. Thin-gauge (0.05 mm) polyethylene bags are used for carriers to be gamma-irradiated. Carriers to be autoclaved are packaged in polypropylene bags of the same gauge. The rhizobial broth culture is aseptically injected into the packaged carrier with a manually operated motorized syringe.

Inoculants based on sterile carriers are usually of higher quality than the nonsterile carrier type. The number of viable rhizobia per gram can be between 10^9–10^{10} cells in inoculants produced with sterilized carriers. In nonsterile carriers, as in nonsterile peat, the initial number of viable rhizobia tend to be lower by at least one log after curing. The number of rhizobia added to most sterile carriers remain high during shelf life or storage because there are no other microorganisms in the carrier competing with the rhizobia. The quality of such inoculants may still be acceptable after 6–12 months, depending on the temperature during storage.

Although producing inoculants based on sterile carriers is more costly than nonsterile carrier-based inoculants, mainly due to the need for sterilization facilities and labor-intensive production operations, using the dilution technique can substantially lower the production cost. Here, the broth culture is aseptically diluted with sterile water up to 1000-fold before incorporation into the sterile carrier as demonstrated in sterile peat. The low cell population in the diluted culture will multiply to the same level as with undiluted cultures during the maturing time of 5–7 days.

Inoculants are cured for about 2 weeks at 25–30°C to gain maximum numbers in excess of 10^8 and 10^9 cells g^{-1} for nonsterile and sterile carrier-based inoculants, respectively. Thereafter, inoculants are stored in a refrigerated or air-conditioned environment, protected from direct light. Most inoculants are stored at 4°C and tend to survive best at this temperature. However, there are inoculant strains, including CB 627 from *Desmodium*, CB 1923 from *Centrosema*, and CB 82 from *Stylosanthes*, that have very poor survival at 4°C, but have good survival at 26°C after 12 months.

The final moisture content of the peat inoculant should be 40–60% on a wet-weight basis for inoculants produced with presterilized peat. A lower moisture content (30–40%) is preferred for better rhizobial survival in nonsterile peat.

QUALITY CONTROL

Quality control begins with selecting of the rhizobial strains for inoculant production. Only authenticated, and pure cultures of the strains are used. Purity is tested by growing the cultures on different test media. Growth reactions must agree with those characteristics known for the culture. Samples of the starter broth culture are tested for purity by Gram stain and by serological methods if antisera of the intended strains are available.

The broth culture must be protected from other microorganisms during mass culture in a fermentor or other culture vessel. Contaminants such as fungi or other bacteria compete with rhizobia and prevent the maximum growth of rhizobia during mass culture. Contaminants usually have fast growth rate, uncharacteristic odor, and cause excessive foaming. The broth culture is checked frequently for any abrupt change in pH (usually lower pH), which indicates contamination. Prior to incorporation with the carrier material (peat, coal, charcoal, etc.), the fully grown culture is once again checked by Gram stain and/or by serological methods for culture purity.

Before leaving the production plant, batches of inoculants are usually sampled to check their quality. (*Quality of a solid carrier-based inoculant* may be defined as the number of viable and infective cells of the specified rhizobial strain in 1 g of moist inoculant.) In the case of inoculants based on sterile carriers, viability is tested by the plate count, but infectivity is tested on the appropriate host legume in the growth room or greenhouse. When nonsterile carrier material is used, inoculant quality is commonly tested by a plant infection test, as contaminants would interfere with the plate-count methods. A modification of this technique, the preenriched plant infection technique (PEPI), has been claimed as a more accurate method for enumerating rhizobia in the presence of other microorganisms. Plant infection tests can also be valuable for detecting lost infectivity in a culture. Unfortunately, quality check by the plant infection methods requires about 3 weeks, conflicting with a need for quick distribution of the inoculants.

Serological methods may also be used for the quality control of inoculants. The enzyme-linked immunosorbent assay (ELISA) and the membrane-filter immunofluorescent technique (MFIF) have been applied. They give much quicker results than the plant infection test. However, the MFIF procedure must be used with caution because it is a direct counting procedure that does not distinguish between viable and nonviable rhizobia. Another disadvantage is the need for equipment, chemicals, and biological reagents, which are costly and not easily available in some locations.

Inoculants must bear an expiration date and comply with quality standards regarding numbers of rhizobia present. These standards vary with different countries. Generally, a minimum of 10^9 viable rhizobia g^{-1} is required for sterile carrier-based inoculants and 10^8 viable rhizobia g^{-1} for nonsterile carrier-based inoculants. In the United States, credit for returned inoculants that have passed their expiration date is an essential facet of inoculation technology acceptance by farmers, and is the standard policy of the more reputable inoculant producers.

SOIL AND SEED INOCULATION

The essence of legume inoculation is the placement of a large population of a compatible, highly effective N_2-fixing rhizobial strain in close proximity to the emerging radicle of the legume so that the majority of the nodules will be formed by the inoculant strain. It is important to assess the need to inoculate a particular legume at a specific site. Sometimes an adequate population of effective native strains will ensure ample nodulation without inoculation. Alternatively, an inoculant strain may not survive in adequate numbers or be sufficiently competitive against the native rhizobia, and yield benefits are unlikely from inoculation.

The most common means of introducing rhizobia to the soil is as a seed-applied inoculant. In its simplest (and least satisfactory) form, peat inoculant is mixed with water to form a slurry and mixed with the seeds. Better results are obtained when the inoculant is coated on the seed with an adhesive. An adhesive increases the amount of inoculant that will adhere to the seed. A good inoculant adhesive must be nontoxic to the rhizobia and provide protection during planting and in the soil. Gum arabic has these properties, but it is expensive to farmers and not readily available at many locations. Other adhesives used successfully include methylethylcellulose, sucrose solutions, and vegetable oils. An additional coating of calcium carbonate, rock phosphate, or other pelleting material can enhance the success of inoculation. This is often done when adverse weather conditions prevent immediate sowing of inoculated seeds, as protection against insects in the soil, when the soil is hot and dry or very acidic, or as protection against pesticides.

Seed coating is not always the best way to inoculate. Some inoculants are designed to be placed into the soil. Soil implant or granular inoculants are generally made from peat granules with a particle size of 0.5–1.5 mm. Other types of granular inoculants are made by spraying a suspension of peat inoculant on an inert substance. Soil inoculants may also be improvised by suspending peat inoculant in water or by mixing peat inoculant with sand. Soil inoculant is generally placed into the furrow under or alongside the seed. This is usually done when very large numbers of rhizobia are needed to overcome an ineffective native population, when planting in hot and dry soil is unavoidable, when seed inoculation has failed, with crops in which the seed tends to lift the seed coat out of the soil with the cotyledons, and with seed-applied pesticides or chemical seed protectants.

COMPATIBILITY WITH PESTICIDES AND FERTILIZERS

When seeds are precoated with pesticides or herbicides, they should not be seed inoculated because these chemicals are toxic to rhizobia. Soil inoculation is recommended before sowing so that the pesticides will not harm the inoculant. Most phosphate and calcium carbonate fertilizers do not harm the inoculants. However, direct contact with acid superphosphate can seriously affect rhizobial survival. Calcium hydroxide may be used before or after sowing inoculated seeds or soil inoculation. Avoid direct contact with the inoculant.

25

Producing Broth Cultures in Simple Glass Fermentors

A small glass fermentor of 2–3-liter capacity can be used to produce broth cultures to be applied as liquid inoculants in field experiments. It is also suitable for producing starter cultures for medium-sized commercial fermentors. In this chapter, a glass fermentor is assembled and then used for small-scale production of broth cultures. The broth cultures are monitored periodically for quality control, including checks for contamination and the progress of growth and cell multiplication.

KEY STEPS/OBJECTIVES

1. Initiate starter broth cultures.
2. Assemble small fermentor units.
3. Sterilize fermentors.
4. Become familiar with operation details.
5. Inoculate the fermentors.
6. Take broth samples periodically for cell count and check for contamination.
7. Test for contamination.
8. Perform total counts and optical density measurements.
9. Perform viable counts by the spread-plate method on the presumptive test media.
10. Perform agglutination tests with the homologous antisera.

a. Inoculating Starter Cultures (Key Step 1)

Prepare four 50-ml flasks or tubes each containing 25 ml of yeast-mannitol broth (YMB). Obtain slant, lyophilized or bead preserved cultures of a strain of *Bradyrhizobium* (e.g., *B. japonicum* strain TAL 102) or a strain of *Rhizobium* (e.g., *Rhizobium* sp. strain TAL 1145 from *Leucaena leucocephala*). Inoculate two flasks with each rhizobial strain and aerate at 25–30°C. These will serve as starter cultures for inoculating the YMB in the fermentors. Other cultures of *Bradyrhizobium* spp. and *Rhizobium* spp. may be substituted for the suggested rhizobia.

b. Assembling Simple Fermentors (Key Steps 2 and 3)

Set up two fermentors (one for each strain) as shown in Figure 25.1. The main fermentation vessel is a slightly modified 4-liter Erlenmeyer flask with a sampling port (glass tubing 4-mm i.d.) fitted close to its base. Fill each fermentor with 2–3 liters of YMB. Connect the cotton-packed filters to prevent the entry of contaminants via the air lines. All rubber stoppers and tubings must be autoclavable. Insert the large rubber stopper, which holds the air inlet and outlet tubes with their respective filters, firmly into the neck of the flask.

Connect the air inlet tube to an aquarium pump. Activate the pump and check the air inlet and outlet filters for air resistance. Air should flow freely through both filters while bubbling through the broth, and simultaneously aerating and agitating the me-

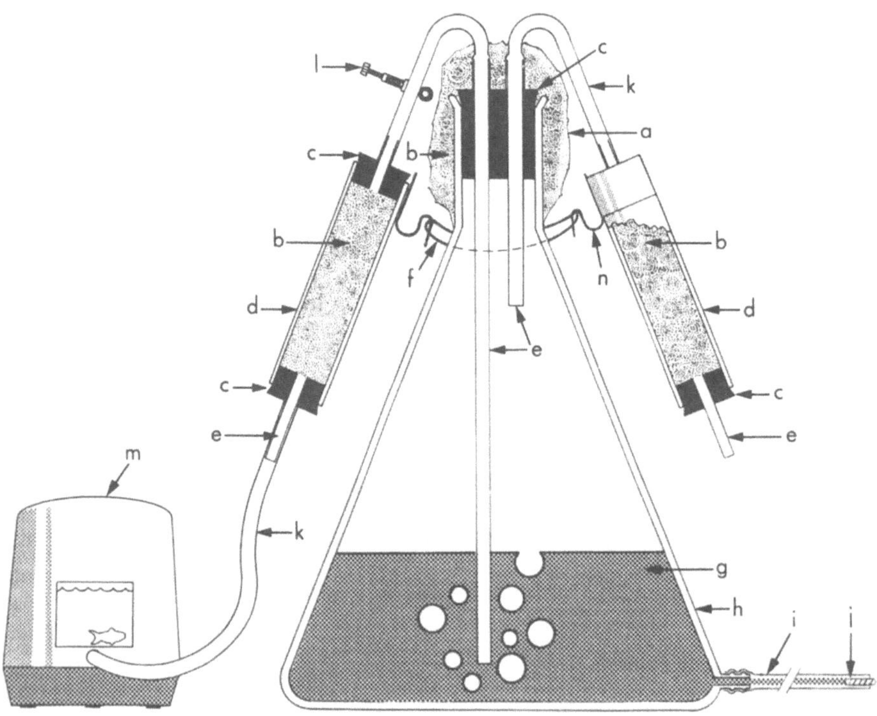

FIGURE 25.1 Scheme of a simple fermentor unit. a, aluminum foil; b, nonabsorbent cotton; c, autoclavable stopper; d, filter unit; e, glass tubing; f, wire ring; g, growth medium; h, flask; i, sampling tube; j, plug; k, latex tubing; l, hose clamp; m, aquarium pump; n, wire hook.

dium. The cotton in the filters should be packed uniformly, but loosely. Overpacking the air inlet filter can cause resistance to incoming air and lead to poor aeration. Overpacking the outlet filter can lead to poor air escape and pressure buildup in the fermentor.

Disconnect the fermentor from the pump and prepare it for autoclaving. Make sure that the stopper which holds the air tubes is still firmly seated. The air-supply system must be well protected to prevent entry of contaminants. Wrap the top of each flask with a wide band of nonabsorbent cotton and secure it with a string. Then add a protective wrapper of aluminum foil (Figure 25.2). Close the air inlet tube with a clamp at the spot indicated in Figure 25.1 to prevent the broth from leaving the flask due to pressure buildup in the flask during autoclaving. Pressure relief during autoclaving occurs through the air outlet tube, which must be left open. The filters should remain connected to the fermentor during autoclaving. To provide a convenient place for them, make an oversized wire ring to fit snugly around the neck of the fermentor vessel and twist it to obtain an eyelet or loop on each side. Each filter may then also be fitted with a piece of wire ending in a small hook. Hook the filters onto the eyelet (Figure 25.2). Sterilize the assembly for 40 min, if it contains approximately 2 liter of broth. Adjust the sterilization time according to the volume of liquid; increase time by 10 min for each additional liter.

FIGURE 25.2 Simple fermentor in operation.

After the fermentor has cooled, remove the clamp from the air inlet tubing. Connect the air supply to check for proper aeration once again and for leaks in the system. Various types of air systems have been used to aerate small fermentors, including compressors, compressed air in tanks, aspirators, and aquarium pumps. The latter have been very satisfactory for small units and are inexpensive, silent, and dependable. Although a pressure relief valve may be desirable, it is not really necessary. Most aquarium pumps generate only low pressure, sufficient however, for several (four) fermentor units that may be connected to one aquarium pump using a manifold.

c. Operating the Glass Fermentors (Key Step 4)

General Operation

If, after autoclaving, the fermentor has been inspected and found to function properly, it is ready for inoculation with the starter culture. If an aquarium pump is used, and more than one fermentor is attached, adjust the air to achieve an equal flow to each fermentor. For other air-supply systems, adjust the air flow on the bypass, which may be installed between the pump and the air inlet filter (not shown).

The glass fermentor is inoculated through the latex air inlet tubing (at a point just above the main stopper) with a sterilized syringe fitted with an 18-gauge needle. Care must be taken that no contaminants are introduced. Twenty milliliters of the starter culture are removed aseptically from its flask. The air inlet tubing is swabbed with 70% alcohol (or 3% hydrogen peroxide) about 1 in (2.5 cm) above its connection to the glass tube. The needle is inserted downwards into the tubing and the culture is injected. The airstream will facilitate speedy entry and incorporate the starter inoculum into the YMB. The culture is incubated at 25–30°C under continuous aeration.

Sampling Procedures

Aseptically, with a sterile syringe, withdraw culture broth from the fermentor through the sampling tubing attached to the sampling port. Swab the tubing with 70% alcohol or 3% hydrogen peroxide. Insert the needle into the sterilized portion of the tubing and withdraw the desired amount of culture broth. For quality control purposes (such as Gram stain, pH measurements, optical density measurements, the total count, and plate counts), 5–10 ml of culture are sufficient and may be withdrawn by using a 5- or 10-ml syringe fitted with a 22-gauge needle.

For injecting the broth culture into bags of sterile carrier (peat), 40-ml samples are usually withdrawn with a sterile 50-ml syringe fitted with an 18-gauge needle (Figure 25.3). Alternatively, an automatic motorized syringe equipped with a 16-gauge needle may also be used to withdraw broth culture if large numbers of bags are to be injected.

d. Producing Broth Inoculum (Key Steps 5–10)

When the starter cultures have reached the end of their log phase of growth (7 days for *Bradyrhizobium* spp. and 5 days for *Rhizobium* spp.), they are ready to be used for

FIGURE 25.3 Inoculant production with broth culture from a glass fermentor.

inoculating the fermentor. Inoculate one fermentor with *B. japonicum* (e.g., TAL 102) and the other with *Rhizobium* sp. (e.g., TAL 1145). Take a 10-ml sample from each fermentor at the end of the growth period of each strain and conduct the following tests:

1. pH tests: A contamination problem is usually evident when the pH of the broth decreases toward acidity, especially with bradyrhizobia. This is because *Bradyrhizobium* spp. cultured in fermentors do not decrease the pH of the medium. However, with *Rhizobium* spp. the pH test is less helpful because most contaminants, like the *Rhizobium* spp., are usually acid producers. Test the broth pH of both the rhizobia by adding two drops of bromthymol blue (BTB) (0.5% w/v in alcohol) in 1 ml of broth. With most strains of bradyrhizobia tested, the pH does not change during mass culturing. A yellow coloration indicates acidity (presence of contaminants) and a green to blue coloration indicates alkalinity (absence of contaminants).

2. Gram stain (Chapter 3).

3. Peptone glucose test (Chapter 3).

4. Total count with Helber or Petroff-Hausser counter (Chapter 5).

5. Optical density measurement (Chapter 5).

6. Spread-plate count on yeast-mannitol agar (YMA) containing congo red (CR) and on YMA containing BTB (Chapter 5).

7. Agglutination with the homologous antiserum: This test should be done just before harvesting when the culture has no less than 1×10^9 cells ml^{-1}. Dilute 2 ml of the cell suspension with 2 ml of saline. Mix well and heat in boiling water for 30 min. After cooling, pipette 0.5 ml into an agglutination tube and add 0.5 ml of a 1:50 dilution of the homologous antiserum, which should have a titer of at least 800. Perform the agglutination test as described in Chapter 9.

The broth cultures may be incorporated into carrier material when the total count indicates a cell concentration of more than 1×10^9 cells ml^{-1} and purity of culture has been established.

REQUIREMENTS

a. Inoculating Starter Cultures

Agar slant cultures of *B. japonicum* strain TAL 102 and *Rhizobium* sp. strain TAL 1145
Transfer chamber
Platform shaker
Inoculation loop, flame
Erlenmeyer flasks or screw-capped tubes of 50-ml capacity each containing 25 ml of YMB

b. Assembling Simple Fermentors

Large autoclave, aquarium pumps or compressor
Cork borer, small glass file, Bunsen burner
For each fermentor: Erlenmeyer flask, 4 liters. [This flask is modified by the addition of an outflow tube (4-mm i.d.) at its base. These modified flasks are not available commercially, but any glassblower should be able to attach the short 3–5-ml glass tube.]
No. 12 autoclavable stopper
Glass tubing (4-mm i.d.), approximately 120 cm
Glass tubing (30-mm i.d.), two pieces of 10-cm length for making air filters. Barrels of 50-ml syringes may be cut to size and used instead.
Rubber stoppers no. 4, autoclavable, four pieces
Hose clamps (two) air bypass (T piece with short latex tube and clamp)
Surgical rubber tubing (4-mm i.d.), approximately 150 cm
Glass wool; cotton wool, nonabsorbent; aluminum foil

Sampling tubes (4-mm i.d.)
YMB (2–3 liters per fermenter)

c. Operating the Glass Fermenters

Sterile syringes, 30 ml, with 18-gauge needles
70% alcohol, cotton swabs or tissue paper
Broth culture of *B. japonicum* TAL 102
Broth culture of *Rhizobium* sp. TAL 1145
Sterile syringes, 10 ml, with 22-gauge needles and 18-gauge needles
Sterile test tubes (for samples)

d. Producing Broth Inoculum

Spectrophotometer; cuvettes; transfer chamber
Antisera of TAL 102 and TAL 1145
Plates of peptone glucose agar
Plates of YMA containing BTB
Plates of YMA containing CR
Syringes, 20-30 ml with 22-gauge needles and 18-gauge needles
Sterile test tubes (for samples)
Sterile pipettes, 1 and 10 ml
Tubes containing 9 ml of sterile diluent, rack
Sterile Pasteur pipettes, calibrated
Solution of BTB (0.5% w/v in ethanol)
Materials and supplies for Gram stain (Appendix 3)

KEY REFERENCE

Thompson, J.A. 1980. Production and quality control of legume inoculants. pp. 489–533. *In* F.J. Bergersen (ed.) Methods for Evaluating Biological Nitrogen Fixation. John Wiley & Sons, New York.

26

Producing Inoculum in a Steel Fermentor

The mass culture of rhizobia in large volumes is a major process in commercial inoculant production. Mass culture requires large capacity fermentors for commercial production. Fermentors must be simple enough to easily sterilize the growth medium, and provide access for inoculating, sampling, aerating, and cleaning. The fermentor should be made of stainless steel for strength and corrosion resistance. This chapter describes the procedure for mass culturing rhizobia in a 141-liter stainless steel fermentor, designed by the NifTAL Project.

KEY STEPS/OBJECTIVES

1. Prepare starter cultures.
2. Inspect the fermentor for use and function.
3. Test the fermentor for absence of air leaks.
4. Prepare air filters.
5. Test the fermentor for proper function.
6. Install the air inflow filter.
7. Prepare growth medium.
8. Operate the fermentor for growth medium sterilization.
9. Check the growth medium for absence of contamination.
10. Inoculate the fermentor.
11. Operate the fermentor for broth culture production.
12. Perform quality control on the fermentor broth culture.

a. Inoculating Starter Cultures for the Steel Fermentor (Key Step 1)

The commercial-size steel fermentor of 100-liter working capacity requires 1–2 liters of starter culture. The glass fermentors described in Chapter 25 are ideal for this purpose.

Set up two glass fermentors as described in Chapter 25 and inoculate each fermentor with 50 ml of starter cultures of a *Bradyrhizobium* spp. (e.g., *B. japonicum* TAL 102) and a *Rhizobium* spp. (e.g., *Rhizobium* sp. TAL 1145 from *Leucaena leucocephala*).

b. Becoming Familiar with the Fermentor (Key Step 2)

Use the diagram in Figure 26.1 to become familiar with the fermentor. The body of the fermentor is a pressure vessel with a 141-liter total capacity. It has a working capacity of 20–100 liters. It is domed at the top and bottom, and held upright by a welded-on stainless steel skirt. The top has a centrally positioned oval opening with a snap-type closure, which uses a Viton O-ring seal. Encircling the opening are the following accessories: a steam pressure gauge, a pressure relief valve, and an aeration system with filters for the intake and exit of sterile air. Inlet and outlet ports for water passage through the builtin stainless steel cooling coil are not shown in Figure 26.1. The inoculation port, thermometer, and the sampling port are positioned on the vessel wall. The fermentor is situated on a sturdy steel support that houses a 98,000 Btu gas burner. The air inlet hose is connected to a prefilter, which is attached to a regulated pressurized manifold (not shown in Figure 26.1). When in operation, the air manifold is kept at constant pressure.

c. Getting the Fermentor Ready for Operation (Key Steps 3 and 4)

To safeguard against contamination of the growth medium during mass culture, the fermentor must be leak proof. Assuming the fermentor is fully assembled, close all valves and turn on the air compressor for air supply. Now open the air inlet valves slightly (i.e., first the ball valve above the air inlet filter and the ball valve below the air inlet filter). Allow the air pressure to build up to 5lb/in^2, as indicated by the pressure gauge. If the line pressure is higher than 5 lb/in^2, open the air outflow filter slightly so that 5 lb/in^2 of pressure is maintained. Now check all valves and connections for air leaks by applying soap solution (0.5% detergent in water). Using a wash bottle filled with soap solution, apply a small amount of the soap solution to all screw-in connections and valves. Start with the hose connection of the air inlet hose and end with the sampling port valve at the bottom of the fermentor. Whenever bubbles form after the application of the soap solution, indicating leak(s), tighten the joints or connections to remove the leak. Once the fermentor has been established to be free of leaks, release the pressure via the air outlet filter.

Remove the air inlet and the air outlet (exit) filters. The air inlet filter has a ball valve on top that connects to the air hose with a snap fitting. The bottom of the filter is attached to the fermentor inlet valve via a union coupling. Loosen the union coupling

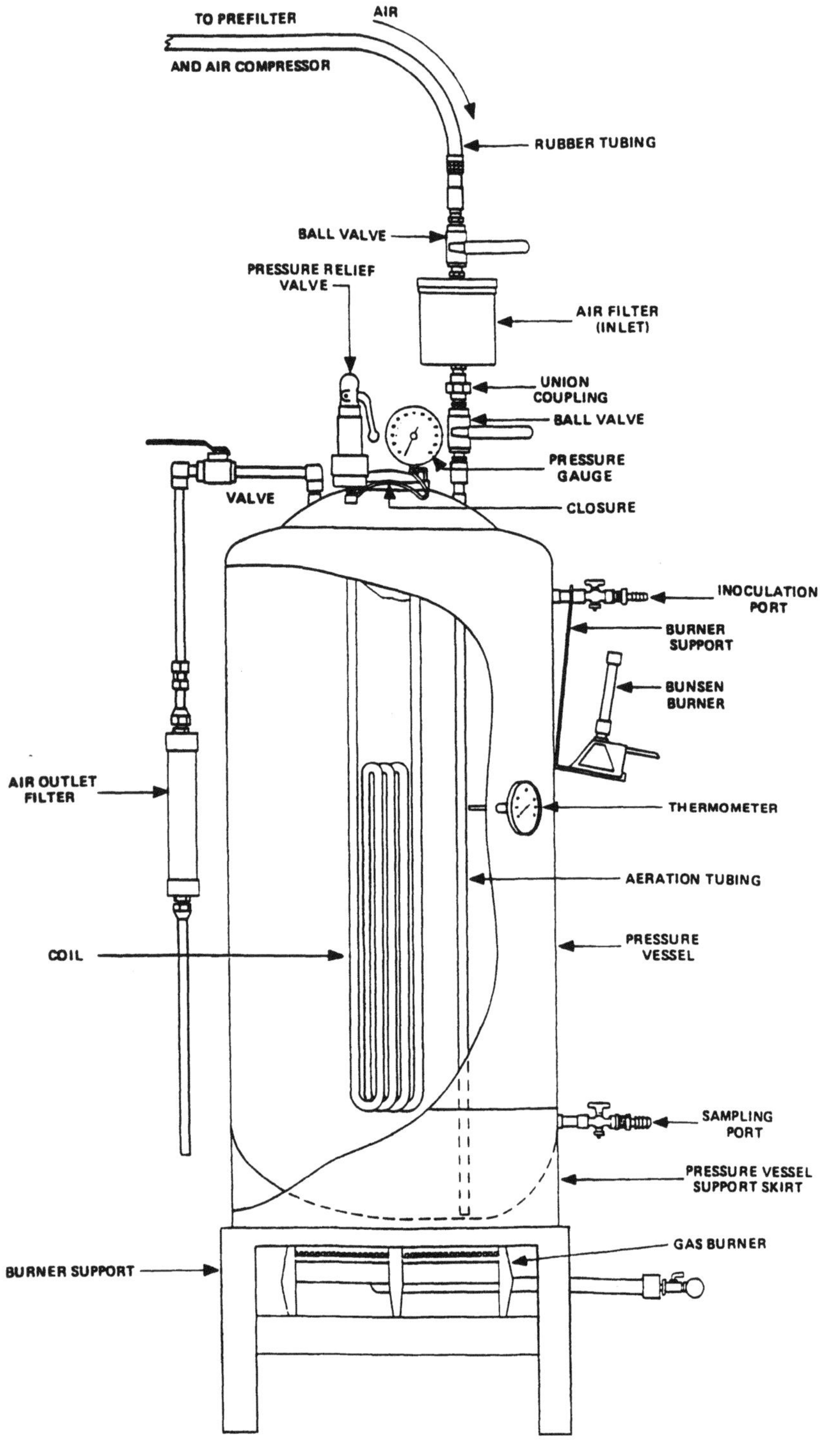

FIGURE 26.1 Steel fermentor.

with an adjustable wrench and remove the air inlet filter unit. Unscrew and remove the air inlet filter cap. Pack the filter with layers of nonabsorbent cotton. The cotton must be packed uniformly and loosely to prevent the air from channeling while under pressure. Screw the filter cap back on and wrap the whole filter unit with aluminum foil. Autoclave the filter unit at 121°C and 15 lb/in^2 for 1 h. Store the sterilized air inlet filter unit in a clean environment until needed. The in-line air outlet filter need not be autoclaved. Remove it with a wrench and pack it loosely with fine, high-grade glass wool. Do not use the coarser insulation grade material. Reattach the filter.

d. Pretesting the Fermentor (Key Steps 5 and 6)

A pretest is necessary if the fermentor is used for the first time or if it has not been used for a while. The fermentor is fully assembled, but the air inflow filter is not yet attached. Through the main opening, fill the vessel with 40–100 liters of water. Close the opening with the snap-type closure. The snap-type closure with the O-ring in place must be immersed in water to obtain a proper seal. Wet the O-ring before closing. Close all valves except the air outlet valve, which should be left open to allow air and steam to escape when the growth medium is boiling. Light the burner and bring the growth medium to boiling under maximum heating. When the boiling point is reached, turn the burner to the lowest flame level at which boiling can be maintained. Adjust the air outlet valve to allow steam to escape slowly in order to sterilize the glass wool packed in the outlet filter. After 15 min, close the outlet filter completely to allow pressure to build up inside the vessel. The relief valve will discharge to control the pressure when the pressure reaches more than 15 lb/in^2. Allow the relief valve to discharge twice or more. As the pressure rebuilds, reduce the flame and open the air outlet valve slightly to maintain a pressure of 15 lb/in^2 at 121°C for 5 min. Now recheck all joint connections and valves for leaks, evidenced by escaping steam. If necessary, tighten joints to stop leaks.

Install the air inlet filter while the fermentor is still hot. (The installation needs to be done carefully and quickly.) Squirt alcohol into the lower half of the union coupling located above the air inflow valve. Ignite it with the flame of a Bunsen burner or gas torch. Just before the flame extinguishes, quickly unwrap the air inlet filter, bring the union ends together, and secure the filter in place by hand-tightening the coupling screw. Heat the union coupling with a flame for approximately 30 s and use a wrench to further tighten the coupling.

The filter packing should be replaced and the filter unit resterilized after each production run. (Although experienced operators have used a filter for as many as 10 runs before repacking and resterilizing.) Turn off the flame and gradually release the tank pressure by opening the air outlet valve. Turn on the water to the cooling coil to cool down the contents of the fermentor.

e. Sterilizing the Growth Medium (Key Steps 7 and 8)

When cool, empty the fermentor with a water pump and hose, and/or by draining it at the sampling port. Using a flashlight to illuminate the interior, inspect the vessel for cleanliness. Make sure the sampling valve is closed and fill the vessel with 90 liters of clean water. In a clean plastic bucket or other suitable container, prepare a concentrate of the growth medium (Appendix 3) in 10 liters of water. Pour the concentrated growth medium into the 90 liters of water in the fermentor. Check the pH and adjust to 6.8 if necessary. Close the fermentor opening and all valves except the air outlet valve. Wrap aluminum foil around the inoculation and sampling ports. Light the burner and bring the medium to a sterilizing pressure and temperature as described in (d).

After a 45-min sterilization period, turn off the burner and slowly release the pressure by opening the air outlet valve, allowing the steam to escape through the outlet filter. When the pressure reaches 10 lb/in^2, slowly turn on the cooling water. Carefully control the air outlet valve so the pressure decreases slowly. A rapid drop in tank pressure may cause a partial vacuum in the vessel; this should be avoided. Increase the flow of the cooling water when the tank temperature has reached 90°C. When the temperature has reached 30°C, shut off the cooling water, open the air outlet valve completely, and allow the medium to equilibrate to ambient temperature (overnight). If the growth medium is not completely sterilized, surviving contaminants (e.g., spore-forming bacteria) will grow during this period.

f. Mass Culturing Rhizobia (Key Steps 9–12)

The next step is to check the sterility of the growth medium. Spray the sampling port with alcohol and thoroughly flame it with a torch. Open the port with a valve tool or adjustable wrench and allow a small amount of broth to flow out without being sampled. Then quickly and aseptically obtain a 50-ml sample in a sterile flask. Perform the following tests:

1. Smell: The medium should have the odor of the yeast-extract mineral salts medium.
2. Clarity: A clear medium indicates the absence of contaminants. However, if the water used is rich in minerals, precipitation may cause turbidity, usually associated with contamination.
3. pH: A near-neutral pH (6–7) is expected in a sterile medium.
4. Gram stain: Perform a Gram stain (Chapter 1) if any turbidity is detected.

If the medium is found free from contamination, inoculate the fermentor. Use the starter culture of *B. japonicum* (TAL 102) from (a) for this purpose. Use the starter culture of *Rhizobium* sp. (TAL 1145) for inoculating a second fermentor.

Remove the aluminum foil wrapping and spray the inoculation port of the steel fermentor with alcohol and flame thoroughly. Allow the port to cool. Remove the protective cover from the glass fermentor latex sampling tubing. Close the latex tubing with a clamp 2–4 cm closer to the flask. Briefly spray the tubing with alcohol and flame. Allow the alcohol to burn off without melting the latex tubing. Quickly cut the latex tubing with sterile scissors. Slip the cut end of the tubing over the flame-sterilized inoculation port and release the clamp on the latex tubing. Open the inoculation port valve and allow the starter culture (inoculum) to flow into the steel fermentor (Figure 26.2). Close the inoculation port and disconnect the latex tubing. Flame the inoculation port until remaining inoculum inside the inoculation port has burned off. Replace the aluminum foil wrapping.

With the air outlet valve open, turn on the air pump. Open the top ball valve attached to the air inlet filter first. Next, open the second ball valve that is closest to the tank. Watch the pressure gauge. Adjust air inlet and outlet valves so there is no internal pressure buildup. Regulate the airflow to 3–10 liters of air per hour per liter of medium. The sterile air provides aeration as well as agitation for the rhizobial growth.

Sample the contents of the fermentor about 3 days after the starter culture is inoculated. Aseptically remove the broth through the sampling port as previously described in (f). Monitor the growth of the culture by total count or optical density (OD) measurement (Chapter 5). Perform pH measurements and Gram stain as checks for contamination. The rhizobial population in the culture medium should reach full growth (approximately 1.5–2.5×10^9 cells ml^{-1}) 3 or 6 days after inoculation, depending on the type of rhizobia.

Turn off the air supply (aeration) and perform a final quality test of the culture in

FIGURE 26.2 Inoculating the steel fermentor.

the steel fermentor. Use the measurements or the total count to monitor cell numbers. Use pH measurements and Gram stain to check for contamination. Use agglutination to test for strain identity (Chapter 9).

REQUIREMENTS

a. Inoculating Starter Cultures for the Steel Fermentor

Transfer chamber
Broth culture, 100 ml, of *B. japonicum* TAL 102 in 250-ml Erlenmeyer flasks
Broth culture of *Rhizobium* sp. TAL 1145
Cotton swabs or tissue paper
Alcohol, 70%
Fully assembled glass fermentors, each filled with 2 liters of yeast-mannitol broth (YMB)
Sterile syringes, 30 ml, with 18-gauge needles

b. Becoming Familiar with the Fermentor

NifTAL-designed 141-liter stainless steel fermentor (NifTAL Project, Maui, HI; basic vessel built by Alloy Products Corp., Waukesha, WI)
Air pump and air manifold
Fermentor stand and gas burner
Viton O ring (Alloy Products Corp.)

c. Getting the Fermentor Ready for Operation

All equipment as in (b)
Autoclave
Wash bottle with 1% liquid soap solution
Set of wrenches or an adjustable wrench
Nonabsorbent cotton
Aluminum foil
Glass wool

d. Pretesting the Fermentor

Fully assembled fermentor with stand and pump
Sterilized air inflow filter
Source of water, filler hose
Gas lighter

Valve-opening tool
Set of wrenches or an adjustable wrench
Alcohol in spray bottle
Bunsen burner or torch

e. Sterilizing the Growth Medium

All equipment and supplies as in (d)
Salts and nutrients for fermentor medium (Appendix 3)

f. Mass Culturing Rhizobia

Fermentor containing sterilized medium
Alcohol in spray bottle
Valve-opening tool
Sterile flask, 50 ml
pH meter
Gram stain solutions (Appendix 3)
Microscope slide
Microscope with phase-contrast condenser
Glass fermentor with 1–2 liters of starter culture
Bunsen burner or torch, lighter
Petroff-Hausser counting chamber
Sterile Pasteur pipette with rubber bulb
Saline
Test tube
Antisera specific to TAL 102 and TAL 1145

KEY REFERENCES

Food and Agriculture Organization of the United Nations. 1991. Expert Consultation on Legume Inoculant Production and Quality Control. Rome, 19–21 March 1991. Food and Agriculture Organization of the United Nations, Rome.

Thompson, J.A. 1980. Production and quality control of legume inoculants. pp. 489–533. *In* F.J. Bergersen (ed.) Methods for Evaluating Biological Nitrogen Fixation. John Wiley & Sons, New York.

27

Preparing a Range of Carrier Materials and Producing Inoculants

After rhizobia have been cultured in broth, they are usually incorporated into carrier material where they remain until use. Although peat is generally the most suitable carrier, other materials are used if peat is not available. In any case, the suitability of the carrier needs to be confirmed. In this exercise, carriers for rhizobia are prepared from various materials such as peat, charcoal, and lignite. These carriers are used for the production of granular and powdered inoculants. The quality of these inoculants is tested and compared.

KEY STEPS/OBJECTIVES

1. Select and dry carrier materials.
2. Grind carrier materials.
3. Sift carrier materials and select suitable particle sizes for granular and powdered inoculants.
4. Neutralize carrier materials.
5. Determine water-holding capacity of carriers.
6. Package the carrier materials.
7. Sterilize the carriers.
8. Examine the carriers for sterility after sterilization.
9. Inoculate carriers with broth cultures from fermentors.
10. Plate peat cultures for quality control.
11. Inoculate plants for the plant infection count.
12. Test strain identity serologically.
13. Record and tabulate results. Compare carrier treatments.
14. Apply quality standards.

a. Milling Inoculant Carrier Materials (Key Steps 1, 2, and 3)

Carrier materials are chosen to fill criteria set forth in the introduction to Section IV. For this exercise, select peat, charcoal, and lignite, or three other carrier materials if these are not available. Work with each carrier individually. Weigh 5 kg of each carrier and grind it in a hammer mill. Thoroughly clean the hammer mill with a brush or with an air jet from a compressor before grinding the next carrier.

Stack up a set of sieves in series: 16 mesh (1 mm), 42 mesh (355 μm), 100 mesh (150 μm), and 200 mesh (75 μm). Place this series of sieves on a collecting pan, and clamp the stack and collecting pan to a sieve shaker. Add the milled carrier to the uppermost sieve and activate the shaker for 60 min. Collect the fraction caught on the 42-mesh sieve and the fraction caught in the pan. The remainder should be returned to the mill and ground again. Particles of 16–42 mesh are used for preparing granular carriers (soil implants); particles of 200 mesh and finer make carriers suitable for seed coating.

b. Preparing and Characterizing Inoculant Carriers (Key Steps 4, 5, and 6)

The pH of an inoculant carrier should be around 6.5–7.0. In a 400-ml glass beaker, suspend 10 g of the carrier into 90 ml of water. Stir the mixture on a magnetic stirrer while monitoring the pH with the electrode of a pH meter. If the pH is lower than 6.5, gradually add precipitated, powdered calcium carbonate ($CaCO_3$) until a pH of 6.5 has been reached. Record the amount of $CaCO_3$ needed to neutralize 10 g of the carrier. Add a corresponding amount to the remaining carrier, e.g., if 0.25 g were needed to neutralize 10 g of carrier in the water suspension, add 2.5 g of $CaCO_3$ to every 100 g of dry carrier. Mix well by hand. Repeat the same procedure for all carriers.

The water (moisture)-holding capacity of a carrier determines the maximum amount of liquid inoculum that can be added to it. Carriers vary greatly in their water-holding capacities. Before the water-holding capacity can be measured, the inherent moisture level in the carrier must be determined. This may be done most conveniently on a moisture balance. Use a drying oven if a moisture balance is not available. Weigh 10 g accurately on a foil or glass weighing dish and place it into the oven at 70°C for 24 h. Weigh and return to the oven. Another weighing at 48 h will confirm the endpoint of moisture loss.

Use the following formula to calculate the inherent moisture content on the dry-weight basis.

$$M = \frac{(W_1 - W_2) \times 100}{W_2}$$

where M = Moisture content (%)
W_1 = Weight of carrier before drying
W_2 = Weight of carrier after drying at 70°C

Proceed to determine the moisture-holding capacity of the carrier. Weigh 100 g of oven-dried carrier material into a 500-ml beaker. Add water with continuous stirring until the carrier appears to be saturated. Add additional water to produce a thin slurry. Transfer this slurry to a preweighed measuring cylinder, which has a sieve-covered drainhole on its bottom. Allow the water to drain overnight, then weigh the measuring cylinder with the contents. Give the moisture-holding capacity on the dry-weight basis of the carrier. For example, if 100 g of predried carrier can hold 120 ml of water, then its moisture-holding capacity is 120%.

The amount of inoculum broth to be added to the carrier must be well below the carrier's moisture-holding capacity as the resulting inoculum should be friable in texture. It is, however, desirable to add the largest amount possible while still retaining the desirable texture. A high moisture level is necessary because moisture is lost during storage, and the survival of rhizobia in a carrier is affected by low moisture levels.

Proceed to determine the desirable amount of moisture to be added to the carrier by a trial-and-error method. Prepare six bags (polyethylene 127 × 178 × 0.076 mm) of each neutralized carrier (50 g per bag). To the first bag add an amount of water that is approximately 5 ml less than the carrier's moisture-holding capacity. If this moisture-holding capacity is 60 ml (or 120%), add 55 ml. To the next bag add 5 ml less (50 ml). Continue until each successive bag has received 5 ml less than the preceding one. Thus, bag 6 will receive 30 ml of water. Seal the bags with a bag sealer and incorporate the water into the carrier by kneading. Knead or massage the bags thoroughly until all moisture has been absorbed and the carrier/water mixture appears to be homogeneous.

Examine the bags for total absorption of the water. Check for dry areas in the carrier that can usually be recognized because an unwetted carrier has a lighter color. Allow the six treatments to equilibrate for 2 h, then cut the bags open and sample a few grams of each bag with your hand. A suitable carrier/water mixture should feel moist, but not soggy. It should crumble in your hand (i.e., be friable) and it should not be sticky. From each representative carrier select that treatment which has absorbed a maximum amount of water while still retaining friability. Record the carrier/water ratio and use this information to calculate the recommended moisture level for each carrier. The recommended moisture level is usually given in percent, calculated on the wet-weight basis of the final preparation. Of course, the inherent moisture level of the carrier must be considered. The inoculants total moisture content is the sum of the weights of the broth culture and the inherent moisture of the carrier. Thus, a 90-g package of inoculum with a moisture content of 50%, made from a carrier with an inherent moisture level of 10%, contains 45 g of dry carrier, 5 g of inherent moisture, and 40 g of broth inoculum.

Determine the moisture-holding capacity of all the carriers used (powdered and granular), then prepare them as outlined in Table 27.1. A similar table may be made for the granular carriers. Record the moisture-holding capacities in the last column of Table 27.1.

Gamma-irradiation (5 Mrads) is preferred for peat sterilization over autoclaving. Gamma-irradiated peat is used here in one treatment only since irradiated peat is often unavailable. It serves as a standard because its properties as carrier material for various

TABLE 27.1 Carrier Types, Treatments, and Quantities Required for Inoculant Preparation and Evaluation

Carrier	Sterilization Treatment	Carrier Quantity	Recommended Moisture Level
Peat	Gamma-irradiated	4 bags × 50 g	50%
Peat	Autoclaved	4 bags × 50 g	[1]
Peat	Autoclaved	2 trays × 1 kg	[1]
Peat	Not sterilized	2 trays × 1 kg	[1]
Charcoal	Autoclaved	4 bags × 50 g	[1]
Charcoal	Autoclaved	2 trays × 1 kg	[1]
Charcoal	Not sterilized	2 trays × 1 kg	[1]
Lignite	Autoclaved	4 bags × 50 g	[1]
Lignite	Autoclaved	2 trays × 1 kg	[1]
Lignite	Not sterilized	2 trays × 1 kg	[1]

[1]To be determined.

strains of rhizobia are well known. It is regularly used for inoculant production at NifTAL. It is packaged and sealed in 127- × 178-mm polyethylene bags of 0.076-mm thickness. Weigh 50-g portions of all other carriers into 127- × 178- × 0.076-mm autoclavable (polypropylene) bags. Add 1 ml of water per bag. Make an incomplete heat seal, leaving the bags slightly open. Autoclave the bags in a foil-covered tray. After the bags are cool, completely heat seal in a sterile hood.

For the bulk preparations, place 1 kg of neutralized carrier into each of four autoclavable trays, approximately 46 × 46 cm wide and 10 cm deep. Spread into an even layer and cover with aluminum foil. Set aside two of these trays as nonsterilized treatments. Autoclave the other two trays at 121°C and 15 lb/in^2 for 1 h. Allow to cool in the autoclave overnight. After sterilization, test a representative sample of each autoclaved carrier for sterility as described in Chapter 28.

c. Producing Inoculants (Key Steps 7, 8, and 9)

Prepare inoculants following the treatments and replications outlined in Table 27.1. Obtain the fermenter cultures of *B. japonicum* (TAL 102) and *Rhizobium* sp. (TAL 1145), which were produced in Chapter 25. Use broth culture of TAL 102 to inoculate each 1-kg portion of the autoclaved carriers in trays. Add broth culture according to recommended moisture levels determined in (b). Use your gloved hands to mix the broth into the carrier until its consistency becomes uniform. (Tools are not needed for mixing, but your hands should be covered with sterile gloves to minimize contamination.) Replace the foil cover and allow the inoculant to mature at 25–30°C for 2 weeks. Repeat this procedure with TAL 1145 using the second autoclaved tray of each carrier. Similarly, prepare inoculants by hand-mixing the untreated (nonsterile) bulk carriers with broth cultures of TAL 102 and TAL 1145.

The presterilized carrier materials in sealed bags are aseptically injected with the suitable amount of broth culture using a sterile 50-ml syringe fitted with a sterile 18-gauge needle as follows. Withdraw the desired amount of broth culture from the outlet tubing of the glass fermentor as described in Chapter 25. Sterilize a small area in a corner of the carrier bag with 70% ethanol. Puncture the bag in the sterilized area and insert the needle carefully to avoid piercing the opposite wall of the bag. Inject the desired amount of inoculum, aiming the tip of the needle toward the center of the bag. Seal the puncture hole with plastic labeling tape and write the treatment number, the strain used, and the date of preparation on the tape. Work the broth into the peat by kneading the bags until the liquid inoculum has been uniformly absorbed by the carrier. Incubate at 25–30°C for 2 weeks. Obtain inoculants prepared earlier and stored for 6 months at room temperature. One bag of each will be used in (d).

d. Testing the Quality of the Inoculants (Key Steps 10, 11, and 12)

Rhizobia in the various treatments are expected to reach their maximal population 2 weeks after inoculation. Determine the number of viable rhizobia in all treatments. Inoculants prepared in bagged gamma-irradiated and autoclaved carriers are not expected to contain many contaminants. The usual recommended counting technique is the drop-plate method (Chapter 5).

In this exercise, rhizobia in the presterilized carriers should also be enumerated using the plant-infection technique because the populations in sterile- and nonsterile-based carriers must be compared using the same enumeration methods. Make serial dilutions of duplicate samples of the 2-week-old inoculants and those stored for 6 months. Plate dilutions ranging from 10^{-4}–10^{-7} on yeast-mannitol agar (YMA) + congo red (CR) and on YMA + bromthymol blue (BTB). If proper aseptic procedures are not fully observed, contaminants may be accidentally introduced during injecting the broth culture and during serial dilution and plating. Such contaminants will usually be detectable on these indicator media and their number should also be reported.

The hand-mixed inoculants, especially those based on nonsterilized carriers, can be expected to contain contaminants. The plant infection count will be necessary for a reliable determination. Plate counts on indicator media may be used to give a measure of the contaminants. Set up the plant infection most-probable-number (MPN) count in growth pouches using (*Leucaena leucocephala*) as host for TAL 1145 and soybean (*Glycine max*) for TAL 102 (Chapter 6). Plate dilutions of sterile- and nonsterile-based carriers from 10^{-4}–10^{-7}. Plate both the 2-week-old inoculants and those stored for 6 months from (c).

e. Collecting, Recording, and Analyzing the Data (Key Steps 13 and 14)

Determine the number of viable rhizobial cells in the various carrier treatments as described in (d). Transform the data to $\log_{10}$ and calculate the mean for the replications.

Organize the data in the format as shown in Table 27.2. The experiment with each rhizobial strain is a factorial involving three carriers (peat, charcoal, and lignite), two carrier forms (powdered and granular), and three carrier-sterility conditions. Assistance may be needed for statistical analysis of the data.

Compare the 2-week-old inoculants and note the decline in cell numbers. Compare the different treatments and decide whether the number of cells in the 6-month-old inoculants is sufficiently high to comply with minimum standards of quality. Minimum standards are given for the date of expiration, usually 6 months after manufacture. The minimum standards vary in different countries. In Canada, 10^6 viable rhizobia per gram of peat are acceptable. In the United States, there is no federal regulation for legume inoculant quality. Some of the states, however, have their own standards, as do the inoculant manufacturers. Australia, like NifTAL, requires a minimum of 1×10^9 viable rhizobia per gram at expiration. These inoculants are produced with irradiated peat.

Examine the results critically and contemplate the following questions:

TABLE 27.2 Multiplication of *B. japonicum* (TAL 102) in Inoculants Prepared from Various Carriers and under Different Sterility Conditions

	Log_{10} No. Rhizobia g^{-1} Moist Inoculant					
Carrier	Peat		Charcoal		Lignite	
Treatment	Powdered	Granular	Powdered	Granular	Powdered	Granular
Autoclaved in polypropylene bag	____	____	____	____	____	____
Autoclaved in polypropylene trays	____	____	____	____	____	____
Untreated in polypropylene trays	____	____	____	____	____	____
Irradiated in polyethylene bags	____	ND[1]	ND	ND	ND	ND
Mean	____	____	____	____	____	____

[1]ND, not done.

- Are all treatments well above the NifTAL minimum standard for inoculants at expiration?
- Which level of sterility contributes to the highest cell population?
- How are the carriers affected by the sterilization measures with respect to their ability to support high cell populations?
- Why are different counting methods suggested for different levels of sterility?
- Compare bulk sterilization of carriers in trays with bag sterilization and explain the advantages and disadvantages of each method.

REQUIREMENTS

a. Milling Inoculant Carrier Materials

Hammer mill with collecting tray, bag, or bucket
Screen shaker equipped with 16-mesh (1 mm), 42-mesh (355 μm), 100-mesh (150 μm), and 200-mesh (75 μm) screens (60 × 60 cm or larger)
Balance 1–5000-g capacity
Unground dried peat, 5 kg
Unground charcoal, 5 kg
Unground lignite, 5 kg
Scoop or small shovel
Brush or air jet from compressor for cleaning the hammer mill
Other locally available carrier material (e.g., filter mud, bagasse, coir dust to replace some or all the previously mentioned carrier materials if these are not available, 5 kg of each
Aluminum foil, large roll
Trays to contain 1–5 kg of carrier material

b. Preparing and Characterizing Inoculant Carriers

Transfer chamber
Balance, toploading 0.1–100-g capacity
Magnetic stirrer and 1-in (2.5 cm) stirring bar
pH meter
Moisture balance or drying oven
Bag sealer
Autoclave
Glass beakers, 500 ml
Bottle of pH 7 buffer solution

Beakers, 50 ml, for pH meter calibration
Autoclavable trays, approximately 46 × 46 × 10 cm
Weighing dishes (metal or glass)
Measuring cylinder (250 ml) with drainhole and sieve
Packaged gamma-irradiated peat
Sterile pipettes, 1 ml, one canister
Pipettes, 5 ml
Measuring cylinders, 50 ml
Aluminum foil, large roll
Scissors
Alcohol in spray bottle
Dilution tubes, each containing 9 ml of water
Rubber bulbs, 1 ml (for pipetting)
Calibrated, sterile Pasteur pipettes
Bottle of distilled water, 1 liter
Carrier materials from (a)
YMA plates containing CR
YMA plates containing BTB
Polypropylene and polyethylene bags (127- × 178- × 0.076-mm wall thickness)
Precipitated $CaCO_3$

c. Producing Inoculants

Incubator
Broth culture of *B. japonicum* TAL 102 and *Rhizobium* sp. TAL 1145, approximately 13–15 liters
Four bags each of autoclaved, powdered carriers prepared from three different materials, e.g., peat, charcoal, and lignite from (b)
Four bags each of autoclaved granular carriers of the same materials, also from (b)
Four 50-g bags of neutralized powdered gamma-irradiated peat packaged in polyethylene bags (127- × 178- × 0.076-mm thickness)
Carrier material in trays as prepared in (b)
Powdered and granular peat, two batches of 1 kg each, autoclaved
Powdered and granular peat, two batches of 1 kg each, nonsterile
Powdered and granular charcoal, two batches of 1 kg each, autoclaved
Powdered and granular charcoal, two batches of 1 kg each, nonsterile
Powdered and granular lignite, two batches of 1 kg each, autoclaved
Powdered and granular lignite, two batches of 1 kg each, nonsterile
Three bags each of inoculants of *Rhizobium* sp. (e.g., TAL 1145) and *Bradyrhizobium* sp. (e.g., TAL 102) made from each of the carriers previously listed and stored for 6 months at 26°C
Surgical gloves (one package)

Sterile plastic syringes, 50 ml (one box)
Sterile syringe needles, 18 gauge
Alcohol, 70%
Tissue paper
Labeling tape

d. Testing the Quality of the Inoculants

Plates of YMA + CR
Plates of YMA + BTB
Plates of YMA + brilliant green
Sterile serological pipettes, 1 ml; glass spreaders
Sterile, calibrated Pasteur pipettes
Dilution bottles with 99 ml of sterile diluent
Test tubes containing 9 ml of sterile diluent
Test tube racks
Wrist-action shaker (optional)
Balance, spatula, weighing paper
Growth-pouch racks, growth pouches
Sterile plant nutrient solution
Soybean and *L. leucocephala* seeds
Bottles of sterile water
Chlorine bleach (Chlorox) or hydrogen peroxide for seed sterilization
Concentrated sulfuric acid
Erlenmeyer flasks, 500-ml capacity

e. Collecting, Recording, and Analyzing the Data

Soybean and *L. leucocephala* plants from (d)
Plates with bacterial colonies from (d)

KEY REFERENCES

Parker, F.E., and J.M. Vincent. 1981. Sterilization of peat by gamma radiation. Plant Soil 61: 285.

Roughley, R.J. 1968. Some factors influencing the growth and survival of root nodule bacteria in peat culture. J. Appl. Bacteriol. 31:259–265.

Roughley, R.J., and J.M. Vincent. 1967. Growth and survival of *Rhizobium* spp. in peat culture. J. Appl. Bacteriol. 30:362–376.

28

Preparing Inoculants Using Diluted Cultures of Rhizobia and Presterilized Peat

The production capacity of small-scale inoculant production plants using presterilized peat can be increased by using diluted liquid cultures of rhizobia. In this exercise, fully grown cultures are diluted in water and other diluents of different formulations prior to incorporating of presterilized peat in packages or in polypropylene trays. The multiplication of rhizobia in the inoculants is studied.

KEY STEPS/OBJECTIVES

1. Culture *Rhizobium* sp. and *Bradyrhizobium* sp.
2. Make culture dilution flasks.
3. Prepare diluents in dilution flasks.
4. Prepare and package peat.
5. Sterilize peat in packages and polypropylene trays.
6. Prepare yeast-mannitol broth (YMB) + peat blanks and check for sterility.
7. Examine yeast-mannitol agar (YMA) congo red (CR) plates plated with YMB-peat blanks.
8. Perform viable counts on late-log-phase cultures.
9. Prepare diluted cultures.
10. Inject diluted cultures into peat.
11. Mix diluted cultures with autoclaved peat in trays and package.
12. Perform viable counts on inoculants at 2 weeks.
13. Perform viable counts on inoculants at 8 weeks.
14. Record and analyze the data.

a. Culturing Rhizobia in YMB (Key Step 1)

Prepare 500 ml of YMB in each of two 1-liter Erlenmeyer flasks. Inoculate one flask with *Bradyrhizobium* sp. (e.g., *B. japonicum* TAL 102) and the other with *Rhizobium* sp. (e.g., TAL 1145 from *Leucaena leucocephala*). Both rhizobia should have antisera available for strain recognition and confirming purity (by serology) to be done later in the experiment. Incubate the inoculated flasks at 25–30°C on a shaker. To obtain late-log-phase cultures, allow the fast- and slow-growing rhizobia to grow for 4–7 days, respectively. At the end of the specified growth period, check the purity of the culture by Gram stain and by serology (simple tube agglutination or by the fluorescent antibody (FA) technique as described in Section II).

b. Making a Culture Dilution Flask and Its Operation (Key Step 2)

The culture dilution flask is basically a 2-liter Erlenmeyer flask modified by a short glass-tubing outlet at the base of the flask as shown in Figure 28.1. Seek the assistance of a skilled glassblower for fitting the glass tubing to the base of the flask. Five culture dilution flasks are required per rhizobial strain (four for diluents and one for the undiluted culture as control, Table 28.1).

Attach a piece of surgical latex tubing of suitable size to the glass tubing outlet of each dilution vessel. Close the open end of the latex tubing with a plug made from a short piece of glass rod. Add appropriate diluent, close the flask, and sterilize the entire

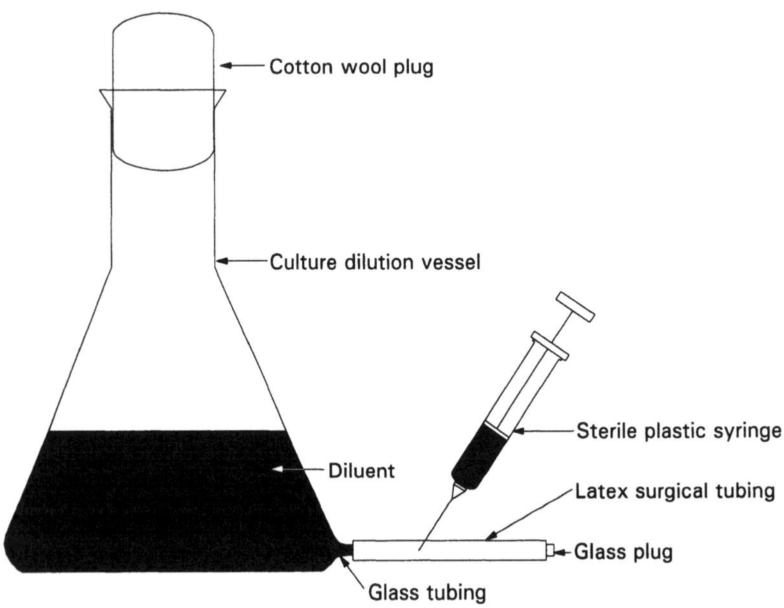

FIGURE 28.1 Apparatus for diluting rhizobial cultures.

TABLE 28.1 Protocol for Preparing Inoculants of TAL 102 and TAL 1145 with the Various Diluents and Sterilized Peat

	Packages of Sterilized Peat Needed per Strain		
Treatment	Gamma-Irradiated	Autoclaved	Total ml of Diluted Culture Needed per Strain
Water	6	6	420
YMB (20%)	6	6	420
YSB (20%)	6	6	420
YW (20%)	6	6	420
Control[1]	6	6	420

[1]Consists of undiluted late-log-phase cultures.

unit by autoclaving. To dilute the culture, aseptically introduce (with a pipette or a hypodermic plastic syringe fitted with a 3.5-cm and 14-gauge needle) the fully grown culture via the mouth of the culture dilution flask. Swirl the flask to ensure proper mixing and dilution of the culture in the diluent. Withdraw the diluted culture for inoculation with a sterile plastic syringe as described for the fermentor in Chapter 25.

c. Preparing the Diluents (Key Step 3)

The late-log-phase cultures of each strain are diluted in 20% (v/v) solutions of YMB, yeast sucrose broth (YSB), yeast water (YW), and distilled (or deionized) water. YSB has the same ingredients as YMB (Appendix 3) except that sucrose (10 g/liter) is substituted for mannitol. YW is prepared by dissolving 0.4 g of yeast extract (Difco Laboratories, Detroit, MI) in 1 liter of distilled or deionized water.

Accurately prepare 500 ml of 20% YMB, YSB, and YW by mixing 100 ml of full-strength media with 400 ml of distilled (or deionized) water in the culture dilution flasks. Prepare each diluent in duplicate since two strains will be used. Sterilize the diluents by autoclaving in the dilution flask. Also, fill two 2-liter Erlenmeyer flasks with 750 ml of distilled (or deionized) water each, and sterilize by autoclaving. These will be used for the bulk inoculants.

d. Preparing Packaged Presterilized Peat and Checking for Sterility (Key Steps 4–7)

Packages containing 40 g of neutralized peat (pH 6.5–6.8) in 3-mil thickness (0.003 in or 0.076 mm) polyethylene and in autoclavable polypropylene bags are needed. Prepare 62 bags of peat in polyethylene bags and heat seal after excluding all air from the bags.

Expose the peat in polyethylene bags to gamma-irradiation (2.5–5.0 Mrads). Alternatively, prepackaged irradiated peat is produced commercially and can be purchased from some commercial inoculant producers. Similarly, package 40 g of neutralized peat in 62 autoclavable polypropylene bags (127 × 178 × 0.076 mm). Pipette 1 ml of water into each bag. (Including water during autoclaving is necessary for proper sterilization.) Follow the procedure described in Chapter 27 on using polypropylene bags for autoclaving carriers.

Autoclave the peat in the polypropylene bags for 45–60 min at 121°C and 15 lb/in². Allow sufficient time for the autoclave to cool before removing the bags. (Avoid rapid release or loss of pressure from the autoclave after sterilization.) Check the sterility of the treated peat by setting up peat + YMB blanks. To set up these blanks, aseptically inject 30 ml of sterile YMB into peat in two polyethylene and two polypropylene bags. Massage the bags to ensure proper incorporation of the YMB into the peat. Incubate the bags at 25–30°C for 1 week.

At the end of the incubation period, aseptically remove a 10-g sample from each bag and transfer into 90 ml of sterile water in dilution bottles. Prepare serial dilutions from 10^{-1}–10^{-4}. Plate 0.1 ml of each dilution in duplicate on plates of glucose peptone agar and YMA + CR. Check the plates daily for 7 days for signs of growth and appearance of microorganisms that survived the sterilization.

If there is growth at any dilution, the sterilization was not complete. (It is not unusual to get growth of contaminants, e.g., actinomycetes, from peat samples that were previously irradiated and stored for a long time.) If there is growth, note the different types of colony morphology the survivors produced. Make wet amounts of colonies picked from the plates and observe under phase-contrast microscopy to establish cell morphology of the survivors (e.g., bacteria, filamentous fungi, yeasts, etc.).

Only sterile peat is recommended for inoculant production by the dilution procedure. However, inoculants have been prepared from peat with surviving contaminants, as long as the contaminants were not detectable at dilutions higher than 10^{-2}. Irradiation sometimes does not provide absolute sterility, but the dilution method still produces high-quality inoculants in irradiated peat carriers.

e. Preparing Presterilized Peat in Polypropylene Trays (Key Step 5)

Obtain two sturdy trays (46 × 46 × 10 cm) made of autoclavable polypropylene. Place 1 kg of neutralized peat in each tray and spread it out to give a layer of even thickness. Cover the tray with aluminum foil. Autoclave both batches of peat at 121°C and 15 lb/in² for 60 min. Allow the autoclave to cool before removing the trays of sterilized peat. Leave the peat to cool in the trays overnight. Do not remove the aluminum foil cover.

f. Preparing Diluted Cultures of Rhizobia (Key Steps 8 and 9)

The various diluents prepared in (c) are used for diluting the late-log-phase cultures of TAL 102 and TAL 1145. Perform serial dilutions for viable counts (Chapter 5) of the late-log-phase cultures of TAL 102 and TAL 1145. Plate on YMA + CR. Use the drop- or spread-plate methods. (Late-log-phase cultures may have 1–5 $\times$ 10^9 cells ml^{-1}.)

Immediately after performing viable counts with the undiluted culture, accurately pipette 1 ml of the broth culture of TAL 102 into 500 ml of the 20% YMB in the dilution flask to obtain a diluted culture. (The diluted culture will contain approximately 2–10 $\times$ 10^6 cells ml^{-1}, based on the assumption that the original undiluted culture had at least 1–5 $\times$ 10^9 viable cells ml^{-1}. The dilution factor is better estimated at a later stage, after actual viable counts of the undiluted culture are obtained.) Finish preparing the diluted cultures of TAL 102 with YSB, YW, and water as diluents. Similarly, prepare diluted cultures of TAL 1145 using the various diluents in the dilution flasks.

g. Preparing Inoculants with Presterilized Peat (Key Step 10)

Aseptically, with a 50-ml plastic syringe, inject 30 ml of the diluted culture into each package of autoclaved peat and 40 ml into the irradiated peat. Inoculate the bags as summarized in Table 28.1. Massage or knead the inoculated bags to work the inoculum into the peat. Label the bags to indicate treatments and dates. Incubate the packages at 25–30°C.

h. Preparing Inoculants with Presterilized Peat in Polypropylene Trays (Key Step 11)

Add 10 ml of the late-log-phase culture of TAL 102 to the 750 ml of sterile water (c). Swirl the flask to ensure proper dilution. (The diluted culture will contain approximately 1.33–6.67 $\times$ 10^7 cells ml^{-1} based on the assumption that the original undiluted culture had 1–5 $\times$ 10^9 cells ml^{-1}.) Add this diluted culture to the autoclaved peat in the tray.

Work the diluted culture into the peat by hand-mixing. Sanitized disposable polyethylene or latex gloves must be worn during the mixing. Hand-mixing without wearing gloves results in high levels of contaminants. Continue mixing until the culture is absorbed by the peat. Break up any lumps that may result during the mixing. Immediately after mixing, weigh out approximately 70-g quantities of the peat inoculant into polyethylene bags and heat seal. Label the packages to indicate treatments and date. Incubate the bags at 25–30°C.

Repeat the procedure to prepare inoculants of TAL 1145. Best results are obtained if mixing and packaging the inoculants are done in simple, but clean rooms (e.g., 5 $\times$ 3 $\times$ 3 m). Rooms of this size can be kept clean and disinfected regularly.

i. Determining Multiplication of the Rhizobia in Peat Inoculants Prepared Aseptically (Key Steps 12 and 13)

The inoculants produced as described in (g) are most unlikely to contain significant numbers of contaminants because they were prepared aseptically by injecting the diluted cultures into presterilized (irradiated and autoclaved) peat. Determine the multiplication of the rhizobia in these inoculants at 2 and 8 weeks of storage. Use three replications of each treatment at each enumeration period. Enumerate the rhizobia in these inoculants by the drop- or spread-plate methods (see Chapter 5). Plate dilutions ranging from 10^{-4}–10^{-7}.

j. Determining the Multiplication of the Rhizobia in Peat Inoculants Prepared by Hand-Mixing in Trays (Key Steps 12 and 13)

Enumerate the rhizobia in these inoculants at 2 and 8 weeks, using three replications of each treatment. The inoculants produced in (h) will contain contaminants since mixing the culture and peat was done without full aseptic precautions. Multiplication of the rhizobia in these inoculants may be determined by plate counts, but more reliably by the plant infection technique (see Chapter 6).

Establish ahead of time seedlings of *L. leucocephala* and soybean (*Glycine max*) for TAL 1145 and TAL 102, respectively, in growth pouches. (Growth tubes with seedling agar or NifTAL tubes may also be used for *Leucaena* sp.) Following the recommendations given in Chapter 6, 50 seedlings will be needed for enumerating the rhizobia in each bag of inoculant. Since three replications of each strain treatment are being enumerated, 150 seedlings of each host are needed. Pregerminate *Leucaena* sp. seeds after acid scarification/sterilization (see Appendix 10).

Prepare serial dilutions of the inoculant ranging from 10^{-2}–10^{-10}. Spread-plate dilutions 10^{-5}–10^{-7} on YMA + CR and YMA + brilliant green for plate counts (Chapter 5). Record the contamination on the plates and quantify, if possible. Inoculate 10^{-1}–10^{-10} dilutions onto plants in growth pouches or in tubes.

k. Collecting, Recording, and Analyzing the Data (Key Step 14)

Determine the number of viable rhizobia in the inoculants prepared by the various diluent formulations, sterilization, and method of preparation. Transform the data to $\log_{10}$ and calculate the mean for the replications. Organize the transformed data in the forms of Tables 28.2 and 28.3. Determine the number of rhizobia in the inoculants prepared in (h) by the most-probable-number (MPN) method. Analyze statistically for differences in the various diluent treatments for both strains and enumeration methods as indicated in Table 28.3.

Because many biological, chemical, and physical factors influence the multiplication and survival of rhizobia in carriers, examine the data and contemplate the following questions:

- Did the inoculants produced with diluted cultures reach maximum populations compared with the undiluted culture control?

- How did water perform as a diluent compared with other diluents?

TABLE 28.2 Multiplication of *B. japonicum* (TAL 102) in Inoculants Prepared with Diluted Cultures and Presterilized Peat

	Log$_{10}$ No. of Rhizobia g^{-1} Moist Inoculant			
	Gamma-Irradiated Peat		Autoclaved Peat	
Diluent	2 weeks	8 weeks	2 weeks	8 weeks
Water				
YMB (20%)				
YSB (20%)				
YW (20%)				
Undiluted culture control				

TABLE 28.3 Multiplication of *B. japonicum* (TAL 102) and *Rhizobium* sp. (TAL 1145) in Inoculants Prepared by Mixing Diluted Cultures and Autoclaved Peat in Trays

	Log$_{10}$ No. of Rhizobia g^{-1} Moist Inoculant			
	TAL 102		TAL 1145	
Enumeration Method	2 weeks	8 weeks	2 weeks	8 weeks
Plant infection (MPN)				
YMA + CR				
YMA + brilliant green				

- Compare the practicality and inoculant quality of the aseptic method of inoculant preparation in prepackaged carriers with that of mixing diluted cultures with autoclaved peat in trays.
- Can you confidently recognize colonies formed by rhizobia on plates in the presence of colonies formed by other microorganisms during plate counts? How did these plate counts agree with the values obtained by the plant infection technique?

REQUIREMENTS

a. Culturing Rhizobia in YMB

YMA-slant cultures of *B. japonicum* (TAL 102) and *Rhizobium* sp. (TAL 1145)
Two 1-liter flasks, each containing 500 ml of sterile YMB

Shaker
Gram-stain reagents (Appendix 3)
Antisera of TAL 102 and TAL 1145 for agglutination (Chapter 9) or for FA (Chapter 13)

b. Making a Culture Dilution Flask and Its Operation

Erlenmeyer flasks, 2 liter (10)
Glass tubing
Surgical latex tubing

c. Preparing the Diluents

Distilled or deionized water, 500 ml
20% YMB, 20% YSB, and 20% YW, 500 ml each
Culture dilution flasks (ten)
Erlenmeyer flasks, 2 liter (two)

d. Preparing Packaged Presterilized Peat and Checking for Sterility

Neutralized peat, approximately 8 kg
Autoclavable polypropylene bags, approximately 65 pieces
Polyethylene bags, approximately 150
Bag sealing machine
Facilities for irradiating peat
Sterile YMB; sterile plastic syringes, 50 ml, fitted with 3.5-m and 14-gauge needles
Incubator
Pipettes and milk dilution bottles containing sterile water
Plates of peptone glucose agar (PGA)
Plates of YMA + CR
Phase-contrast microscope

e. Preparing Presterilized Peat in Polypropylene Trays

Autoclavable polypropylene trays (two)
Aluminum foil

f. Preparing Diluted Cultures of Rhizobia

Plates of YMA + CR
Culture dilution flasks, each containing 20% YMB (two)
Culture dilution flasks, each containing 20% YSB (two)

Culture dilution flasks, each containing 20% YW (two)
Culture dilution flasks, each containing sterile water (two)
Late-log-phase cultures of TAL 102 and TAL 1145

g. Preparing Inoculants with Presterilized Peat

Sterile plastic syringes, 50 ml, fitted with 3.5-cm and 14-gauge needles (five)
Packages of gamma-irradiated and autoclaved peat
Diluted cultures from (f)

h. Preparing Inoculants with Presterilized Peat in Polypropylene Trays

Trays of autoclaved peat (two)
Flasks, each containing 750 ml of sterile water from (c) (two)
Late-log-phase cultures of TAL 102 and TAL 1145
Sterile pipettes, 10 ml, (two)
Sanitized disposable polyethylene or latex gloves
Spatula and weighing balance
Sealing machine

i. Determining the Multiplication of the Rhizobia in Peat Inoculants Prepared Aseptically

Plates of YMA + CR and YMA + brilliant green
Serological pipettes, 1 ml; calibrated Pasteur pipettes; milk dilution bottles with 90 and 99 ml diluents; test tubes containing 9 ml of sterile diluent
Balance, spatula, weighing paper
Wrist-action shaker

j. Determining the Multiplication of the Rhizobia in Peat Inoculants Prepared by Hand-Mixing in Trays

Requirements as in (i)
Seedlings (7 days old) of soybean and *Leucaena* sp.
Growth pouches, N-free plant nutrient solution

k. Collecting, Recording, and Analyzing the Data

Calculators, statistical tables
Statistical assistance

KEY REFERENCES

Somasegaran, P. 1985. Inoculant production with diluted cultures of *Rhizobium* spp. and autoclaved peat: Evaluation of diluents, *Rhizobium* spp., peats, sterility requirements, storage and plant effectiveness. Appl. Environ. Microbiol. 50:398–405.

Somasegaran, P., and J. Halliday. 1982. The dilution of liquid cultures of *Rhizobium* to increase the production capacity of inoculant production plants. Appl. Environ. Microbiol. 44:330–333.

29

Testing the Survival of Rhizobia on Inoculated Seeds

The most common method of legume inoculation is coating the seeds with inoculant. Survival of rhizobia on the seed depends on factors such as seed coat toxicity, application method, stickers used, and storage time and temperature. In this exercise, soybean (*Glycine max*) seeds are coated with several stickers and inoculated by various methods. The seeds are then stored up to 9 days at two temperatures. The survival of the rhizobia on the seed is monitored by the spread-plate method.

KEY STEPS/OBJECTIVES

1. Prepare inoculants.
2. Prepare adhesives.
3. Coat and pellet seeds and glass beads.
4. Plate count rhizobia for viable numbers on seeds and beads.
5. Record and analyze results.

a. Preparing Inoculants for Seed Inoculation (Key Step 1)

Prepare sterile carrier-based inoculants for a *Bradyrhizobium japonicum* strain (e.g., TAL 102) as described in Chapter 27. One week after inoculating the peat, inoculate two 50-ml batches of yeast-mannitol broth (YMB) with TAL 102. Incubate at 25–30°C on the shaker for 7 days. Set this broth culture aside in the refrigerator after maximum turbidity has been reached. This broth culture is to be used as a liquid inoculum for seed coating.

b. Preparing Adhesives (Key Step 2)

The adhesives (stickers) used in this exercise are water, a 40% solution of gum arabic, a 5% solution of methyl ethyl cellulose (Cellofas A), a 15% sucrose (household sugar) solution, and vegetable oil. Water and oil do not require special preparations. To prepare the gum arabic solution, heat 100 ml of distilled water to near boiling. Regulate the heat to prevent boiling. Add granular gum arabic in small lots (1–2 g) while continuously stirring the mixture. Each lot should be completely dissolved between additions until a total of 40 g have been added. The recommended gum arabic has the graininess of normal household sugar. Unlike the powdered form, which is also frequently used, it dissolves easily and without clumping. The solution should be clear and straw colored. Neutralize with 1 N NaOH if the pH of the solution is lower than 6.0. Refrigerate until needed. Prepare the Cellofas A solution by dissolving 5 g in 100 ml of distilled water, adding it in small increments while stirring the solution. Heating may not be necessary.

c. Inoculating and Pelleting Seeds (Key Step 3)

Inoculate soybean seeds in batches of 100 g with inoculant preparations as indicated in Table 29.1. Also inoculate a control group of 400 glass beads, 4 or 5 mm in size, using the same inoculation techniques as in treatment 7 of Table 29.1. If seeds from other legume species are chosen for this exercise, refer to Appendix 20 for recommended amounts of adhesive and inoculant. The glass beads should be approximately the same size as the seeds. Containers, spatulas, and glass rods should be sterile. Glass beads

TABLE 29.1 Amount of Stickers, Inoculant, and Lime Used for Various Inoculation and Pelleting Methods

Treatment	Sticker	Inoculant	Pellet	Inoculation Method
1	—	Broth (2.0 ml)	—	Direct coating
2	Water (1.5 ml)	Peat (0.5 g)	—	Slurry
3	Gum arabic (1.5 ml)	Peat (0.5 g)	—	Slurry (with adhesive)
4	Gum arabic (2.0 ml)	Peat (1.0 g)	$CaCO_3$ (20 g)	Slurry (with adhesive + lime)
5	Sucrose (1.5 ml)	Peat (0.5 g)	—	Slurry (with adhesive)
6	Cellofas A (1.5 ml)	Peat (0.5 g)	—	Slurry (with adhesive)
7	Gum arabic (1.0 ml)	Peat (1.0 g)	—	Two-step method
8	Gum arabic (2.0 ml)	Peat (1.0 g)	$CaCO_3$ (20 g)	Two-step method (with lime)
9	Vegetable oil (1.0 ml)	Peat (1.0 g)	—	Two step

should be washed well with detergent, rinsed with distilled water, and oven dried. Seeds are not surface sterilized for this exercise.

Four inoculant coating methods are shown in Table 29.1. Direct coating (treatment 1), the slurry method (treatments 2–6), the two-step method (treatments 7–9), and seed pelleting (treatments 4 and 8) are used in combination with the slurry method and the two-step method, respectively.

Direct Coating

Direct coating is self explanatory. Place seeds into a 1-liter flask; add 2 ml of inoculant broth and shake for approximately 1 min until all seeds are uniformly wetted. Spread seeds on clean paper and allow to dry.

The Slurry Method

Farmers most commonly use the slurry method. It is the most economic method, using less sticker and inoculant than other methods. When inoculating seeds by this method, mix sticker solution and the peat inoculant at a ratio of 1:3 immediately before use. Place seeds into a 500-ml beaker and add approximately 2 ml of the slurry to the seeds, using a small measuring spoon. Stir continuously until the seeds are uniformly coated. Spread seeds on clean paper to dry.

The Two-Step Method

The two-step method is especially useful when large numbers of rhizobia must be applied to the seed. Approximately 10 times as many rhizobia can be bound to the seed as compared with the slurry method. In this method, the sticker and the inoculant are applied to the seed separately. In the first step, the seeds are uniformly coated with the sticker. In the second step, the inoculant is added to the sticky seeds. The procedure is illustrated in Appendix 20, Figure A20.1 and performed as follows.

Place the preweighed seeds into a plastic bag. Add the sticker and then inflate the bag. Twist the bag shut to trap as much air as possible inside the bag. Swirl the bag for at least 1 min or until all the seeds are uniformly wet. Open the bag, add the inoculant, reinflate the bag, and shake gently. Stop as soon as the seeds are uniformly black. Stop at this stage because prolonged shaking will break down the coating. Again, dry the seeds on clean paper. Gauging the correct quantity of sticker solution is important in this method and is based more on experience than any specific recipe. Satisfactory coating will not occur if there is too little or too much gum.

Seed Pelleting

Seed pelleting is used to provide the inoculant with additional protection for survival. Immediately after seed coating, add $CaCO_3$ to the sticky seeds in the plastic bag. Inflate the bag and gently shake for 1 min or until all seeds are uniformly white. Dry on clean paper. The glass beads are included as a control since their surface is relatively inert. Comparing the various seed inoculation treatments with glass bead controls will help in detecting significant effects of toxic seed coat diffusates. Divide each treatment of inoculated seeds and glass beads into two batches of equal size. Store one batch at 4°C (batch A) in the refrigerator and the other batch (batch B) at room temperature (25–30°C). Petri dishes are recommended as storage containers.

d. Determining the Number of Viable Rhizobia on Seeds (Key Steps 4 and 5)

The number of rhizobia on the seeds and glass beads of each treatment will be determined at 0, 1, 2, 3, 6, and 9 days after inoculation. On each plating day, remove 20 seeds from batch A of each treatment (stored in the refrigerator) and 20 seeds from batch B of each treatment (stored at room temperature). Make four subsamples of five seeds from each.

If all treatments shown in Table 29.1 are chosen (nine treatments—including the control—at two temperatures divided into four subsamples), the total number of samples to be plated on one day will be $9 \times 2 \times 4 = 72$. Transfer each subsample into a screw-capped test tube containing 5 ml of sterile diluent. Shake the test tubes vigorously for 5 min to wash the inoculum off the seeds. One milliliter of the resulting suspension will contain the rhizobia derived from one seed. Make a serial dilution from 10^{-1}–10^{-5} from each subsample as described in Chapter 5.

Plate 0.1 ml of each dilution by the spread-plate method on yeast-mannitol agar (YMA) plates containing brilliant green (1.25 μg/ml) and on YMA plates containing congo red (CR) (25 μg/ml). The brilliant green will suppress fungal growth while CR will aid in detecting contaminants. Count the rhizobial colonies and express the results as number of viable rhizobia per seed basis. Also, convert viable rhizobia per seed to percent of 0-day viability. Enter both these data side by side. Organize the results of all counts as in Table 29.2.

Plot two graphs using the mean counts of each treatment.

1. Mean log of viable rhizobia per seed (y axis) against time (x axis). This graph will show which treatment permits the largest number of cells to be applied to the seed and also which treatment allowed the longest survival of the applied inoculum.
2. Percent viable rhizobia per seed (y axis) against time (x axis). This graph will indicate the percent decline of the applied inoculum in relation to the initial number of viable cells.

REQUIREMENTS

a. Preparing Inoculants for Seed Inoculation

Transfer chamber
Platform shaker, incubator, refrigerator

TABLE 29.2 Percentage Viability of Rhizobia per Seed as Affected by Different Inoculation Methods

	Viable Rhizobia per Seed After (days)					
Treatment	0	1	2	3	6	9
1						
2						
3						
4						
5						
6						
7						
8						
Glass beads (control)						

Inoculation loop, bunsen burner
Requirements for Gram stain (Appendix 3)
Sterile syringe, 50 ml, with 18-gauge sterile needle
Erlenmeyer flasks, 250-ml capacity (20)
YMB, 2 liters
Plates of YMA and bromthymol blue (BTB), plates of YMA and CR
Solution of BTB, 0.5% in alcohol
Spreader; beaker of alcohol, 95%; spray bottle of alcohol, 70%
Sterile peat, sealed bag of 50 g
Sterile pipettes, 1 ml; sterile test tube

b. Preparing Adhesives

Refrigerator, balance
Hot plate (unit that includes a magnetic stirrer if possible), stirring bar
Beakers, 100-ml capacity
Weighing paper, spatula
$CaCO_3$ (precipitated powder)
Methyl ethyl cellulose (Cellofas A, Sigma Chemical Co., St. Louis, MO), gum arabic, sugar
Distilled water

c. Inoculating and Pelleting Seeds

Refrigerator, balance
Spatula, Bunsen burner, weighing paper

Glass stirring rods, glass beads (400)
Pipettes, 5 ml; pipettes, 10 ml (wide mouth)
Beakers, 100 ml
Plastic bags, 2-liter capacity
Distilled water, alcohol in spray bottle
Gum arabic solution from (b), Cellofas A solution from (b), sugar solution
$CaCO_3$ powder
Paper towels (to place coated seeds on for drying)
Petri dishes
Peat inoculant from (a), broth culture from (a)
Soybean seeds, 800 g

d. Determining the Number of Viable Rhizobia on Seeds

Transfer chamber
Incubator, Bunsen burner, Refrigerator
Sterile pipettes, 1 ml
Tubes, 20 ml, screw-capped with 5 ml of sterile diluent
YMA plates containing brilliant green
YMA plates containing CR
Coated seeds from (c)

KEY REFERENCES

Hoben, H.J., N.N. Aung, P. Somasegaran, and U.G. Kang. 1991. Oils as adhesives for seed inoculation and their influence on the survival of *Rhizobium* spp. and *Bradyrhizobium* spp. on inoculated seed. World J. Microbiol. Biotechnol. 7:324–330.

Vincent, J.M. 1970. A Manual for the Practical Study of Root Nodule. IBP Handbook no. 15, Blackwell Scientific Publications, Oxford.

Vincent, J.M., J.A. Thompson, and K.O. Donovan. 1962. Death of root nodule bacteria on drying. Aust. J. Agric. Res. 13:258–270.

Additional References and Recommended Reading

Bezdicek, D.F., D.W. Evans, B. Abede, and R.E. Witters. 1978. Evaluation of peat and granular inoculum for soybean yield and N-fixation. Agron. J. 70:865–868.

Boonkerd, N., and R.W. Weaver. 1982. Cowpea rhizobia: Comparison of plant infection and plate counts. Soil Biol. Biochem. 14:305–307.

Brockwell, J. 1963. Accuracy of a plant infection technique for counting populations of *Rhizobium trifolii*. Appl. Microbiol. 11:377–383.

Burton, J.C. 1967. *Rhizobium* culture and use. pp. 1–33. *In* H.J. Peppler (ed.) Microbial Technology. Van Nostrand Reinhold Co., New York.

Burton, J.C. 1975. Methods of inoculating seeds and their effect on survival of rhizobia. pp. 175–189. *In* P.S. Nutman (ed.) Symbiotic Nitrogen Fixation, International Biological Programme, Vol. 7, Cambridge University Press, Cambridge, UK.

Burton, J.C., and R.L. Curley. 1966. Compatibility of *Rhizobium japonicum* and sodium molybdate when combined in a peat carrier medium. Agron. J. 58:327–330.

Deschodt, C.C., and B.W. Strijdom. 1974. Effect of prior treatment of peat with ethylene oxide or methyl bromide on survival of rhizobia in inoculants. Phytophylactica 6:229–234.

Deschodt, C.C., and B.W. Strijdom. 1976. Suitability of a coal-bentonite base as carrier of rhizobia in inoculants. Phytophylactica 8:1–6.

Diatloff, A. 1970. The effects of some pesticides on root nodule bacteria and subsequent nodulation. Aust. J. Exp. Agric. Anim. Husb. 10:562–567.

Graham, P.H., G. Ocampo, L.D. Ruiz, and A. Duque. 1980. Survival of *Rhizobium phaseoli* in contact with chemical seed protectants. Agron. J. 72:625–627.

Hiltbold, A.E., D.L. Thurlow, and H.D. Skipper. 1980. Evaluation of commercial soybean inoculants by various techniques. Agron. J. 72:675–681.

Hoben, H.J., and P. Somasegaran. 1982. Comparison of the pour-, spread-, and drop-plate methods for the enumeration of *Rhizobium* in peat inoculants. Appl. Environ. Microbiol. 44:1246–1247.

Kremer, R.J., and H.L. Peterson. 1983. Effects of carrier and temperature on survival of *Rhizobium* spp. in legume inocula: Development of an improved type of inoculant. Appl. Environ. Microbiol. 44:1790–1794.

Kremer, R.J., J. Polo, and H.L. Peterson. 1982. Effect of suspending agent and temperature on survival of *Rhizobium* in fertilizer. Soil Sci. Soc. Am. J. 46:539–542.

McLeod, R.W., and R.J. Roughley. 1961. Freeze dried cultures as commercial legume inoculants. Aust. J. Exp. Agric. Anim. Husb. 1:29–33.

Odeyemi, O., and M. Alexander. 1977. Use of fungicide resistant rhizobia for legume inoculation. Soil Biol. Biochem. 9:247–251.

Paczkowski, M.W., and D.L. Berryhill. 1979. Survival of *Rhizobium phaseoli* in coal-based legume inoculants. Appl. Environ. Microbiol. 38:612–615.

Philpotts, H. 1976. Filter mud as a carrier for *Rhizobium* inoculants. J. Appl. Bacteriol. 41:277–281.

Roughley, R.J. 1970. The preparation and use of legume seed inoculants. Plant Soil 32:675–701.

Roughley, R.J., and J.A. Thompson. 1978. The relationship between the numbers of rhizobia in broth and the quality of peat based legume inoculants. J. Appl. Bacteriol. 44:317–319.

Singleton, P.W., P. Somasegaran, P. Nakao, H. Keyser, H.J. Hoben, and P. Ferguson. 1990. Applied BNF Technology: A Practical Guide for Extension Specialists. NifTAL, Paia, HI.

Skipper, H.D., J.H. Palmer, J.E. Giddens, and J.M. Woodruff. 1980. Evaluation of commercial soybean inoculants from South Carolina and Georgia. Agron. J. 72:673–674.

Somasegaran, P., V.G. Reyes, and H.J. Hoben. 1984. The influence of high temperatures on the growth and survival of *Rhizobium* spp. in peat inoculants during preparation, storage, and distribution. Can. J. Microbiol. 30:23–30.

Sparrow, S.D., and G.E. Ham. 1983. Survial of *Rhizobium phaseoli* in six carrier materials. Agron. J. 75:181–184

Strijdom, B.W., and H.J. van Rensburg. 1981. Effect of steam sterilization and gamma irradiation of peat on quality of *Rhizobium* inoculants. Appl. Environ. Microbiol. 41:1344–1347.

Toomsan, B., O.P. Rupela, S. Mittal, P.J. Dart, and K.W. Clark. 1984. Counting *Cicer-Rhizobium* using a plant infection technique. Soil Biol. Biochem. 6:503–507.

van Rensburg, H.J., and B.W. Strijdom. 1974. Quality control of *Rhizobium* inoculants produced from sterilized and non-sterile peat in South Africa. Phytophylactica 6:307–310.

van Shreven, D.A. 1970. Some factors affecting growth and survival of *Rhizobium* spp. in soil peat cultures. Plant Soil 32:113–130.

Weaver, R.W. 1979. Adsorption of rhizobia to peat. Soil Biol. Biochem. 11:545–546.

Weaver, R.W., and L.R. Frederick. 1972. A new technique for most probable number (MPN) counts of rhizobia. Plant Soil 36:219–222.

Wilson, D.O., and K.M. Trang. 1980. Effects of storage temperature and enumeration method on *Rhizobium* spp. numbers in peat inoculants. Trop. Agric. (Trinidad) 57:233–238.

SECTION V

Genetic Techniques for Rhizobia

The advances in microbiology along with the wealth of refinements in modern molecular biology techniques in the last several years have significantly influenced the research on the genetics of rhizobia. The emphasis and application of a range of molecular genetic techniques and nucleic acid hybridization-based assays have been instrumental in the physical location, cloning, and analysis of the genes involved in the symbiotic interaction between rhizobia and legumes. Also, with the application of molecular biology, a better and clearer picture is now beginning to emerge on the taxonomy and classification of the rhizobia. Restriction enzyme digests of rhizobial genomic DNA and restriction fragment length polymorphism (RFLP) analysis utilizing specific gene probes have become useful in studying genetic diversity and in strain identification for ecological studies.

NUCLEIC ACIDS

There are two types of nucleic acids found in microorganisms. These are deoxyribonucleic acid (DNA) and ribonucleic acid (RNA). Native DNA is double-stranded helix and is composed of two sugar phosphate backbones with pairs of nucleotide bases held between them by hydrogen bonds. DNA is composed of the nucleotides adenine (A), thymine (T), guanine (G), and cytosine (C). During base pairing, adenine pairs with thymine and guanine with cytosine. Unlike DNA, RNA is single stranded. RNA is very similar in chemical composition to DNA with the exception that thymine is replaced by uracil (U).

DNA molecules are among the largest known and the mass of DNA is expressed in daltons (Da) or in kilobase (kb) units. For conversion purposes, 1 kb double-stranded (ds) DNA equals 6.6×10^5 Da. (The mass of a single hydrogen atom is defined as 1 Da). Fragments of DNA have a constant charge/length ratio due to the negative charge of the phosphate backbone. This causes the DNA to migrate to the anode or positive electrode during electrophoresis.

The physical properties of DNA itself form the main basis of nucleic acid hybridization technology. A unique physical property of the double-stranded DNA is that, under certain conditions (high temperature or high pH), the complementary strands will dissociate (denature). When the resulting single-stranded DNA is then subjected to a lower temperature (slow cooling) and a higher salt concentration, the complementary strands reassociate (renature) to form double-stranded DNA that are very similar, if not identical, to the native DNA. It follows that if the denatured DNA of one microorganism is mixed with the denatured DNA of another where there are complementary base sequences, pairing will take place, resulting in stable DNA-DNA hybrids.

Perfect matches hybridize readily and withstand high temperatures in the hybridization and washing reactions. These complexes also form in the presence of low salt concentrations. The high temperatures and low salt concentrations are referred to as stringent conditions. Less-than-perfect matches do not tolerate these stringent conditions and hybridization either never occurs or is disrupted during the washing steps. Stringency conditions are critical in order to avoid nonspecific or false positive reactions.

PLASMIDS AND SYMBIOTIC GENES

Plasmids are extrachromosomal DNA found in a variety of bacterial species. They are double-stranded and range from 1 to >200 kb in molecular weight. Plasmid DNA can be covalently closed circular (CCC); open circular (OC), where the strands are relaxed; and the linear conformation, which is generated by breakage of the double-stranded molecule. Bacterial plasmids confer several phenotypes, including resistance to antibiotics, production of antibiotics, colicins, enterotoxins, restriction endonucleases, and degradation of complex organic compounds. The plasmids can also harbor N_2 fixation genes.

One to six large indigenous plasmids (molecular weight 90–350 $\times$ 10^6 Da) are found in rhizobia, especially in the genus *Rhizobium*. Indigenous plasmids of *Rhizobium* sp. (*Leucaena* sp.) and *R. leguminosarum* bv. viceae are shown in Figure V.1. In most species investigated in the genus *Rhizobium* (e.g., *R. leguminosarum* bv. trifolii, *R. leguminosarum* bv. viciae, *R. leguminosarum* bv. phaseoli, *R. meliloti*, *Sinorhizobium fredii*), the genes that control nodulation (*nod*), host range specificity (*hsn*), and N_2 fixation (*nif*) have been located on the large plasmids called *symbiotic plasmids*, commonly abbreviated as pSym. In the *Rhizobium* spp. that nodulate tree legumes such as *Acacia melanoxylon*, *A. cyanophylla*, *Prosopis chilensis*, *Sophora chrysophylla*, and *Leucaena leucocephala*, the genes for N_2 fixation have also been located on large plasmids. Plasmids can account for about 25% of the total DNA in some strains of *Rhizobium*.

It has been shown in *R. leguminosarum* bv. trifolii and *R. leguminosarum* bv. viceae, that when the entire pSym is cured using high temperatures, then infection, nodulation, and N_2 fixation are not possible. However, these symbiotic functions can be restored after pSym is reintroducted. Also, if the pSym of *R. leguminosarum* bv. trifolii is introduced into another species, for example, into a pSym cured *R. leguminosarum* bv. viceae cell environment, the *R. leguminosarum* bv. viceae will then nodulate clover (*Trifolium* spp.)

Genetics of the bradyrhizobia are relatively less studied. Most of the studies have focused on *Bradyrhizobium japonicum*. In *B. japonicum*, many *nif*, *fix*, and *nod* genes

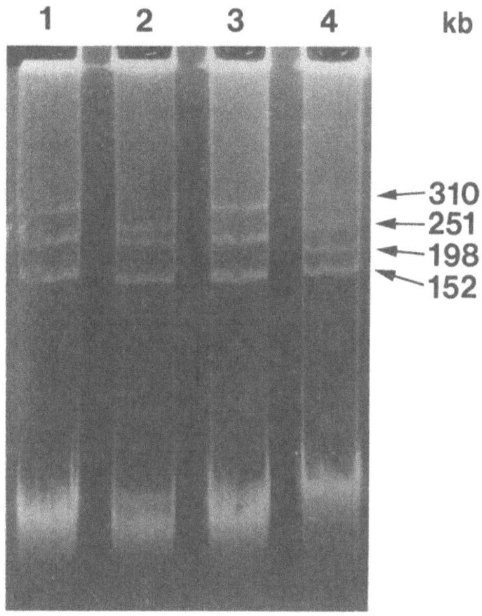

FIGURE V.1 Lanes 1 and 3, *Rhizobium* sp. (*Leucaena* sp.) strain TAL 1145 (CB 3060); lane 2, nonnodulating variant of TAL 1145 with a deletion on the largest (Sym) plasmid; lane 4, *R. leguminosarum* bv. viceae, strain 6015 (pJB5JI).

have been identified, but these genes are located as clusters on the bacterial chromosome. Though plasmids have been detected in *B. japonicum*, it has not been shown that these carry symbiotic genes. In recent years, the genetics of N_2 fixation has received much attention. More than 50 symbiotic genes of *Rhizobium*, *Bradyrhizobium*, and *Azorhizobium* species have been identified, cloned, and analyzed.

The *nif* genes code for the proteins of the enzymes in the N_2-fixing system and the regulation of their synthesis. The *fix* genes code for different functions with a specific role in supporting N_2 fixation. Some of the genes of *Rhizobium* and *Bradyrhizobium* that are involved in the N_2-fixation process are structurally similar to the *nif* genes of *Klebsiella pneumoniae*, which is a free-living N_2 fixer. Genes that are essential for nodulation (*nod*) show a considerable level of homology between the different species of *Rhizobium* and *Bradyrhizobium*. These *nod* genes, which are functionally conserved in all rhizobia, are referred to as the common *nodABC*. The expression of the nodulation genes is controlled by regulatory *nodD* genes. The products of the *nodD* genes are activated by flavonoids excreted by the legume roots.

GENE PROBES FOR RHIZOBIA

Serological and antibiotic-resistance marker identification techniques make use of the properties of complex macromolecular gene products and not the genes themselves. Because DNA-DNA hybridization detects the genes, regardless of serological cross-reactions, it is much more specific in strain identification, and also facilitates the study of genetically unmodified rhizobia.

Gene probes are the key to strain identification using nucleic acid hybridization. Generally, probes are pieces of DNA or RNA labeled with a ^{32}P-containing nucleotide. Nonradioactive biotinylated nucleotide probes have now become increasingly popular. Both radioactive and nonradioactive labels are commonly incorporated into the probe DNA via an enzymatic process called *nick translation*. The probe must recognize a complementary sequence to be effective. A summary of the steps involved in identifying a specific sequence using DNA-DNA hybridization is presented in Figure V.2.

Generally, probes are designed to identify a specific microorganism of interest. In the case of rhizobia, an ideal and desirable gene probe should identify all and only rhizobia. Probes may be further developed specifically to distinguish between different genera of rhizobia, species within a genus, and strains within a species. The application of gene probes in research with rhizobia is still developing, though much has been done to identify and characterize the genes involved in specificity, infection, and nodulation.

The *nif* and *nod* gene probes have been well studied and are frequently used for

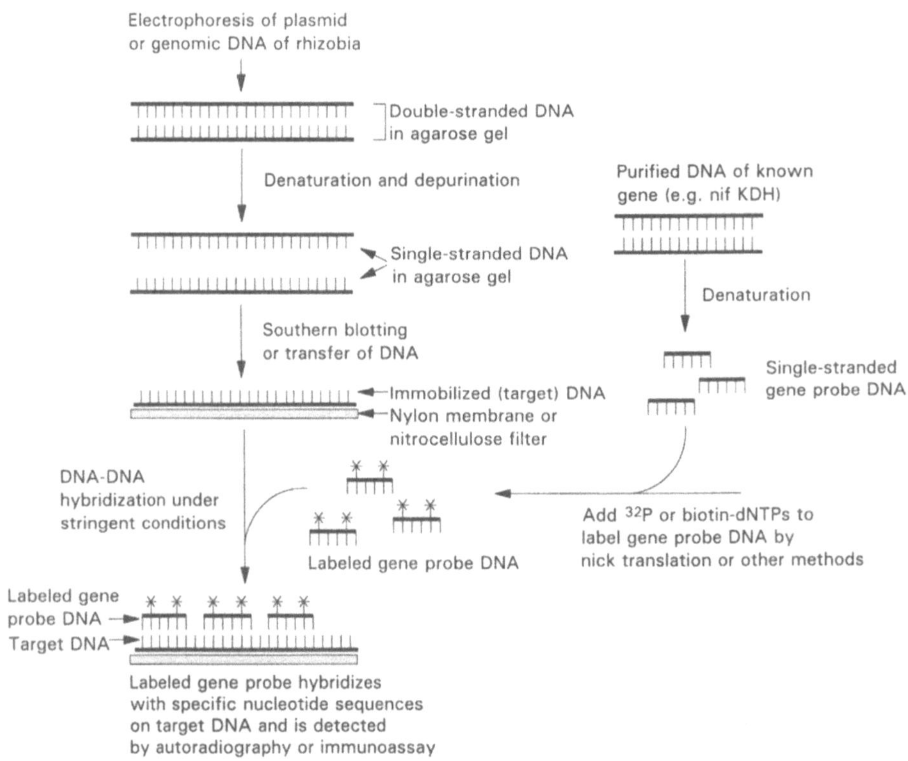

FIGURE V.2 Summary of steps in identifying a specific nucleotide sequence through DNA-DNA hybridization between gene probe DNA and target DNA.

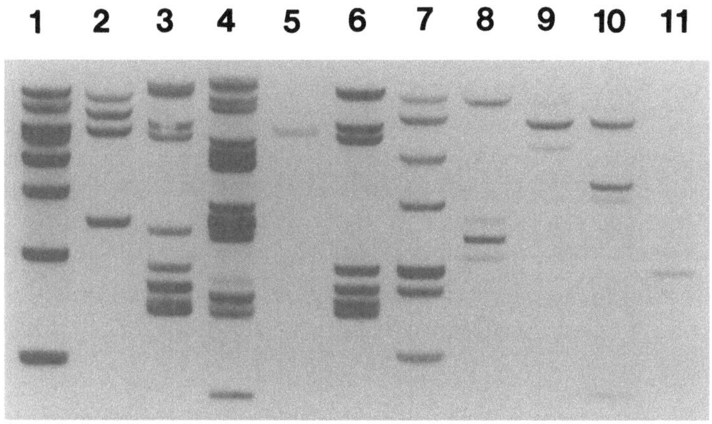

Figure V.3 Insertion sequence (IS) probes can be used to study RFLP patterns in *Rhizobium* spp. The *Hind*III digested genomic DNA of *Rhizobium* sp. (*Leucaena* sp.) (lane 1), *R. meliloti* (lanes 2–6), and *Rhizobium* sp. (*Sophora chrysophylla*) (lanes 7–11) were probed with an IS isolated from *Rhizobium* sp. (*Leucaena* sp.) strain TAL 1145.

genetic analysis of rhizobia. When these or other probes are used on restriction-enzyme-digested rhizobial plasmid or genomic DNA that have been transferred and immobilized (Southern blotting) on nitrocellulose or nylon membranes, specific RFLP hybridization patterns emerge. Each pattern is a fingerprint of the rhizobial strain and can be used in classifying rhizobia, in identifying strains, and in maintaining culture collections. Another highly conserved region of the nodulation genes is the *nod*-box. This has been shown to be species-specific as in the case of the clover rhizobia. Recently, it has been shown that insertion sequences (IS), a class of transposable elements or mobile DNA, can be used as positive strain identification probes for *R. meliloti*. Figure V.3 illustrates applying an IS probe to show the relationship between *Rhizobium* sp. (*Leucaena* sp.) and other *Rhizobium* spp. IS probes can be useful to monitor genetic changes in the genome of inoculant rhizobia that have persisted in the soil for many years (Figure V.4). In contrast to traditional classification based on phenotypic characteristics, gene probes offer precision in distinguishing between superficially similar rhizobia or determining phylogenetic relationships among groups of rhizobia.

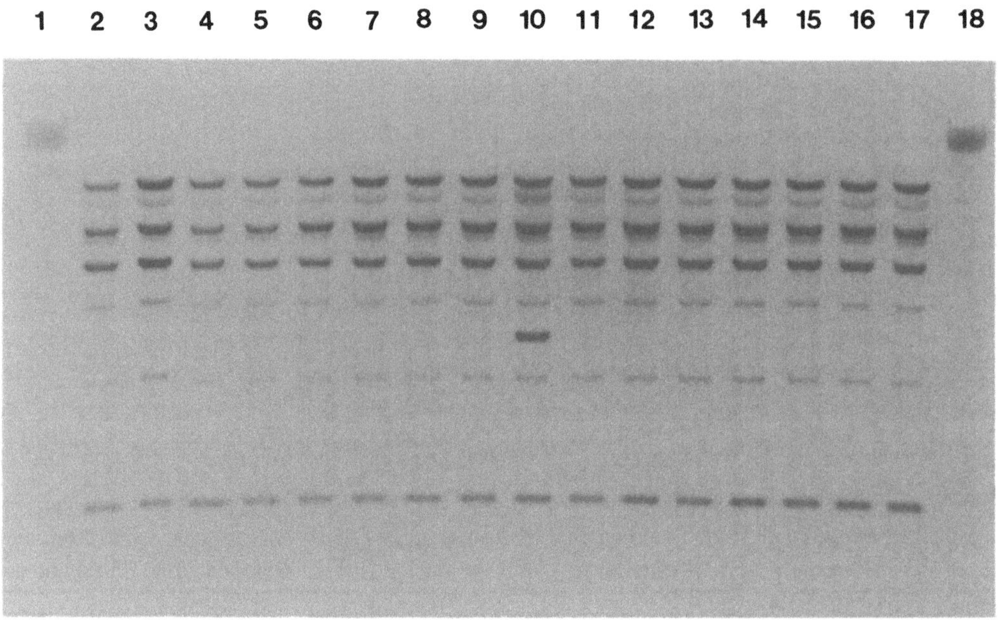

FIGURE V.4 Genetic change in an inoculant *Rhizobium* sp. (*Leucaena* sp.) strain TAL 1145 persisting in an Oxisol. Genomic DNA was digested with *Hin*dIII and probed with an IS isolated from TAL 1145. Lanes 2–17 were TAL 1145 nodule isolates that reacted positively with the fluorescent antibody of TAL 1145. Note that isolate in lane 10 has acquired an extra band, indicating a genetic change.

30

Analyzing Plasmid Profiles of Rhizobium *spp. by a Modified Eckhardt Vertical Gel Electrophoresis Procedure*

Rhizobia have one to several copies of indigenous plasmids. Plasmid copy number and the distinct migration patterns in agarose gel during electrophoresis can be valuable in recognizing a specific rhizobial strain and also for detecting deletions and loss of plasmids. In the Eckhardt procedure, the cells of the rhizobia are lysed in situ in the wells of the agarose gel to release the chromosomal and supercoiled plasmid DNA. The in-well lysis is achieved by treating the cells with lysozyme and sodium dodecyl sulfate (SDS). The specified concentration of agarose used in preparing the gel forms a molecular sieve of a desired pore size that allows the supercoiled plasmids and smaller DNA fragments to migrate through the gel while the larger molecular weight bacterial chromosomes are retained in the wells. The molecular size of the plasmids can be estimated by comparison with known molecular weight (MW) standards. In this exercise, the plasmid profiles of several *Rhizobium* spp. are analyzed by vertical gel electrophoresis.

KEY STEPS/OBJECTIVES

1. Culture rhizobia.
2. Assemble electrophoresis apparatus.
3. Pour agarose gel.
4. Centrifuge and wash cells.
5. Load cells.
6. Perform electrophoresis.
7. Stain DNA in gel.
8. Photograph DNA bands in gel.

a. Culturing Rhizobia (Key Step 1)

Young 24-h cultures are needed for plasmid analysis. Inoculate cultures stored on agar slants into 2–3 ml of tryptone-yeast (TY) medium (Appendix 3) in tubes. Use vigorous shaking (200–300 rpm) on a rotary shaker to aerate. Cultures of *Rhizobium* spp. should reach maximum turbidity in 3 days at room temperature. Inoculate four drops (or 50 μl) of each turbid culture into 5.0 ml of fresh TY medium in tubes and allow to grow for 24 h.

Up to nine cultures may be tested at a time in a single Eckhardt gel. Strain *R. leguminosarum* bv. viceae 6015(pJB5JI) may be used as the standard because it has at least four plasmids whose MWs are known. The MWs are 310, 250.8, 197.6, and 152 kilobases (kb). The suggested test strains (see Appendix 24) are as follows: TAL 182 and TAL 1797 (*R. leguminosarum* bv. phaseoli), TAL 634 and TAL 1400 (*R. leguminosarum* bv. viceae), TAL 380 and TAL 1372 (*R. meliloti*), and TAL 1145 and TAL 82 (*Rhizobium* sp. from *Leucaena leucocephala*). Rhizobia other than suggested here may also be used.

b. Assembling the Electrophoresis Apparatus (Key Step 2)

The apparatus used in this exercise is a commercially manufactured vertical slab (gel) unit whose specifications are given in the requirements list of this chapter. Handling and assembly instructions are provided with the purchase of the apparatus. Familiarize yourself with the various parts, accessories, and operational details of the vertical slab unit. Ensure that the two glass plates (one clear and the other with a frosted surface) for casting the gel are clean. Assemble the glass plate sandwich unit (i.e., cast) and secure it in casting position for pouring the gel. Level the apparatus.

c. Pouring the Gel (Key Step 3)

Use sterile glassware and wear latex gloves at all times. The presence of nucleases on your hands could contaminate glassware and working materials. Weigh out 0.7 g of agarose in a sterile 250-ml Erlenmeyer flask. Add 100 ml of 1x Tris-borate-EDTA (TBE) buffer (Appendix 5). Microwave or heat to dissolve the agarose completely. (Avoid moisture loss during heating because this would change the concentration level.)

Place the liquid gel in a 50°C water bath until needed. When ready, slowly pour the gel with a sterile 25-ml glass pipette into the cast. Allow the gel to flow down the sides of the spacers to the bottom of the cast. (Avoid chipping the edges of the glass plates with the pipette tip.) Continue pouring until the gel level is about 1 cm from the top. Insert a 3-mm comb into the gel without trapping any air bubbles. Using a Pasteur pipette, bring the gel level to the top by filling the spaces between the teeth of the comb with gel. Allow the gel to solidify and cool for at least 1 h.

To remove the comb, loosen two or three clamp screws at the top end of each clamp. Hold the gel cast with both hands and push the comb forward with the thumbs, applying just enough pressure to break the seal between the comb and the gel. Slowly slide out and remove the comb while continuing forward pressure on the comb. Bubbles may result between the gel and the glass plates, but this will not affect the electrophoresis. Retighten the clamp screws. Transfer the gel cast to the lower buffer chamber and secure it in place. (Note that removal of the comb results in the formation of 10 wells.) Fill each well with 250 µl of 1x TBE buffer using a micropipette for dispensing.

d. Centrifuging and Washing the Cells (Key Step 4)

Transfer 1000 µl of the young 1-day-old culture (10^7–10^8 cells ml^{-1}) to a sterile microfuge tube. Centrifuge the tube at 14,000 rpm (16,000 × g) for 3 min. (Observe the volume of the pellet, which is approximately 30 µl.) Decant the supernatant. Pipette 500 µl of sarkosyl-TEN solution (0.1% Sarkosyl in TEN buffer) into the microfuge tube. Vortex the pellet into suspension and centrifuge at 14,000 rpm for 3 min. Decant the supernatant and place the microfuge tube in a container containing crushed ice. Following similar procedures, process the rest of the strains.

e. Loading Wells with Lysed Cells (Key Step 5)

Begin this step by working with strain *R. leguminosarum* bv. viceae 6015 (pJB5JI). Vortex the pellet in the small amount of liquid remaining in the microfuge tube. Add 40 µl of Eckhardt solution A (cell lysis solution) and gently mix in the cell suspension by drawing in and out through the pipette tip several times.

Using the same pipette tip, load 40 µl of the cell mixture into the well. To load, introduce the tip of the pipette into the buffer (placed in the wells previously) as close as possible to the base of the well and hold it against the side of the well. Carefully and slowly depress the plunger of the micropipette to deliver the cell mixture to the base of the well. The cell mixture forms a layer under the buffer. (Release the plunger only after the pipette tip is withdrawn completely out of the well.)

Following the procedure described, treat the rest of the strains with solution A and load the wells. After all wells have been loaded with cell mixtures, add 40 µl of Eckhardt solution B to each well. Finally, add 100 µl of Eckhardt solution C to each well. Seal the wells with molten agarose, kept aside for this purpose. Dispense the agarose using a Pasteur pipette. Allow the agarose to solidify and trim off the overflow, if necessary, with a sharp scalpel blade.

f. Performing Electrophoresis (Key Step 6)

Fit the upper buffer chamber to the top of cast and secure it in position. Fill the upper and lower buffer chambers, each with 250–300 ml of 1x TBE buffer. Cover the upper buffer chambers with the safety lid. Check for leaks from the upper buffer chamber.

Turn on the power source and allow it to warm up for 10–15 min. Plug in the power cables. Set the current to 8 mA and allow electrophoresis to proceed for 1 h. After 1 h, increase the current to 40 mA and run for 3 h. (If two gels are run with the same power source, set the current to 16 mA for the initial 1 h, followed by 80 mA for the extended 3-h run.)

Periodically, during electrophoresis, observe the migration of the blue band of the tracking dye (bromphenol blue), a component in solution A. When the blue band migrates close to the bottom of the gel, it indicates that the electrophoresis may be terminated. Reduce the current to zero, turn off the power, and unplug the electrophoretic unit from the power source.

g. Staining and Photographing the DNA in the Gel (Key Step 7)

Prepare the staining solution by adding 50 μl of stock ethidium bromide (EtBr) (Appendix 5) into 500 ml of 1x TBE buffer. **(Caution: EtBr is a powerful mutagen. Wear gloves when handling.)**

Carefully remove the upper buffer chamber from the top of the gel cast and pour off the buffer. Remove the gel cast and dismantle. Slide the clear glass plate off of the gel and remove the spacers on each side. With a scalpel, cut off a small piece of the bottom corner of the gel on the same side as well 1. (This cut corner is a marker and will help in orienting the gel during handling.) The gel adheres firmly to the frosted surface of the glass plate. Using one of the spacers, gently push the gel off the frosted surface of the glass plate and into the staining solution. Allow the staining to continue for 20 min. At the end of the staining period, slide a suitable piece of low-flexibility plastic sheet (slightly larger in area than the gel) under the gel and carefully lift it out of the staining solution. Slide the gel off of the plastic sheet and onto the screen of the transilluminator. Move the gel to a central position on the screen.

Wear a protective face shield to protect the eyes from harmful UV rays and turn off the lights. Switch on the transilluminator. The plasmids are visible as fluorescent bands on the gel. Note the number of plasmid bands in each lane and record the information.

Photograph the gel on the transilluminator at f 5.6 after an exposure of approximately 1–3 s.

As soon as the gel has been photographed, overlay a plastic transparency on the gel and trace the plasmid bands and well locations. Switch off the transilluminator.

With a suitable marker pen, indicate on the photograph the identity of the strain in each lane.

Carefully return the gel to the staining solution. Save the gel to use later in another experiment. Clean transilluminator's screen using alcohol and soft tissue paper.

REQUIREMENTS

a. Culturing Rhizobia

Cultures of the following *Rhizobium* spp. on yeast-mannitol agar (YMA) slants are suggested for this experiment: *R. leguminosarum* bv. viceae strains 6015 (pJB5JI), TAL 634, and TAL 1400; *R. leguminosarum* bv. phaseoli strains TAL 182 and TAL 1797; *Rhizobium* sp. (*Leucaena leucocephala*) strains TAL 82 and TAL 1145; *R. meliloti* strains TAL 380 and TAL 1372

TY medium (Appendix 3) in screw-cap tubes
Sterile Pasteur pipettes
Rotary or reciprocating shaker
Transfer chamber/laminar flow hood, Bunsen burner

b. Assembling the Electrophoresis Apparatus

Sturdier Slab Gel Electrophoresis Unit (Model SE 400) (Hoefer Scientific Instruments, San Francisco)
Instruction manual for unit assembly

c. Pouring the Gel

Agarose (Type 1, Low EEO) (Sigma Chemical Co., St. Louis)
Latex gloves, balance, spatula
Erlenmeyer flasks, 250 ml; sterile pipette, 25 ml
Microwave oven
Teflon comb (3 mm, 10 teeth)
Water bath (70°C)
Micropipette, 0–500 or 0–1000 μl, and tips
1x TBE buffer (Appendix 5)

d. Centrifuging and Washing the Cells

Micropipette, 0–500 or 0–1000 μl, and tips
Sterile microfuge tubes, 1.5 ml; microfuge tube rack
Microcentrifuge, Vortex mixer
Crushed ice
Sarkosyl-TEN buffer solution (Appendix 5)

e. Loading Wells with Lysed Cells

Micropipettes, 0–100 μl and 0–500 μl, and tips
Microfuge tubes containing cell pellet from (d)
Gel cast from (c)
Eckhardt solutions A, B, and C (Appendix 5)

f. Performing Electrophoresis

Completely assembled electrophoresis unit
Power source for electrophoresis
1x TBE buffer, 500–600 ml

g. Staining and Photographing the DNA in the Gel

EtBr stock solution (Appendix 5)
Plastic container for staining solution
Transilluminator, UV face shields or goggles
1x TBE buffer, low-flexibility plastic sheet
Micropipette, 0–100 μl, and tips
Camera set-up (Kodak or other)

KEY REFERENCES

Eckhardt, T. 1978. A rapid method for the identification of plasmid deoxyribonucleic acid in bacteria. Plasmid 1:584–588.

2. Nuti, M.P., A.M. Ledboer, A.A. Lepidi, and R.A. Schilperoot. 1977. Large plasmids in different *Rhizobium* species. J. Gen. Microbiol. 100:241–248.

3. Priefer, U. 1984. Characterization of plasmid DNA by agarose gel electrophoresis. pp. 26–37. In A. Pühler and K.N. Timmis (ed.) Advanced Molecular Genetics. Springer-Verlag KG, Berlin.

Isolating and Purifying Genomic DNA of Rhizobia Using a Large-Scale Method

The DNA of bacteria are found on the chromosomes and plasmids, which together make up the genome. Isolating and purifying the total genomic DNA constitute the first step in the procedure for the genetic fingerprinting of a rhizobial strain. Total genomic DNA can also be used for constructing the gene library of a rhizobial strain. In the isolation procedure, the cells are first treated with lysozyme to digest much of the cell wall. This is followed by adding a mixture of an ionic detergent (sarkosyl) and proteolytic enzyme (pronase), which causes cell lysis to release the total cellular DNA and simultaneous digestion of many native proteins (e.g., endogenous nucleases). The cell lysate can then be treated with phenol to dissociate the DNA from the cellular proteins. The dissociated DNA is precipitated by adding isopropanol or ethanol. In this exercise, a young culture of a strain of *Rhizobium* spp. or *Bradyrhizobium* spp. is processed to isolate, purify, and quantify total genomic DNA employing a large-scale method.

KEY STEPS/OBJECTIVES

1. Culture rhizobial strains.
2. Prepare Sarkosyl/pronase mixture and lysozyme solution.
3. Centrifuge and wash cells.
4. Treat cells with lysozyme.
5. Lyse cells, denature, and digest cell proteins.
6. Deproteinize and purify DNA by phenol extraction.
7. Remove phenol by chloroform extraction.
8. Precipitate DNA with isopropanol.
9. Spool DNA and wash in ethanol.

10. Dry DNA and dissolve in buffer.

11. Quantify DNA spectrophotometrically.

a. Culturing the Rhizobial Strain (Key Step 1)

Obtain a pure culture of *Rhizobium* sp. (TAL 1145) or other rhizobia. Inoculate a loopful of culture into 5 ml of yeast-mannitol broth (YMB) in a screw-cap tube. Incubate for 3 days at 28–30°C on a rotary shaker to obtain a turbid culture. Aseptically transfer 1 ml of the turbid YMB culture into 40 ml of tryptone yeast (TY) medium in a 125-ml Erlenmeyer flask. Incubate the culture at 28–30°C on a rotary shaker overnight (24 h) to obtain a vigorously growing young culture.

b. Preparing Sarkosyl/Pronase Mixture and Lysozyme Solution (Key Step 2)

Prepare the sarkosyl/pronase mixture and lysozyme solutions in tubes and incubate in a water bath set at 37°C. These preparations need to be done at least 1 h ahead of time. Carefully dissolve 1 g of sarkosyl in 10 ml (10% solution) of Tris-EDTA (TE_{25}) buffer. Add 10 mg of pronase (final concentration 5 mg ml^{-1}) to 2 ml of the sarkosyl solution in another tube. Prepare a 2 mg ml^{-1} lysozyme solution by dissolving 4 mg of lysozyme in 2 ml of TE_{25} buffer.

c. Centrifuging and Washing Cells (Key Step 3)

The young 24-h culture needs to be washed to remove extracellular polysaccharide before it can be treated with lysozyme to digest the cell wall. Place 30 ml of the young culture in a 50-ml plastic centrifuge (Oak Ridge) tube. Centrifuge the cells at 6000 rpm (5000 $\times$ g) for 10 min. Pour off the supernatant and resuspend (vortex) the pellet in 25 ml of 1.0 M NaCl solution. Partially immerse the centrifuge tube with its contents in a 1-liter beaker containing ice. Place the beaker of ice on a rotary shaker for 30 min.

Centrifuge the suspension again at 6000 rpm for 10 min. Discard the supernatant and resuspend the cells in 20–25 ml of N-Tris(hydroxymethyl) methyl-2-aminoethanesulfonic acid (TES) buffer. Complete the washing by centrifuging the cells once again at 6000 rpm for 10 min. Pour off the supernatant. Finally, resuspend the pellet in 5 ml of TE_{25} buffer.

d. Lysing Rhizobial Cells (Key Step 4)

Lysing the rhizobial cells is carried out in two steps. Cells are first exposed to the lysozyme and then to the sarkosyl/pronase mixture. Add 0.5 ml of the lysozyme solution to the cell suspension. Mix by gently inverting and rotating the tubes. Incubate for 15 min in a water bath set at 37°C. Add 0.6 ml of sarkosyl/pronase mixture and incubate at 37°C (water bath) for 1–2 h. At the end of the incubation time the lysate will become viscous. The incubation with sarkosyl/pronase may also be done overnight without any loss in quality and yield of DNA.

e. Extracting Proteins with Phenol (Key Step 5)

Note the total volume of the lysate (6.1 ml). Add an equal volume (6.1 ml) of phenol. (**Phenol is toxic**. Wear gloves and safety eyeglasses or a face shield, and perform the operation in a fume hood. Use a pipette filler (rubber bulb) to draw phenol into the pipettes.)

Upon adding the phenol, mix by gently inverting and rotating the tubes. An emulsion will form. Briefly incubate the mixture at 37°C in a water bath or centrifuge at low speed to allow phase separation. The top aqueous clear phase contains the DNA. Carefully and slowly remove the top aqueous phase with a wide-bore pipette. Note the volume and empty the contents of the pipette into a fresh Oak Ridge tube. Repeat the phenol extraction of the proteins in the aqueous phase once or twice. Note the volume of the aqueous phase each time.

f. Removing Phenol Using Chloroform Extraction (Key Step 6)

Extract the last aqueous phase by adding an equal volume of chloroform. Perform the operation in a fume hood. Mix by gently inverting and rotating, allowing the phases to separate. With a wide-bore pipette, draw up the aqueous phase. Note the volume and transfer the contents into a fresh tube.

g. Precipitating the DNA (Key Steps 7–10)

DNA can be precipitated by adding an equal volume of ethanol or isopropanol to an aqueous solution with 0.1–0.5 M monovalent cations (acetate or chloride). Ammonium acetate, sodium acetate, and NaCl are commonly used as counter ions for ethanol or isopropanol precipitation.

Based on the volume of the aqueous phase obtained in (f), add one-ninth that volume of 3 M sodium acetate solution. Mix well by gently inverting and rotating the tube. Add 0.6 volume of isopropanol and invert several times to ensure complete mixing. At this point the DNA will precipitate as a mass of white threads that will float to the top because of trapped air bubbles. Set aside the precipitating DNA.

Take a Pasteur pipette and heat seal the narrow end. Allow the sealed end to droop into a hook. Air cool the pipette and spool the precipitated DNA with the hooked end. Wash the spooled DNA by immersing into 1 ml of 70% ethanol. Dry the DNA on the glass rod for 5–10 min. Dissolve the DNA by pipetting it into 1 ml of TE buffer in a 1.5-ml microfuge tube. Place the tube at 37°C in a water bath for 1 h and then transfer to 4°C (refrigerator). Usually it takes several hours to dissolve genomic DNA at 4°C.

h. Determining the DNA Yield (Key Step 11)

When DNA is completely dissolved, measure its absorbance (optical density) using a spectrophotometer set at 260 nm. Zero the instrument with 1x TE buffer. Make a 1:50 or 1:100 dilution of the DNA in TE buffer. (Example: 1:100 dilution is made by dissolving 5 μl of DNA solution in 495 μl of TE buffer.) Double-stranded DNA at a concentration of 50 μg ml^{-1} has an absorbance (A) of 1 at 260 nm, therefore:

$$\text{DNA concentration of sample } (\mu\text{g ml}^{-1}) = A_{260} \times \text{Dilution factor} \times 50$$

REQUIREMENTS

a. Culturing the Rhizobial Strain

YMB, 5 ml, in screw-cap tubes
Rotary/reciprocating shaker
Inoculating loop; pipettes, 1 ml
TY medium (Appendix 3) in 125-ml Erlenmeyer flasks
Transfer chamber, Bunsen burner
Rhizobium sp. (TAL 1145) or other rhizobia

b. Preparing Sarkosyl/Pronase Mixture and Lysozyme Solution

TE_{25} buffer (Appendix 5)
10% sarkosyl (sodium-n-lauroylsarcosine) solution
Pronase, 5 mg ml^{-1} in TE_{25} buffer
Lysozyme, 2 mg ml^{-1} in TE_{25} buffer
Water bath (37°C)

c. Centrifuging and Washing Cells

Overnight culture of TAL 1145 or other rhizobia
Oak Ridge centrifuge tubes for ss-34 rotor; centrifuge vortex; beaker, 1 liter; crushed ice; rotary shaker
NaCl solution (1 M), TES buffer (Appendix 5)

d. Lysing Rhizobial Cells

Lysozyme solution and sarkosyl/pronase mixture from (b)
TE_{25} buffer; Eppendorf micropipettes, 0–1000 μl, with tips
Water bath (37°C), pipettes

e. Extracting Proteins with Phenol

Gloves, face shield or safety eyeglasses
Fume hood, water bath (37°C), centrifuge (optional)
Phenol (Appendix 5)
Wide-bore pipettes, 10 ml; pipette fillers
Oak Ridge tubes

f. Removing Phenol Using Chloroform Extraction

Chloroform
Fume hood
Wide-bore pipette
Oak Ridge tubes

g. Precipitating the DNA

Isopropanol, 3 M sodium acetate solution (sterile)
Pasteur pipettes, microfuge tubes
Ethanol, 70%; TE buffer (Appendix 5)
Water bath (37°C), refrigerator

h. Determining the DNA Yield

Spectrophotometer with quartz cuvettes, 0.5–1.0-ml size
Eppendorf micropipettes, 0–20 μl and; 0–500 μl, with disposable tips
Microfuge tubes, TE buffer

KEY REFERENCE

Maniatis, T., E.F. Fritsch, and J. Sambrook. 1982. Molecular cloning. pp. 86–96. *In* A Laboratory Manual. Cold Spring Harbor Laboratory, Cold Spring Harbor, N.Y.

32

Isolating and Purifying Genomic DNA of Rhizobia Using a Rapid Small-Scale Method

When Gram-negative bacteria are exposed to lysozyme, the integrity of the outer cell wall is affected by hydrolytic cleavage of complex polysaccharides. Some of the cell wall remains and the cell becomes spherical, protected mainly by the cell membrane. These cells are called *spheroplasts* and can be lysed to release the DNA by adding ionic detergents such as sodium dodecyl sulfate (SDS) or sodium N-lauroylsarcosine (sarkosyl). This is followed by phenol extraction to remove the proteins from the nucleic acids. Phenol denatures proteins readily, but does not completely inhibit ribonuclease activity. Guanidine isothiocyanate disrupts the cell membrane and causes rapid protein denaturation. Also, the guanidinium cation and the isothiocyanate anion inactivate the ribonucleases by disrupting their tertiary structure.

In this chapter, guanidine isothiocyanate cell lysis method is used in preparing small quantities (30–40 μg) of genomic DNA of rhizobia. The entire procedure is performed in microcentrifuge (microfuge) tubes.

KEY STEPS/OBJECTIVES

1. Culture rhizobial strains.
2. Obtain cell pellet.
3. Wash cells with acetone.
4. Lyse the cells.
5. Precipitate DNA.
6. Wash DNA.
7. Determine DNA concentration.

a. Selecting Rhizobia (Key Step 1)

The genomic DNA isolated and purified in this chapter will also be used later in other chapters. Therefore, it will be helpful to select specific rhizobia at this point to make later studies on the DNA meaningful. Rhizobia may be selected to study diversity among the various species belonging to the different cross-inoculation groups; differences among strains within the same species, but from various geographical locations; diversity among strains of the same species, but from different sero-groups; or effective and ineffective strains of the same species and/or other aspects. Select 12–15 cultures representing the various species in the genera *Rhizobium* and *Bradyrhizobium* (see Table 24.1).

b. Culturing Rhizobia (Key Step 2)

Obtain pure cultures of *Rhizobium* spp. and *Bradyrhizobium* spp. Inoculate a loopful of each culture into 5 ml of yeast-mannitol broth (YMB) in screw-cap tubes. Incubate at 28–30°C for 3–6 days on a rotary shaker to obtain turbid cultures. Examine the cultures for purity by simple Gram staining (Chapter 3). When cultures are turbid and confirmed to be pure, inoculate 1 ml of each culture into 5 ml of tryptone yeast (TY) medium in screw-cap tubes. Incubate the culture at 28–30°C for 24 h with vigorous shaking to obtain actively growing young cultures.

c. Pelleting and Washing Cells (Key Step 3)

A pellet volume of 25–30 μl in a microfuge tube is needed. Pipette 1.5 ml of the young TY culture into a sterile microfuge tube. Centrifuge at 16,000 × g (or 14,000 rpm) for 1 min. Pour out the supernatant. Resuspend cells in 0.75 ml of 50 mM Tris-HCl buffer (pH 7.2). Repeat the centrifugation step to obtain a pellet. Discard the supernatant and vortex (mix) to resuspend the cells in the small amount of residual Tris buffer.

d. Washing the Cells with Acetone (Key Step 4)

Add 0.75 ml of ice-cold acetone to the cell suspension and vortex immediately to prevent the cells from clumping. Place the tube in ice for 5 min. Centrifuge the acetone-washed cells at 14,000 rpm (16,000 × g) for 1 min. Pour off the acetone and aspirate any remaining liquid using a micropipette tip connected to a vacuum-pump. (Exercise extreme care to prevent pellet loss during the aspiration.) Air dry the contents for 5 min.

e. Lysing the Cells (Key Step 5)

Prepare lysozyme solution (16.7 mg ml^{-1}) by dissolving the lysozyme in 50 mM Tris-HCl, 1 mM EDTA buffer (pH 8.0) containing 25% (w/v) sucrose. Add 40 μl of Tris-EDTA (TE) buffer (pH 8.0) to the acetone-washed cells and vortex. Add 60 μl of lysozyme and mix by inverting. Incubate the mixture for 10 min at room temperature (20–25°C). (During incubation, microfuge tubes can be held in place in holes made on a flat piece of Styrofoam of suitable thickness.)

Lyse the spheroplasts (cells without cell walls) by adding 200 μl of 5 M guanidine isothiocyanate in 0.1 M EDTA (pH 7.0). Mix the lysate by inverting or by gentle pipetting. Add 150 μl of 7.5 M ammonium acetate. (The lysate turns cloudy.) Emulsify the cloudy lysate by adding an equal volume (500 μl) of isoamyl alcohol-chloroform (1:24, v/v) and vortex. Separate the phases by centrifuging for 4 min at 14,000 rpm.

f. Precipitating the DNA (Key Step 6)

Using a micropipette, transfer 350 μl of the aqueous phase to a fresh microfuge tube. Add 0.54 volume (189 μl) of isopropanol. Set the tube aside for 10 min to allow the DNA to precipitate. Pellet the precipitated DNA by centrifuging at 14,000 rpm for 10 min. Wash the DNA pellet twice with 1 ml of 76% ethanol in 10 mM ammonium acetate. Dry the DNA pellet in a vacuum or desiccator.

g. Determining DNA Concentration (Key Step 7)

Dissolve the DNA pellet in 100 μl of TE buffer. The DNA must be completely dissolved before proceeding with measurements. Invert the tube gently several times or leave the tube in a refrigerator (4°C) overnight to allow the DNA to dissolve. Dilute 5 μl of DNA in 1 ml of TE buffer in a quartz cuvette. Zero the spectrophotometer with 1x TE buffer. Measure the absorbance at 260 and 280 nm. Calculate the 260:280 nm ratio. The ratio should fall between 1.7 and 1.9 for a pure solution of double-stranded DNA. Calculate the DNA concentration in the sample as described in Chapter 31.

REQUIREMENTS

a. Selecting Rhizobia

Catalog of rhizobia or see Table 24.1

b. Culturing Rhizobia

YMB, 5 ml, in screw tubes
Rotary or reciprocating shaker
Inoculating loops; pipettes, 1 ml
TY medium (Appendix 3) in screw-cap tubes
Transfer chamber, Bunsen burner

Strains of *Rhizobium* spp. and *Bradyrhizobium* spp.
Gram-staining reagents.

c. Pelleting and Washing Cells

Sterile microfuge tubes, 1.5 ml
Microcentrifuge
Vortex mixer
Micropipette, 0–1000 μl, and tips
50 mM Tris-HCl buffer (pH 7.2) (Appendix 5)

d. Washing the Cells with Acetone

Acetone (ice cold), ice bath
Vortex mixer
Micropipette, 0–1000 μl, and tips
Microcentrifuge
Vacuum pump with water trap

e. Lysing the Cells

Lysozyme in 50 mM Tris-HCl, 1 mM EDTA sucrose 25% (w/v)
TE buffer (Appendix 5)
5 M guanidine isothiocyanate in 0.1 M EDTA (pH 7.0) (Appendix 5)
7.5 M ammonium acetate; isoamyl alcohol-chloroform (1:24, v/v)
Micropipette, 0–200 μl and 0–500 μl; pipette tips
Microcentrifuge, vortex mixer

f. Precipitating the DNA

Microcentrifuge, microfuge tubes (sterile), microfuge tube rack
Micropipette, 0–200 μl and 0-500 μl; pipette tips
Isopropanol (2-propanol)
Ethanol, 76%, in 10 mM ammonium acetate (Appendix 5)
Vacuum desiccator

g. Determining DNA Concentration

Spectrophotometer, quartz cuvettes
Micropipettes, 0–20 μl, 0–200 μl, and 0–1000 μl; pipette tips
TE buffer (Appendix 5)

KEY REFERENCE

Saunders, N.A. 1989. Analysis of restriction fragment length polymorphisms for epidemiological tracing of bacteria using nonradioactive probes. Focus 11:47–49.

33

Digesting Genomic DNA of Rhizobia with Restriction Endonucleases

Like many other microorganisms, most, if not all, rhizobia have regions of their genomes that are highly variable. In parts of the rhizobial genome, certain DNA sequences tend to vary from species to species and strain to strain, but are stable for a strain. This variability may be seen when the purified genomic DNA of various species and strains of rhizobia are cleaved or digested into numerous small fragments by site-specific enzymes called *restriction endonucleases*. These enzymes make double-stranded breaks or cuts at specific recognition sequences in the DNA. The small linear DNA fragments that result from the digestion can be separated by electrophoresis according to their molecular size and it is often possible to recognize differences in the banding pattern by visual examination. The characteristic DNA banding pattern of each strain is its fingerprint and may be used in strain identification. In this experiment, the genomic DNA isolated and purified from selected rhizobia will be digested by three different restriction endonucleases prior to separation according to molecular size by horizontal gel electrophoresis.

KEY STEPS/OBJECTIVES

1. Prepare restriction enzyme buffers.
2. Select DNA for digestion.
3. Set up protocol.
4. Digest the DNA.
5. Stop the digestion.

a. Preparing Restriction Endonuclease Buffers (Key Step 1)

Purchase or prepare high and medium salt buffers as described in Appendix 5. The high salt buffer is to be used with the restriction endonucleases *Eco*RI and *Bam*HI while the medium salt buffer is used with *Hin*dIII. It is recommended that the restriction buffers be purchased together with the enzyme since the manufacturer optimizes buffer preparation. Most enzyme suppliers provide the buffers with the purchase of the enzyme. Restriction enzyme buffers can also be prepared in the laboratory as described in Appendix 5.

b. Selecting DNA for Digestion (Key Step 2)

Genomic DNA from different species of rhizobia in the genera *Rhizobium* and *Bradyrhizobium* were isolated, purified, and the DNA concentration determined in Chapter 31. Select the purified DNA of only three species in the genus *Rhizobium*. For the rhizobia selected, determine well ahead of time the volume (microliters) needed to provide 1–2 μg of genomic DNA for the digestion.

c. Setting Up the Experimental Protocol (Key Step 3)

To facilitate proper execution of the experiment, a protocol needs to be setup. The following protocol illustrates a typical setup for the digestion of genomic DNA of one hypothetical rhizobial species or strain against three different restriction enzymes. Tube 4 is the control because no restriction enzyme is added to digest the DNA in this tube. Also, it is assumed that 3 μl of the purified DNA would provide 2 μg of DNA for the digestion.

Component	Tube 1	Tube 2	Tube 3	Tube 4
Sterile water (μl)	14	14	14	15
High-restriction buffer (μl)	2	2	0	0
Medium-restriction buffer (μl)	0	0	2	2
Genomic DNA (μg)	3 μl	3 μl	3 μl	3 μl
EcoRI (μl)	0	1	0	0
*Bam*HI (μl)	1	0	0	0
HindIII (μl)	0	0	1	0
Total volume (μl)	20	20	20	20

Write up a protocol for the three species of rhizobia selected for the analysis in this exercise. Note that the volumes of the restriction buffers, enzymes, and the total volume of the mixture are fixed in this exercise. The volume of restriction buffer is usually one-tenth of the final volume of the reaction mixture so that the final concentration of the buffer during digestion is 1x. The volume of the sterile water will change according to the volume of DNA added. Therefore, calculate the volume of sterile water and the quantity of DNA needed, and enter them into the protocol before performing the digestion.

d. Digesting the DNA (Key Step 4)

Restriction enzyme digests are performed in sterile microfuge tubes. It is essential to protect the tubes from contamination with proteases and nucleases present on our fingertips. Therefore, before handling the microfuge tubes, wear a pair of clean, disposable latex gloves.

Restriction enzymes are expensive and very labile. Follow storage and handling instructions for the enzymes that the manufacturer recommends. When the enzyme is taken out of storage for use, immediately immerse it in a container of ice. The enzyme should not be exposed to room temperature. Set up a water bath at 37°C. Make a simple microfuge tube holder out of a flat piece of polystyrene. Make holes in the polystyrene to fit the tubes snugly. This tube holder (and float) will be used later for holding the microfuge tubes during incubation in the water bath.

Four microfuge tubes are needed for the digestion of DNA of each strain against three enzymes. A total of 12 tubes will be needed for the three strains selected. Label and arrange the tubes in a microfuge tube rack. The components of the digestion mixture are to be added to the microfuge tube following the order indicated in the protocols (i.e., sterile water; restriction enzyme buffer; test DNA; and, finally, the restriction enzyme).

Follow the protocol and set up the digestion mixture for the analysis. Leave the tubes until the enzyme is added. Use a micropipette (0–20 μl) to deliver the enzyme. For each enzyme, use a fresh and sterile pipette tip to deliver the enzyme to the reaction mixture. Close the tubes when the enzyme addition has been completed. Pulse each tube for 2–3 s in a microcentrifuge so that all the liquid collects at the bottom of the tube. Transfer the tubes to the polystyrene tube holder and place in the water bath set at 37°C. Allow the digestion to proceed for 1–2 h, or overnight at the same temperature.

e. Stopping the Digestion (Key Step 5)

Stop the reaction by adding 5 μl of stop solution (0.12 M EDTA, pH 8.0) to the digestion mixture and incubate in a water bath at 60°C for 10 min to inactivate the endonucleases. (Note that the volume of the digestion mixture has increased to 25 μl when the stop solution is added.) Store the tubes in a refrigerator if not immediately needed for use.

REQUIREMENTS

a. Preparing Restriction Endonuclease Buffers

Prepare high and medium salt buffers (Appendix 5) if buffers are not provided or purchased with the enzyme

b. Selecting DNA for Digestion

Obtain purified DNA from Chapter 31

c. Setting Up the Experimental Protocol

No special requirements

d. Digesting the DNA

Sterile microfuge tubes and rack
Micropipettes, 0–20 μl and 0–200 μl, with sterile tips
Sterile glass distilled or double-deionized water
High and medium salt restriction endonuclease buffers (Appendix 5)
Purified genomic DNA from Chapter 31
Disposable latex gloves
Container of crushed ice
Water bath (37°C)
Polystyrene microfuge tube holder
Microcentrifuge
Restriction endonucleases (*Eco*RI, *Bam*HI, and *Hin*dIII)

e. Stopping the Digestion

EDTA, 0.12 M, pH 8.0
Micropipettes, 0–20 μl, with sterile tips
Microfuge tube rack
Water bath (55–60°C)

KEY REFERENCES

Maniatis, T., E.F. Fritsch, and J. Sambrook. 1982. Molecular cloning. pp. 98–106. *In* A Laboratory Manual. Cold Spring Harbor Laboratory, Cold Spring Harbor, NY.

Silhavy, T.J., M.L. Berman, and L.W. Enquist. 1984. Experiments with gene fusions. pp. 183–185. *In* Cold Spring Harbor Laboratory, Cold Spring Harbor, NY.

Separating Restriction Fragments of Genomic DNA by Horizontal Agarose Gel Electrophoresis

Electrophoresis is a separation method that functions because charged particles or molecules migrate through a solution under the influence of an electric field. DNA fragments are charged molecules and bear a net negative charge because of the phosphate backbone. Therefore, DNA fragments will migrate towards the positive electrode (anode). The viscosity provided by the agarose allows the DNA to separate according to the molecular size and conformation during electrophoresis. DNA fragments ranging in size from 0.5–30 kilobases (kb) can be separated using agarose electrophoresis. Agarose gels are widely used in the electrophoretic separation of DNA fragments. Agarose is a highly purified and uncharged polysaccharide derived from agar. It dissolves in boiling water and remains liquid at temperatures over 40°C and becomes a stable gel at lower temperatures. The concentration of the agarose can be varied to obtain the suitable pore size. The higher the agarose concentration, the smaller the pore size. An agarose concentration of 0.6–0.7% is generally used in the electrophoretic analysis of DNA. In this experiment, the smaller DNA fragments resulting from the restriction endonuclease digestion of genomic DNA of rhizobia are separated by horizontal gel electrophoresis. The DNA fragments are stained with ethidium bromide (EtBr) and the banding patterns of the rhizobia are visualized by 302 nm UV illumination.

KEY STEPS/OBJECTIVES

1. Assemble electrophoretic apparatus.
2. Prepare agarose gel.
3. Cast agarose gel.
4. Load digested DNA.
5. Carry out electrophoresis.
6. Stain and photograph gel.

a. Assembling the Electrophoretic Apparatus (Key Step 1)

Many commercial and homemade variations in the design and construction of the apparatus for horizontal electrophoresis are available. The apparatus used here is a commercially manufactured unit and the address of the vendor is given in the requirements list at the end of this chapter. Full description of the apparatus and its operation are provided with the purchase of the unit.

The gel is cast in a casting unit consisting of a gel running plate (19 × 15 cm) and a gel casting tray. Clean the gel running plate and the gel casting tray. Assemble the casting unit by placing the gel running plate into the gel casting tray. Place the casting unit on a level surface (e.g., a leveling table, if available). Place a comb (1.5-mm thickness with 15 teeth) at one end of the casting unit. Adjust the comb height to obtain a gap of at least 1 mm between the lower ends of the teeth of the comb and the surface of the gel running plate. (The screw adjustments provided on the comb backing will facilitate this operation.)

b. Preparing the Gel (Key Step 2)

Use sterile glassware and wear latex gloves at all times to avoid contaminating glassware and working materials with nucleases present on the skin surface of your hands and fingers. Prepare 0.7% agarose gel by dissolving 1.05 g of electrophoretic grade agarose in 150 ml of 1x Tris-borate-EDTA (TBE) buffer in a 250-ml screw-cap Erlenmeyer flask. Microwave or heat to dissolve the agarose completely. (If heat is applied to dissolve the agarose, moisture loss should be minimized.) Upon dissolving the agarose, place the flask in a water bath (50°C) until needed.

c. Casting the Gel (Key Step 3)

Pour the dissolved agarose (150 ml) into the casting unit. Allow the agarose to cool and set for at least 1 h. A 5-mm thick gel should result. Carefully remove the comb by lifting one end of the comb. The 15 wells formed by the teeth of the comb should be visible. Each well will have a 42.4-μl capacity, based on the comb thickness, number of teeth, and well depth.

Carefully lift the gel running plate (with the casted gel on it) and transfer it to the central platform of the electrophoretic unit. Pour sufficient 1x TBE buffer into the two tanks of the electrophoretic unit until a 1–1.5-mm layer of buffer covers the gel. (Approximately 1100 ml of buffer are needed to achieve this with the unit used in this experiment. The unit must be level to obtain a uniform layer of the buffer over the gel.)

d. Loading the Digested DNA (Key Step 4)

The restriction enzyme digested DNA of the three selected species of rhizobia in Chapter 32 will be used in this experiment. Four wells are needed for the DNA of each rhizobial species. Prepare a protocol indicating a well number assignment for each of the samples to be analyzed. Assign the first and last wells for a suitable molecular weight marker. Usually 0.5 μg of *Hind*III-digested lambda DNA in a 10–20-μl volume is used as the molecular weight marker.

Remove the digested samples from storage in the refrigerator. Handle one digested DNA sample at a time. Mix each sample (original volume 25 μl) with one-tenth its volume of 10 × loading buffer (Appendix 5). Since the volume of loading buffer needed is 2.5 μl, use a 0–20-μl capacity micropipette. The final volume of the sample will be 27.5 μl. Use a micropipette (0–100 μl) set at 27.5 μl. Draw up the sample and carefully load into the assigned well. Ensure that when loading, the pipette tip is in the well but not touching the bottom. (Note that the well is not filled to its full capacity. Filling to full capacity causes smearing.) Complete loading of all the samples. Cover the electrophoretic unit with its lid.

e. Carrying Out Electrophoresis (Key Step 5)

Plug the electrical terminals of the electrophoretic unit to the power supply. The terminal closest to the wells must be plugged to the negative (cathode) of the power supply since the negatively charged DNA will migrate to the anode. Set the control knob of the power supply to 0 V, turn on the power supply, and allow it to warm up for 15 min. Set the power supply to 50 V and allow to run overnight. The progress of the separation can be monitored by the migration of the tracking dye (bromphenol blue) in the loading buffer. Terminate the run when the tracking dye has migrated to about 1–2 cm away from the end of the gel.

f. Staining and Photographing the Gel (Key Step 6)

Prepare the staining solution by adding 50 μl of stock EtBr (Appendix 5) into 500 ml of 1x TBE buffer. (**Caution: EtBr is a powerful mutagen. Wear gloves when handling.**) Remove the lid and carefully lift the gel running plate out of the central platform of the electrophoretic unit. Place the running plate with the gel on it into the staining solution for a few minutes. Gently dislodge the gel into the staining solution and remove the running plate. Allow the gel to stain (30 min) on a rotary shaker by gently agitating.

Transfer the gel into another container of deionized water (500 ml) for destaining (20 min). To facilitate transfer, slide a suitable piece of low-flexibility plastic sheet (slightly larger than the gel) under the gel and carefully lift it out of the staining solution. Slide the gel off the plastic sheet onto the transilluminator. Move the gel to a central position on the screen.

Wear protective face shields to protect the eyes from harmful UV rays and turn off the lights to darken the room. Switch on the transilluminator to view the fluorescing DNA bands. Switch off the transilluminator. Place a Polaroid camera (with 22A Wratten film) in position over the gel. Once again darken the room, switch on the transilluminator, and take the photograph. Return the gel to the water in the destaining container and save it for use in another experiment.

REQUIREMENTS

a. Assembling the Electrophoretic Apparatus

A complete, large horizontal unit for agarose electrophoresis (Model HE 99, Hoeffer Scientific Instruments, San Francisco)
Gel casting unit (gel running plate and casting tray)
Comb (1.5-mm thickness with 15 teeth and comb backing)

b. Preparing the Gel

Disposable latex gloves
Agarose (electrophoretic grade), weighing balance
1x TBE buffer (Appendix 5)
Measuring cylinder, 100 ml; Erlenmeyer flask, 250 ml
Microwave oven or heater
Water bath (50°C)

c. Casting the Gel

Assembled electrophoretic unit
Gel casting unit and leveling table, if available
Dissolved agarose from (b)
1x TBE buffer

d. Loading the Digested DNA

Restriction enzyme-digested DNA from Chapter 32
Casted gel ready for loading from (c)
Loading buffer (Appendix 5)
Micropipettes, 0–20 μl and 0–100 μl, and pipette tips
Molecular weight marker

e. Carrying Out Electrophoresis

Power supply unit

f. Staining and Photographing the Gel

EtBr stock solution (Appendix 5)
Micropipette, 0–100 μl, and pipette tips
Plastic containers for preparing staining solution
1x TBE buffer, deionized or distilled water
Rotary shaker, flexible plastic sheet, gloves
Transilluminator, protective face shields or goggles
Polaroid camera, 22A Wratten film

KEY REFERENCES

Maniatis, T., E.F. Fritsch, and J. Sambrook. 1982. Molecular cloning. pp. 149–172. *In* A Laboratory Manual. Cold Spring Harbor Laboratory, Cold Spring Harbor, NY.

35

Transferring Electrophoretically Separated DNA from Agarose Gels to a Membrane by Southern Blotting

To detect or probe for DNA containing complementary sequences to other DNA or RNA sequences, the DNA separated on the gel must first be transferred and immobilized on a solid support such as nitrocellulose or nylon membranes. The Southern blotting procedure, originally demonstrated by Southern in 1975, accomplishes this. In this procedure, the double-stranded DNA duplexes are depurinated with acid, then denatured by treatment with an alkali solution while still within the gel. Alkali treatment produces single-stranded DNA, which binds to the membrane while double-stranded DNA does not. The gel is neutralized and placed on top of a layer of filter paper (wick) soaked in a high-salt buffer. A membrane is then placed on the gel followed by a thin layer of filter paper on the membrane. This is followed by a stack of paper towels on top of which a weight is placed. This arrangement creates a moisture gradient and draws the high-salt buffer solution upwards by capillary action, through the filter paper, gel, and paper towels. This flow of buffer by capillary action transfers the DNA to the membrane, but large fragments and supercoiled plasmid DNA do not transfer efficiently. However, efficient transfer is achieved after the DNA is depurinated with acid and the depurinated sites cleaved by alkali treatment. On completing the transfer, the membrane is baked to bind (immobilize) the DNA, almost permanently, on the membrane. In this form, the immobilized DNA can be probed for specific DNA sequences of interest. In this experiment, supercoiled plasmid DNA or restriction endonuclease digested genomic DNA fragments separated by agarose electrophoresis are transferred onto nitrocellulose or nylon membranes by Southern blotting.

KEY STEPS/OBJECTIVES

1. Prepare solutions.
2. Cut filter paper and membrane.
3. Prepare the membrane.

4. Prepare the wick.
5. Depurinate the DNA.
6. Denature the DNA.
7. Set up Southern blot/transfer.
8. Immobilize the DNA on the membrane.

a. Preparing Solutions (Key Step 1)

Prepare 0.2 M HCl (depurination solution), denaturation and neutralization solutions, and 20x sodium chloride/sodium citrate (SSC) solution following recipes described in Appendix 5.

b. Cutting the Membrane and Filter Paper (Key Step 2)

Measure the exact length and width of the gel from which the DNA is to be transferred. These measurements will be used to cut pieces of filter paper and membrane for setting up the Southern blot. Cut one piece of a selected membrane type (nitrocellulose or nylon) and three pieces of 3MM Whatman filter paper with the length and width dimensions of the gel. (Wear disposable latex gloves when handling the nitrocellulose of nylon membranes.) Label or make a mark with a pencil on one corner of the membrane to aid proper orientation and identification in subsequent steps in the experiment.

c. Preparing the Membrane (Key Step 3)

Thoroughly wet the membrane in a tray of distilled or deionized water. Remove the membrane from the water and soak it in another tray containing 20× SSC. Ensure that no dry spots exist after the wetting and soaking steps.

d. Preparing the Wick (Key Step 4)

Prepare a wick by placing two layers of Whatman 3MM paper over a glass-plate platform as illustrated in Figure 35.1. Pour several hundred milliliters of the 20x SSC into a large tray. Place four 50-ml beakers (upside down and in rectangular formation) on the tray. Obtain a glass plate with length and width dimensions slightly greater than that of the gel to be blotted and place the glass plate in a flat position on the beakers. The level of the 20× SSC in the tray should be approximately 2–3 cm below the glass plate platform. Place two layers of the Whatman 3MM filter paper on the glass-plate platform so the two ends of the filter paper dip into the 20x SSC to form a wick. Roll a pipette over the wick to remove any trapped air bubbles.

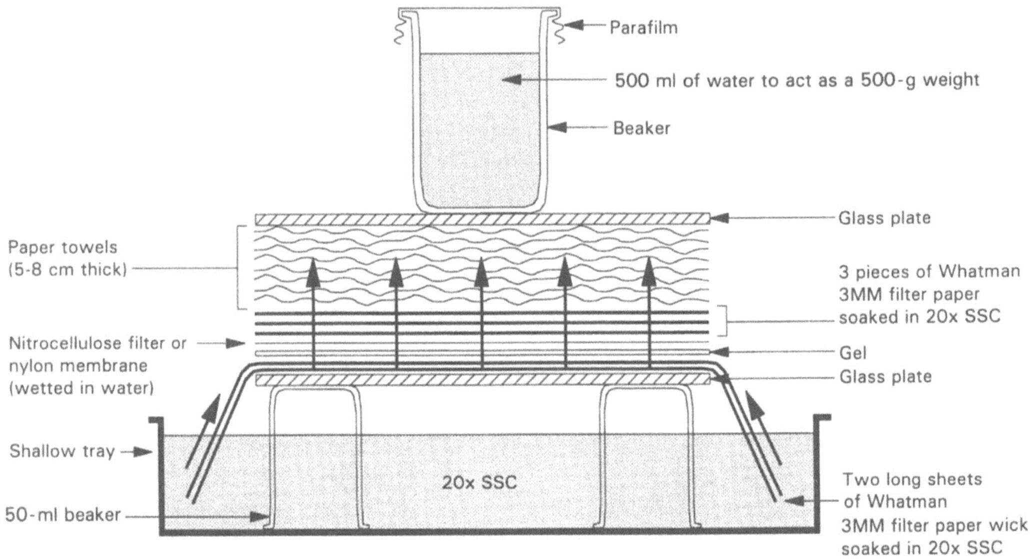

FIGURE 35.1 Arrangement for the transfer of DNA to Nitrocellulose filter or nylon membrane by the Southern blotting procedure. (Heavy arrows indicate the upward flow of the high-salt 20x SSC buffer).

e. Depurinating and Denaturing the DNA (Key Steps 5 and 6)

Place the gel in a tray containing 250–500 ml of 0.2 M HCl and agitate on a rotary shaker. Allow 8–10 min for gels containing restriction endonuclease-digested DNA or 20 min for gels containing plasmid DNA. (Too long in the acid will cause the DNA to cleave into very small fragments that do not bind to the membrane.) Note that the bromphenol blue (tracking dye) turns yellow because of the acidity.

Decant the acid solution and rinse several times with deionized water. Add 250–500 ml of denaturation solution to the gel in the tray. Gently agitate the tray for 30–40 min at room temperature. Note that the bromphenol turns blue, indicating neutralization and denaturation of the DNA. Decant the denaturation solution and add 250–500 ml of neutralization solution and gently agitate on a rotary shaker for 30 min.

f. Setting Up the Southern Blot/Transfer (Key Step 7)

Lift the gel out of the neutralizing solution and carefully lay it (upside down) over the Whatman 3MM filter paper wick prepared on the glass plate platform. Ensure that no air bubbles are trapped between the gel and the filter paper. Cut off a small corner of the gel corresponding to the labeling or marking made on the membrane.

Remove the membrane soaked in the 20x SSC solution and lay it on top of the gel such that the marking occupies the same position as the cut off corner of the gel. Place the membrane to fit exactly on the gel. Do not move the membrane once it is placed on the membrane because DNA transfer begins almost immediately.

Take one of the three pieces of the previously cut filter paper and wet it in 20x SSC and lay it on the membrane. Similarly, wet the other two pieces of filter and place them one at a time over the first piece. Cut paper towels of slightly smaller dimensions than the membrane and stack them over the filter papers. Stack paper towels to a height of at least 5–8 cm and place a glass plate on top of the towels. Finally, position a 500-g weight (500 ml of water in a beaker or Erlenmeyer flask) on the glass plate.

Allow blotting or transfer to proceed for 5–24 h at room temperature. Replace wet paper towels whenever necessary. Remove the paper towels and the layers of filter paper. Transfer the gel and the membrane together (gel side up) as a single unit onto a dry sheet of Whatman 3MM filter paper. Using blunt forceps, peel off the gel (now paper thin) and place the membrane in a 5x SSC solution for 1–2 min. Rehydrate the gel by soaking in deionized water or in 1x TBE. Remove the membrane from the 5x SSC and air dry on a piece of Whatman 3MM filter paper. Examine the rehydrated gel on the transilluminator for efficiency of transfer of the DNA to the membrane.

g. Immobilizing the DNA on the Membrane and Storage (Key Step 8)

Place the dried membrane between two sheets of filter paper and bake for 2 h at 80°C in a vacuum oven. The baked membrane can be stored in a desiccator at 4°C for 6 months or longer.

REQUIREMENTS

a. Preparing Solutions

Weighing balance, weighing paper
Recipes for preparing 0.2 M HCl, denaturation, neutralization, and sodium chloride/sodium citrate (20x SSC) solutions (Appendix 5)

b. Cutting the Membrane and Filter Paper

Whatman 3MM filter paper
Ruler, scissors, disposable latex gloves, soft lead pencil
Nylon or nitrocellulose membranes (Schleicher & Schuell, Inc., Keene, NH)

c. Preparing the Membrane

Nylon or nitrocellulose membrane from (b)
Plastic or glass (baking) trays

Distilled or deionized water, measuring cylinder
20x SSC

d. Preparing the Wick

Whatman 3MM filter paper, scissors
Glass plate, plastic or glass (baking) trays
Measuring cylinders; beakers, 50 ml; pipette, 10 ml
20x SSC

e. Depurinating and Denaturing the DNA

Gel (from Chapter 30 or 34)
Plastic or glass (baking) trays
0.2 M HCl, denaturation, and neutralization solutions (Appendix 5)
Distilled or deionized water
Rotary shaker

f. Setting Up the Southern Blot/Transfer

Denatured gel from (e)
Wick prepared in (d)
Membrane soaked in 20x SSC from (c)
Whatman 3MM filter paper pieces from (a), 20x SSC
Paper towels, scissors, Whatman 3MM filter paper
Forceps, plastic tray
Deionized or distilled water, or 1x TBE
Transilluminator, UV face shields or goggles

g. Immobilizing the Membrane and Storage

Vacuum oven, desiccator, refrigerator
Whatman 3MM filter paper

KEY REFERENCES

Maniatis, T., E.F. Fritsch, and J. Sambrook. 1982. Molecular cloning. pp. 382–389. *In* A Laboratory Manual. Cold Spring Harbor Laboratory, Cold Spring Harbor, NY.

Southern, E.M. 1975. Detection of specific sequences among DNA fragments separated by gel electrophoresis. J. Mol. Biol. 98:503–517.

36

Preparing a DNA Probe for Detecting the nif *Genes on Symbiotic Plasmids of* Rhizobium *spp.*

The large indigenous plasmids of most fast-growing rhizobia (*Rhizobium* spp.) are the sites of the structural genes concerned with the N_2-fixing system. These genes are collectively known as the *nif* genes. They code for the synthesis and activity of the nitrogenase enzyme complex. Some of the nitrogenase structural genes of symbiotic and asymbiotic bacteria have been cloned and specific molecular probes prepared. These molecular probes can seek out and bind to their complementary DNA sequences on the test DNA if there is homology. The *nif* structural genes (K, D, and H) of *Rhizobium meliloti* carried on plasmid pRmR2 have been cloned into the vector plasmid pACYC184 and then introduced into the *Escherichia coli* strain HB101 by transformation. In this exercise, the plasmid pRmR2 carried by the *E. coli* strain HB101 will be extracted and purified by a large-scale method. The extraction of the plasmid DNA will involve alkaline lysis of the cells followed by polyethylene glycol (PEG) purification.

KEY STEPS/OBJECTIVES

1. Culture *E. coli* strain HB 101.
2. Harvest the culture.
3. Lyse the cells.
4. Centrifuge the plasmid DNA.
5. Precipitate the plasmid DNA.
6. Recover the plasmid DNA.
7. Precipitate RNA.

8. Remove contaminating RNA.
9. Purify plasmid DNA by precipitation with PEG.
10. Determine DNA concentration.

a. Culturing the *E. coli* Strain HB101 (Key Step 1)

Inoculate a loopful of strain HB101 to 10 ml of Luria-Bertani (LB) broth (Appendix 3) containing tetracycline (10 μg/ml). Grow the culture for 18 h with vigorous shaking in an incubator shaker set at 37°C. Transfer 5 ml of the LB broth culture into 500 ml of Terrific Broth (TB) medium (Appendix 3) contained in a 2800-ml capacity Erlenmeyer flask. Inoculate a second flask of TB medium. Grow the culture with vigorous shaking for 18–24 h in an incubator shaker set at 37°C.

b. Harvesting the Culture (Key Step 2)

Fill a 500-ml capacity plastic centrifuge bottle with 500 ml of TB culture from one flask. Fill another centrifuge bottle with the culture from the other flask. Balance the bottles and place them in a Beckman JA-10 rotor and centrifuge (4°C) at 4000 rpm (2830 × g) for 15 min. Carefully discard the supernatant and invert the open bottles onto paper towels to drain residual supernatant. Using a Pasteur pipette aspirator, remove traces of supernatant adhering to the wall of the centrifuge bottles.

Add 5 ml of ice-cold sodium chloride Tris EDTA (STE) solution (Appendix 5) to the pellet and vortex to get the pellet into solution. Finally, add 95 ml of STE solution and vortex to completely resuspend the cells. Place the bottles in the Beckman JA-10 rotor. Centrifuge as described earlier in (b). Discard the supernatant, invert the bottles over paper towels, and aspirate residual moisture.

c. Lysing the Cells by Alkali Treatment (Key Step 3)

The washed cells are first treated with lysozyme to degrade the cell wall and then with a mixture of sodium dodecyl sulfate (SDS) and sodium hydroxide solutions to disrupt the cytoplasmic membrane. Centrifugation helps to pellet the larger aggregates of denatured chromosomal DNA, RNA, and cellular protein leaving the plasmid DNA in solution.

Add 6 ml of solution I to each pellet and vortex. Then add 30 ml of solution I again and vortex. To each suspension add 4 ml of freshly prepared lysozyme solution. Incubate on a shaker for 5 min at room temperature. Then add 80 ml of freshly prepared solution II to each bottle. Cap each bottle and mix the contents by inverting the bottles several times. Allow the contents to stand for 15 min at room temperature. Add 40 ml of solution III (at room temperature) to each bottle. Cap the bottles and shake to mix. Place the bottles in ice for 15 min during which time a flocculent white precipitate (chromosomal DNA, RNA, and cellular protein) forms.

d. Centrifuging the Plasmid DNA (Key Step 4)

Remove the bottles from the ice and place them in the Beckman JA-10 rotor. Centrifuge at 10,000 rpm (17,700 × g) for 15 min and allow the rotor to stop without braking. The supernatant contains the plasmid DNA. Construct a filter by placing four layers of cheesecloth in a clean, glass filter funnel. Filter the supernatant through the cheesecloth into a 250-ml capacity plastic centrifuge bottle. Visually estimate the volume of the filtered supernatant.

e. Precipitating the Plasmid DNA (Key Step 5)

Based on the estimated volume of the supernatant in (c), add 0.6 volume of isopropanol to each bottle and mix well. Allow the precipitation to continue for 10 min at room temperature.

f. Recovering the Plasmid DNA (Key Step 6)

Place the bottles containing the precipitated DNA in a Beckman JA-10 rotor and centrifuge at 5000 rpm (4420 × g) for 15 min at room temperature. (The 250-ml bottles need to be contained in the proper sleeves before placement in the Beckman JA-10 rotor.) Carefully pour away the supernatant and invert the bottle on a layer of paper towels. Rinse the DNA pellet and wall of the bottle with 70% ethanol kept in storage at −20°C. Pour away the ethanol and remove all traces of liquid adhering to the wall by aspirating with a Pasteur pipette attached to a vacuum pump.

Dissolve the plasmid DNA pellet in each bottle by adding 6 ml of Tris-EDTA (TE) buffer (Appendix 5) and store at 4°C overnight. If the DNA is not completely dissolved at 4°C, place the bottles in a water bath kept at 48–50°C to completely dissolve the DNA. Centrifuge the bottles at 4000 rpm (2830× g) for 10 min. Transfer the supernatant (DNA in solution) in each bottle to a separate 50-ml polypropylene tube.

g. Precipitating RNA (Key Step 7)

The DNA solution contains high molecular weight RNA that needs to be precipitated out. Lithium chloride (LiCl) is used to precipitate the high molecular weight RNA. Add 6 ml of ice-cold LiCl solution (5 M) to the DNA solution in each tube and mix well. Incubate on ice for 10 min. Centrifuge at 10,000 rpm (17,700 × g) for 10 min at 4°C.

Transfer the supernatant containing the DNA from each tube to polysulfone tubes. Estimate the volume of the supernatant and add an equal volume of isopropanol. Mix well and incubate for 10 min. Recover the precipitated DNA by centrifuging (at room temperature) at 10,000 rpm (17,700 × g) for 10 min.

Decant the supernatant and invert the tubes over a layer of paper towels. Use 70% ethanol (−20°C) to rinse the walls of each tube. Drain off the ethanol and remove any traces of residual liquid from the wall of the tubes by aspirating. Keep the tubes in the inverted position on the paper towels for at least 15–30 min to completely evaporate the alcohol. Dry briefly in a vacuum to completely evaporate the alcohol. Dissolve the DNA pellet in each tube in 1 ml of TE buffer. Incubate in a water bath (48°C) if the DNA does not dissolve.

h. Removing Contaminating RNA (Key Step 8)

The dissolved DNA will contain low molecular weight contaminating RNA that needs to be digested away using RNase. RNase efficiently removes contaminating RNA from plasmid preparations. Briefly centrifuge the DNA solution in the tubes. Add 10 μl of RNase (Appendix 5) to each tube and mix. Incubate at room temperature for 30 min. Incubate the tubes in a water bath at 48°C if the precipitate does not dissolve. Centrifuge at 9500 rpm (16,000 × g) for 10 min to remove impurities. Transfer the supernatant to two fresh microfuge tubes.

i. PEG Purification (Key Step 9)

To the DNA solution in the microfuge tubes [from step (h)], add 500 μl (or equal volume) of 1.6 M NaCl containing 13% (w/v) PEG. Mix well by gently inverting the microfuge tubes several times and centrifuge at 12,000 × g (14,000 rpm) for 5 min at 4°C to recover the DNA. (The microcentrifuge should be set up in a walk-in refrigerator ahead of time.)

Carefully remove the supernatant by aspirating. Dissolve the DNA by adding 400 μl of TE buffer (pH 8.0) and incubate at 48–50°C. If the DNA does not dissolve, add another 300 μl of TE buffer and incubate. Extract the aqueous DNA solution twice with phenol-chloroform and once with chloroform as follows. To the 700-μl sample in each microfuge tube, add 700 μl of phenol-chloroform. Centrifuge at 14,000 rpm for 1 min. Remove 600 μl of the sample to a fresh tube and add an equal volume of phenol-chloroform and centrifuge again. Next, transfer 500 μl of the sample to a fresh tube and add 500 μl of isoamyl alcohol-chloroform (1:24, v/v) and centrifuge as before.

Finally, remove 400 μl of the aqueous phase into a fresh microfuge tube. Add 200 μl of 7.5 M ammonium acetate. Add an equal volume (600 μl) of isopropyl alcohol and store for 10 min at room temperature. (The sample can be stored at −20°C overnight.) Centrifuge the contents at 12,000 x g for 5 min at room temperature. Remove the supernatant by aspirating. Wash the DNA pellet in 1 ml of ethanol (76% ethanol in 10 mM ammonium acetate). Remove the ethanol by aspirating. Dry the pellet in vacuo. Dissolve the DNA in 500 μl of TE buffer (pH 8.0).

j. Determining DNA Concentration (Key Step 10)

When the DNA is completely dissolved, make a 1:50 or 1:100 dilution of the DNA in TE buffer. Measure the absorbance (*A*) of the DNA solution using a spectrophotometer set at 260 nm. Calculate the DNA concentration as follows:

$$\text{DNA concentration } (\mu\text{g/ml}) = (A_{260}) \times \text{Dilution factor} \times 50$$

REQUIREMENTS

a. Culturing *E. coli* Strain HB101

E. coli strain HB101
LB broth (containing 10 μg/ml of tetracycline), and TB medium (Appendix 3)
Flasks, 2.8-liter capacity
Incubator shaker (37°C)

b. Harvesting the Culture

Plastic centrifuge tubes, 500 ml; balance
Beckman centrifuge (Model J2-21) (Palo Alto, CA), Beckman JA-10 rotor or equivalent
Pasteur pipette aspirator, paper towels
Ice-cold STE solution (Appendix 5)
Vortex mixer

c. Lysing the Cells by Alkali Treatment

Solutions I, II, and III (Appendix 5)
Lysozyme solution, 10 mg/ml in 10 mM Tris-HCl (pH 8.0)
Vortex mixer, rotary shaker
Bucket of ice

d. Centrifuging the Plasmid DNA

Beckman Centrifuge, Beckman JA-10 rotor
Filter funnel, cheesecloth
Plastic centrifuge bottle, 250 ml

e. Precipitating the Plasmid DNA

Isopropanol

f. Recovering the Plasmid DNA

Beckman centrifuge, Beckman JA-10 rotor
Plastic centrifuge bottles, 250 ml, with sleeve adapters
Paper towels, freezer (−20°C)
Ethanol, 70%, kept at −20°C, TE buffer (Appendix 5)
Pasteur pipette attached to vacuum pump (aspirator)
Refrigerator, water bath (48–50°C)
Polypropylene tubes, 50 ml

g. Precipitating RNA

Polysulfone tubes
Ice-cold 5 M LiCl solution
Bucket of ice; isopropanol; ethanol, 70% (−20°C)
Beckman centrifuge
Pasteur pipette aspirator, paper towels
Water bath (48°C)
TE buffer

h. Removing Contaminating RNA

Micropipettes, 0–20 μl and 0–1000 μl, and tips
RNase solution (Appendix 5)
Beckman centrifuge and rotor
Water bath (48°C)
Microfuge tubes

i. PEG Purification

PEG, 13%, w/v, in 1.6 M NaCl
Microcentrifuge (kept refrigerated), microfuge tubes
Aspirator, water bath (48°C)
TE buffer, micropipettes, 0–1000 μl, and tips
Phenol-chloroform, isoamyl alcohol, chloroform, 1:24, v/v
Ammonium acetate, 7.5 M; isopropyl alcohol
Ethanol, 76%, v/v, in 10 mM ammonium acetate (Appendix 5)

j. Determining DNA Concentration

Spectrophotometer, quartz cuvettes
Micropipettes, 0–20 μl and 0–1000 μl, and tips
TE buffer

KEY REFERENCES

Sambrook, J., E. Fritisch, and T. Maniatis. 1989. Molecular cloning. pp. 1.21–1.41. *In* A Laboratory Manual, 2nd ed. Cold Spring Harbor Laboratory Press, Cold Spring Harbor, NY.

Tartof, K.D., and C.A. Hobbs. 1987. Improved media for growing plasmid and cosmid clones. Focus 9:12.

37

Incorporating a Nonradioactive Label into a DNA Probe by Nick Translation

After a gene or nucleic acid probe has been prepared, the probe has to be labeled or tagged to facilitate its direct or indirect detection once it has found and hybridized with its homologous or target DNA. A gene probe can be labeled by the widely used enzymatic technique known as *nick translation*, which efficiently incorporates radioactively or nonradioactively labeled deoxynucleotide triphosphates (dNTPs) into the double-stranded DNA probe. The nick translation reaction involves the simultaneous action of two enzymes, namely pancreatic deoxyribonuclease I (DNase I) and *E. coli* DNA polymerase I (DNA pol I). DNase I acts by creating free 3′ hydroxyl and 5′ phosphate ends called *nicks* along each strand of the unlabeled probe DNA. Now DNA pol I, which has a 5′-3′ exonuclease activity, catalyzes reactions at the site of the nicks by progressively removing nucleotides from the double-stranded probe starting at the free 5′-end. Then the same DNA pol I, because of its other 5′-3′ polymerase activity, successfully incorporates a new labeled nucleotide at the position where the preexisting nucleotide was cleaved. Thus, the initial nick is sequentially translated along the DNA backbone and the net effect produces a uniformly labeled probe DNA. In this experiment, molecules of a nonradioactive biotinylated nucleotide (biotin-7-dATP) are incorporated into the purified preparation of the *nif* structural gene probe DNA. A commercially available nick translation kit is used.

KEY STEPS/OBJECTIVES

1. Purchase a nick translation kit.
2. Label the probe DNA.
3. Precipitate the labeled probe DNA.

a. Purchasing a Nick Translation Kit (Key Step 1)

Several different nick translation kits are available commercially with variations in their protocols for the nick translation reaction. Purchase a complete nick translation kit, including the protocol and the biotin-labeled nucleotide (biotin-7-dATP), from a vendor such as the Bethesda Research Laboratories Life Technologies Inc., or Amersham Life Science Corp. (see Requirements section for locations).

b. Labeling the DNA Probe by Nick Translation (Key Step 2)

Biotin-7-dATP can be efficiently incorporated into the probe DNA by nick translation in the presence of dCTP, dGTP, and dTTP.

Pipette 1.0 μg of probe DNA (Chapter 36) into a fresh 1.5-ml microfuge tube. Make up the volume to 37.5 μl by adding sterile distilled water. Place the tube on ice.

Add 5 μl of solution A to the tube. (Solution A contains only the unlabeled nucleotides dCTP, dGTP, and dTTP.)

Add 2.5 μl of 0.4 mM biotin-7-ATP (solution B). Close the tube and mix the contents briefly.

Add 5 μl of solution C (contains DNase I and DNA pol I.) Close the tube, and mix gently but thoroughly. Pulse briefly in a microcentrifuge. Incubate the contents at 15°C for 1–2 h.

Stop the reaction by adding 5 μl of stop buffer (solution D) and mix briefly.

Note that the final volume of the contents adds up to 55 μl. The final volume is useful in computing the reagents used in subsequent steps.

c. Precipitating the Labeled Probe DNA (Key Step 3)

The biotin-labeled probe DNA needs to be separated by ethanol precipitation from the unincorporated nucleotides and salts. Also, if sodium acetate (or ammonium acetate) is used, the precipitation efficiency can be further improved.

Add 6 μl (1/9 volume) of 3 M sodium acetate (pH 5.2) and 165 μl (3 volumes) of ethanol. Mix well and place on ice for 10 min. Centrifuge for 20 min at 12,000 × g in a microcentrifuge. Carefully remove and discard the supernatant with a micropipette. Centrifuge briefly again to remove any residual supernatant.

Resuspend the pellet in 50 μl of Tris-EDTA (TE) buffer and reprecipitate the DNA by adding 5 μl of 3 M sodium acetate and 150 μl of ethanol. Place the tube in ice for 10 min followed by centrifuging for 20 min at 12,000 × g. Discard the supernatant.

Dry the pellet in vacuum for 10 min and dissolve the pellet [37°C for 30 min or in a refrigerator (4°C) overnight] in 50 μl of TE buffer. Store the labeled DNA probe in a freezer at −20°C.

REQUIREMENTS

a. Purchasing a Nick Translation Kit

A nick translation kit from Bethesda Research Laboratories Life Technologies Inc., Gaithesburg, MD or Amersham Life Science Corp., Arlington Heights, IL.

b. Labeling the DNA Probe by Nick Translation

Purified *nif* KDH probe from Chapter 36
Microfuge tubes, microcentrifuge
Distilled water, bucket of ice
Micropipette, 0–20 μl, and tips
Complete DNA labeling kit, biotin-7-dATP, dNTP-labeling mixture, enzymes, buffers, and other reagents from supplier

c. Precipitating the Labeled Probe DNA

3 M sodium acetate solution; TE buffer; ethanol, 95%
Microfuge tubes, microcentrifuge
Micropipette, 0–20 μl, and tips
Bucket of ice, vacuum desiccator
Water bath (37°C)

KEY REFERENCES

Kessler, C. 1992. Nonradioactive labeling methods for nucleic acids. pp. 29–92. *In* L.J. Kricka (ed.) Nonisotopic DNA Probe Techniques. Academic Press, San Diego, CA.

Maniatis, T., E.F. Fritsch, and J. Sambrook. 1982. Molecular cloning. pp. 107–148. *In* A Laboratory Manual. Cold Spring Harbor Laboratory, Cold Spring Harbor, NY.

38

Using a Nonradioactively Labeled nifKDH Gene Probe to Locate Complementary Sequences of Rhizobial DNA Immobilized on Membranes

In the Southern blotting procedure, the DNA separated on the agarose gel is denatured into single strands prior to transfer and immobilization on special nitrocellulose (NC) filters or nylon membranes. Because of the single-stranded nature of the immobilized DNA on the membrane, sequences complementary to the gene probe will lead to DNA-DNA hybridization between the membrane immobilized DNA and the denatured (single stranded) gene probe. The sites of hybridization on the membranes can be visualized using different procedures for the detection, depending on whether a radioactive or nonradioactive label was used on the probe. Signals from radioactively labeled probes are captured by exposing the NC filter or nylon membrane to X-ray films (autoradiography). Biotin-labeled probes are usually detected by the enzyme-linked immunoassay using an enzyme conjugate streptavidin-alkaline phosphatase (SA-AP). The streptavidin (a biotin-binding protein) part of the conjugate binds to the biotin. Adding a chromogenic substrate initiates a color reaction because cleavage by the enzyme produces a colored product. Colored bands become visible on the NC filters or nylon membranes at the sites of DNA-DNA hybridization. In this experiment, Southern blots of genomic or plasmid DNA on NC filters or nylon membranes are probed with a biotinylated *nif*KDH gene sequence and a color reaction indicates hybridization sites. A commercial kit for the nonradioactive labeling and hybrid DNA detection is used.

KEY STEPS/OBJECTIVES

1. Prehybridize the blot.
2. Prepare single-stranded probe DNA.

3. Hybridize probe DNA.
4. Wash the blot.
5. Use immunoassay to detect hybrids.
6. Record results.
7. Store blot.

a. Prehybridizing the Blot (Key Step 1)

The prehybridization treatment serves to block sites on the blot (*blot* refers to the NC filter or nylon membrane containing DNA transferred by Southern blotting) where the free probe can bind nonspecifically. The prehybridization solution contains reagents that will saturate these "sticky" sites. These reagents are Ficoll, polyvinylpyrrolidone, and bovine serum albumin (BSA) in Denhardt's solution, salmon sperm DNA, and sodium dodecyl sulfate (SDS).

Wear latex gloves at all times when handling the blot. Measure the length and width of the blot to calculate the volume of the prehybridization solution, which is usually used at the rate of 50–100 $\mu l\ cm^{-2}$ of the blot. Prepare the prehybridization solution and the denatured salmon sperm DNA as described in Appendix 5. Soak the blot in 2X sodium chloride/sodium citrate (SSC) for 5–10 min. Transfer the blot to a sealable plastic bag (polyethylene). Pipette the required volume of prehybridization solution into the bag. Add the denatured salmon sperm DNA, remove trapped bubbles, and heat seal the bag. Place the bag in a shallow plastic tray and incubate with gentle shaking on a shaking incubator at 65°C for 1–2 h.

b. Preparing Single-Stranded Probe DNA (Key Step 2)

The double-stranded *nif*KDH probe has to be denatured to obtain single strands. This is required for hybridization with the single-stranded test DNA immobilized on the blot. Obtain the biotinylated DNA probe in the microfuge tube (1 μg in 50 μl TE) and add 150 μl Tris-EDTA (TE) buffer to bring the volume to 200 μl. Cap the tube and place in boiling water for 10 min and cool. At the end of this time, remove the tube and immediately immerse it in ice.

c. Hybridizing Probe DNA to the Test DNA on the Blot (Key Step 3)

DNA-DNA hybridization reactions are carried out at high-salt concentration (e.g., 5x SSC) because nucleic acids are more stable under this condition. Prepare the desired volume (in a 10–15-ml plastic tube) of hybridization solution (Appendix 5), which is needed at the rate of 50–100-$\mu l\ cm^{-2}$ blot area. Pipette the denatured *nif*KDH probe preparation into the hybridization solution and mix. Cut open a corner of the plastic bag. Pour out the prehybridization solution. Pipette in the mixture containing the hybridization solution and the *nif*KDH probe into the bag containing the blot. Remove trapped bubbles and heat seal the bag. Place the bag in a shallow tray and incubate at 65°C overnight with gentle shaking to distribute the probe evenly.

d. Washing the Blot (Key Step 4)

Once the hybridization is completed, the blot is washed in low-salt solutions to select for the more stable hybrids and minimize the nonspecific background. After the hybridization has been completed, cut open the bag and pour out the probe-hybridization mixture into a tube. Carefully remove the blot and place it in a shallow plastic tray containing 50–100 ml of 2x SSC, 0.1% SDS. Perform this wash with gentle shaking at room temperature for 5 min. Repeat this wash. Next, wash the blot in 50–100 ml of 0.2x SSC, 0.1% SDS for 5 min at room temperature. Repeat this wash. Finally, wash the blot in 50–100 ml of 0.2x SSC, 0.1% SDS for 15 min at 50°C. Briefly rinse the blot in 2x SSC at room temperature.

e. Using Immunoassay to Detect the DNA Hybrids (Key Step 5)

The hybridization sites of the biotinylated probe with the target DNA can be detected by enzyme-linked immunoassay. The SA-AP cleaves the substrate BCIP (5-bromo-4-chloro-indolyl-phosphate), producing an insoluble purple precipitate resulting from the dephosphorylation of BCIP and subsequent oxidation by the dye Nitro Blue Tetrazolium (NBT). Detailed descriptions of the buffers, SA-AP, substrates, and other components in the kit, as provided by the manufacturer, are identified in the Requirements section.

Wash or rehydrate the blot in buffer 1 for 1 min. Transfer the blot to buffer 2 and incubate for 1 h at 65°C. Dilute the SA-AP in a polypropylene or siliconized tube to obtain a concentration of 1.0 μg ml^{-1}. (This is done by diluting 1 μl of stock solution in 1.0 ml of buffer 1.) Perform the dilution just before use. Prepare approximately 7.0 ml per 100 cm^2 of blot. Drain off buffer 2 and pipette the diluted SA-AP conjugate into the incubation tray containing the blot. Incubate for 10–15 min with gentle agitation. Decant the solution. Wash the blot in 100 ml of buffer 1 for 15 min and decant. Repeat this washing step two more times. Finally, wash the blot once in buffer 3 for 10 min.

Prepare a fresh dye solution (in a polypropylene or glass tube) at the rate of 7.5 ml per 100 cm^2 of blot. Add 33 μl of NBT to 7.5 ml of buffer 3 and mix gently by inverting the tube. Add 25 μl of BCIP solution and mix gently again. Place the blot in a small shallow tray or in a polyethylene bag. Allow the color development to proceed in low light or in the dark for 30 min to 3 h. Purple bands appear at sites where the probe hybridized with the target DNA. Wash the blot in 20 mM Tris (pH 7.5)/0.5 mM EDTA to terminate the color development reaction.

f. Recording the Bands on the Wet Blot (Key Step 6)

The hybridization bands on the wet blot can be recorded by photography or photocopying. Bands are strongest only on one side of the blot. Photograph the wet blot using a

Kodak no. 5 yellow filter. To photocopy, place the blot on a yellow or blue plastic transparency to obtain an enhanced copy.

g. Storing the Blot (Key Step 7)

Dry the blot by baking at 80°C in a vacuum oven for 1–2 min. For storage, a blot can be placed between two pieces of Whatman 3MM filter paper and kept in a desiccator. (The color fades upon drying, but can be recovered by wetting with buffer 3.)

REQUIREMENTS

a. Prehybridizing the Blot

Southern blotted NC filter or nylon membrane
Prehybridization solution (Appendix 5)
Denatured salmon sperm DNA (Appendix 5)
20x SSC (Appendix 5)
Sealable plastic bags, sealing machine
Shallow trays (plastic or glass)
Shaking incubator (65°C)
Micropipette, 0–20 μl and 0–200 μl

b. Preparing Single-Stranded Probe DNA

Biotinylated *nif*KDH DNA probe from Chapter 37
TE buffer (Appendix 5)
Boiling water bath, bucket of ice

c. Hybridizing Probe DNA to the Test DNA on the Blot

Prehybridized Southern blotted DNA [from (a)]
Clean plastic tubes, shallow trays
Hybridizing solution (Appendix 5)
Scissors, bag sealing machine
Incubator shaker (65°C)

d. Washing the Blot

Hybridized blot [from (c)]
Scissors, shallow plastic trays
2x SSC, 0.1% SDS
0.2x SSC, 0.1% SDS

e. Using Immunoassay to Detect the DNA Hybrids

Complete nonradioactive DNA label-detection kit from Bethesda Research Laboratories Life Technologies Inc., Gaithesburg, MD or Amersham Life Sciences Corp., Arlington Heights, IL
Hybridized blot [from (d)]
Buffer 1: 0.1 M Tris-HCl (pH 7.5), 0.15 M NaCl
Buffer 2: 3% (w/v) BSA, or 3% dry skim milk in buffer 1
Buffer 3: 0.1 M Tris-HCl (pH 9.5), 0.1 M NaCl, 50 mM $MgCl_2$
NBT
BCIP
Stop solution: 20 mM Tris (pH 7.5)/0.5 mM EDTA

f. Recording the Bands on the Wet Blot

Camera, Kodak no. 5 yellow filter
Photocopying machine, yellow or blue plastic transparency

g. Storing the Blot

Vacuum oven (80°C), desiccator
Whatman 3MM filter paper

KEY REFERENCES

Banfalvi, Z., V. Sankanyan, C. Koncz, A. Kiss, I. Dusha, and A. Kondorosi. 1981. Location of nodulation and nitrogen fixation genes on a high molecular weight plasmid of *R. meliloti*. Mol. Gen. Genet. 184:318–325.

Nuti, M.P., A.A. Lepidi, R.K. Prakash, R.A. Schilperoot, and F.C. Cannon. 1979. Evidence for nitrogen fixation (nif) genes on indigenous *Rhizobium* plasmids. Nature (London) 282: 533–535.

Rashtchian, A. 1992. Detection of alkaline phosphatase by colorimetry. pp. 147–165. *In* L.J. Kricka (ed.) Nonisotopic DNA Probe Techniques. Academic Press, San Diego, CA.

Additional References and Recommended Reading

Beringer, J.E., N.J. Brewin, and A.W.B. Johnston. 1982. Genetics. pp. 167–181. *In* W.J. Broughton (ed.) Nitrogen Fixation, Vol. 2, *Rhizobium*. Oxford University Press, New York.

Broughton, W.J., U. Samrey, and J. Stanley. 1987. Ecological genetics of *Rhizobium meliloti*: Symbiotic plasmid transfer in the *Medicago sativa* rhizosphere. FEMS Microbiol. Lett. 40:251–255.

Casse, F., C. Boucher, J.S. Julliot, M. Michel, and J. Denarie. 1979. Identification and characterization of large plasmids in *Rhizobium meliloti* using agarose electrophoresis. J. Gen. Microbiol. 113:229–242.

Corbin, D., G. Ditta, and D.L. Helinski. 1982. Clustering of nitrogen fixation (*nif*) genes in *Rhizobium meliloti*. J. Bacteriol. 149:221–228.

Downie, J.A., and A.W.B. Johnston. 1988. Nodulation of legumes by *Rhizobium*. Plant Cell Environ. 11:403–412.

Flores, M., V. Gonzalez, M.A. Pardo, A. Leija, E. Martinez, D. Romero, D. Pinero, D. Davilla, and R. Palacios. 1988. Genomic instability in *Rhizobium phaseoli*. J. Bacteriol. 170:1191–1196.

Hartmann, A., and N. Amarger. 1991. Genotypic diversity of an indigenous *Rhizobium meliloti* field population assessed by plasmid profiles, DNA fingerprinting, and insertion sequence typing. Can. J. Microbiol. 37:600–608.

Hirsch, P. 1979. Plasmid-determined bacteriocin production by *Rhizobium leguminosarum*. J. Gen. Microbiol. 113:219–228.

Hodgson, A.L.M., and W.P. Roberts. 1983. DNA colony hybridization to identify *Rhizobium* strains. J. Gen. Microbiol. 129:207–212.

Jarvis, B.D.W., H.L. Downer, and J.P.W. Young. 1992. Phylogeny of fast-growing soybean-nodulating rhizobia supports synonymy of *Sinorhizobium* and *Rhizobium* and assignment to *Rhizobium fredii*. Int. J. Syst. Bacteriol. 42:93–96.

Kosslak, R.M., R. Bookland, J. Barkel, H.E. Paaren, and E.R. Appelbaum. 1987. Induction of *Bradyrhizobium japonicum* common *nod* genes by isoflavones isolated from *Glycine max*. Proc. Natl. Acad. Sci. USA 84:7428–7432.

Martinez, E., R. Palacios, and F. Sanchez. 1987. Nitrogen-fixing nodules induced by *Agrobacterium tumefaciens* harboring *Rhizobium phaseoli* plasmids. J. Bacteriol. 169:2828–2834.

Martinez, E., D. Romero, and R. Palacios. 1990. The *Rhizobium* genome. Plant Sci. 9:59–93.

Phillips, D.A. 1992. Flavonoids: Plant signals to soil microbes. Recent Adv. Phytochem. 26:201–231.

Plazinski, J., Y.H. Cen, and B.G. Rolfe. 1985. General method for the identification of plasmid species in fast-growing soil microorganisms. Appl. Environ. Microbiol. 48:1001–1003.

Quispel, A. 1988. Bacteria-plant interactions in symbiotic nitrogen fixation. Physiol. Plant. 74:783–790.

Saano, A., and K. Lindstrom. 1990. Detection of rhizobia by DNA-DNA-hybridization from soil samples: Problems and perspectives. Symbiosis 8:61–73.

Sadowsky, M.J., and B.B. Bohlool. 1983. Possible involvement of a megaplasmid in nodulation of soybeans by fast-growing rhizobia from China. Appl. Environ. Microbiol. 46:906–911.

Sadowsky, M.J., R.E. Tully, P.B. Cregan, and H.H. Keyser. 1987. Genetic diversity in *Bradyrhizobium japonicum* serogroup 123 and its relation to genotype-specific nodulation of soybean. Appl. Environ. Microbiol. 53:2624–2630.

Toro, N., M.A. Herrera, and J. Olivares. 1984. Location of *nif* genes on large plasmids in *Rhizobium* strains isolated from legume tree root nodules. FEMS Microbiol. Lett. 24:113–115

Triplett, E.W. 1990. The molecular genetics of nodulation competitiveness in *Rhizobium* and *Bradyrhizobium*. J. Mol. Plant-Microbe Interactions 3:199–206.

Watson, R.J. 1989. Molecular genetics of *Rhizobium meliloti* symbiotic nitrogen fixation. Biotech. Adv. 7:31–45.

Wheatcroft, R., and R. Watson. 1988. A positive strain identification method for *Rhizobium meliloti*. Appl. Environ. Microbiol. 54:574–576.

Zurkowski, W. 1982. Molecular mechanism for the loss of nodulation properties of *Rhizobium trifolii*. J. Bacteriol. 150:999–1007.

SECTION VI
Appendices

APPENDIX 1

Characteristics of the Subfamilies of Legumes[1]

PAPILIONOIDEAE

According to the *International Rules of Botanical Nomenclature*, it would appear that the correct name for the *Papilionoideae* subfamily is either *Faboideae* or *Lotoideae*. It is sometimes designated Papilionatae. This subfamily has about 480 genera and 12,000 spp. of trees, shrubs, herbs, and climbers, generally distributed throughout the world, with the more primitive woody genera mostly in the tropics and the more advanced herbaceous genera more common in the temperate regions. Due to the very distinctive structure of the flower, members of this subfamily are very homogeneous and are easy to recognize.

Lvs. usually alternate and mostly compound, pinnate, trifoliate or digitate; stipulate; stipules often present at base of individual leaflets. Fls. zygomorphic and typically papilionaceous; mostly hermaphrodite; calyx tubular and usually 5-toothed; petals 5, imbricate with descending aestivation; upper (adaxial) petal exterior, usually largest, forming standard (vexillum); 2 lateral petals more or less parallel with each other forming wings (alae); and lowest 2 petals interior, usually joined by lower margins, to form keel (carina), which enclosed stamens and ovary. Stamens usually 10, monadelphous (all united by filaments) or diadelphous with 9 united by filaments and with upper or vexillary stamen free; rarely all stamens free; mostly all perfect; anthers 2-locullar, usually dehiscing lengthwise by slits. Ovary superior, of 1 carpel, usually 1-locular, sometimes with false septa; ovules 1-many on ventral suture. Fr. usually a legume or pod, splitting along dorsal or ventral sutures or both; sometimes indehiscent; occasionally jointed and breaking into 1-seeded segments. Seeds usually without endosperm.

CAESALPINIOIDEAE

The *Caesalpinioideae* subfamily has 152 genera and nearly 2800 spp. of trees and shrubs, rarely herbs, mostly tropical and subtropical, and most numerous in tropical America. Lvs. nearly always alternate, pinnate, or bipinnate; stipules paired, mostly deciduous;

[1]After Purseglove 1968.

stipels mostly absent. Fls. zygomorphic, often showy, usually hermaphrodite; sepals 5 or 4 by union of 2 upper sepals, mostly free, sometimes much reduced when 2 bracteoles, which are large and calyx-like, cover the bud; petals 5 or fewer with upper petal innermost in bud; stamens 10 or fewer, free to variously connate, dehiscing lengthwise or by terminal pore; ovary superior, 1-locular, 1-many ovules, style simple. Fr. a legume or indehiscent and drupaceous. Seeds sometimes arillate, rarely with endosperm.

MIMODOIDEAE

The *Mimosoideae* subfamily has 56 genera and about 2800 spp. of trees and shrubs, very rarely herbs, mainly confined to the tropics and subtropics, and more numerous in the Southern Hemisphere. Lvs. usually bipinnate, rarely once pinnate, sometimes reduced to phyllodes; stipules present, sometimes spine-like. Fls. actinomorphic, small, usually sessile, and massed in cylindrical spikes or globose heads; sepals usually 5, mostly valvate and united to form a toothed or lobed calyx; petals same number as sepals, valvate, free or connate; stamens often numerous, free or monadelphous; anthers small, versatile, often with apical gland, dehiscing longitudinally; ovary 1-locular superior, style usually filiform, stigma small, and terminal. Fr. dehiscent or indehiscent, sometimes a lomentum.

The floral characteristics typical of *Papilionoideae*, *Caesalpinioideae*, and *Mimosoideae* are illustrated in Figures A1.1, A1.2, and A1.3, respectively. Even though the pods of legumes (Figure A1.4) cannot be used in recognizing the various subfamilies, they can be used in identifying a field specimen as belonging to the *Leguminosae*. Leaves and associated structures of legumes (Figure A1.5) are also useful in recognizing legumes in the field. Representative shapes of leguminous nodules and their distribution are illustrated in Figures A1.6 and A1.7, respectively.

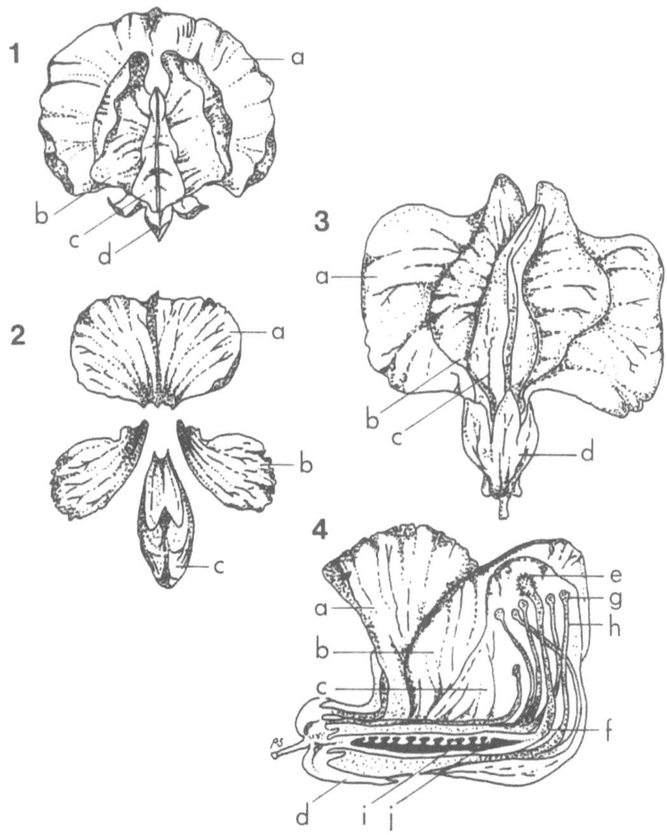

FIGURE A1.1 Subfamily *Papilionoideae.* (1) Front view of flower of *Pisum sativum* (pea); (2) petals of *P. sativum*; (3) flower of *Psophocarpus tetragonolobus* (winged bean) from below; and (4) flower of *Psophocarpus tetragonolobus* in longitudinal section. a, Posterior or standard petal; b, lateral petal; c, keel petals (carina); d, sepals; e, stigma; f, style; g, anther; h, filament; i, ovary wall; and j, ovule.

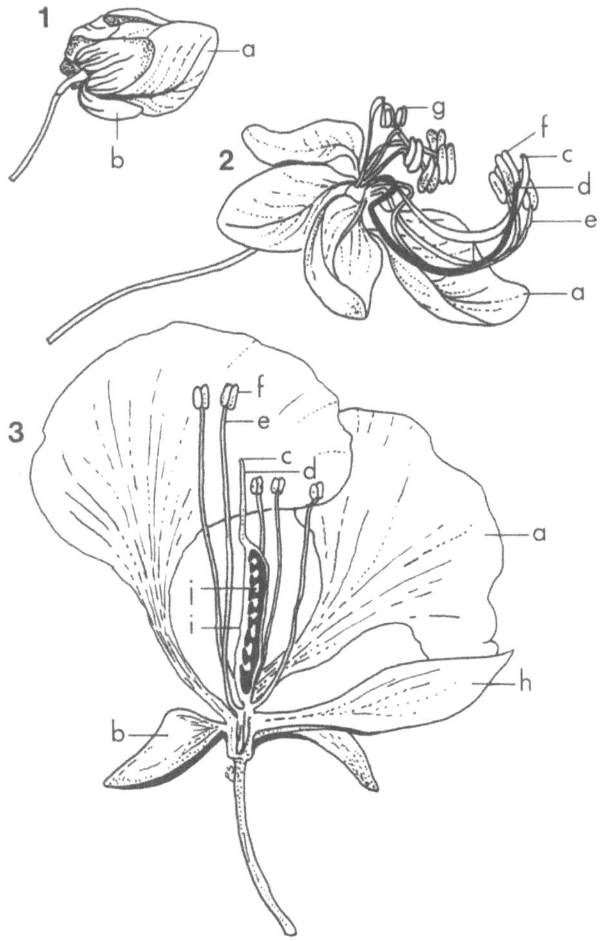

FIGURE A1.2 Subfamily *Caesalpinioideae.* (1) Bud of *Cassia* sp.; (2) flower of *Cassia* sp.; and (3) longitudinal section through flower of *Delonix regia* (Flame of the Forest or Poinciana). a, petal; b, sepal; c, stigma; d, style; e, filament; f, anther; g, anther of staminoid; h, posterior or standard petal; i, ovary wall; and j, ovule.

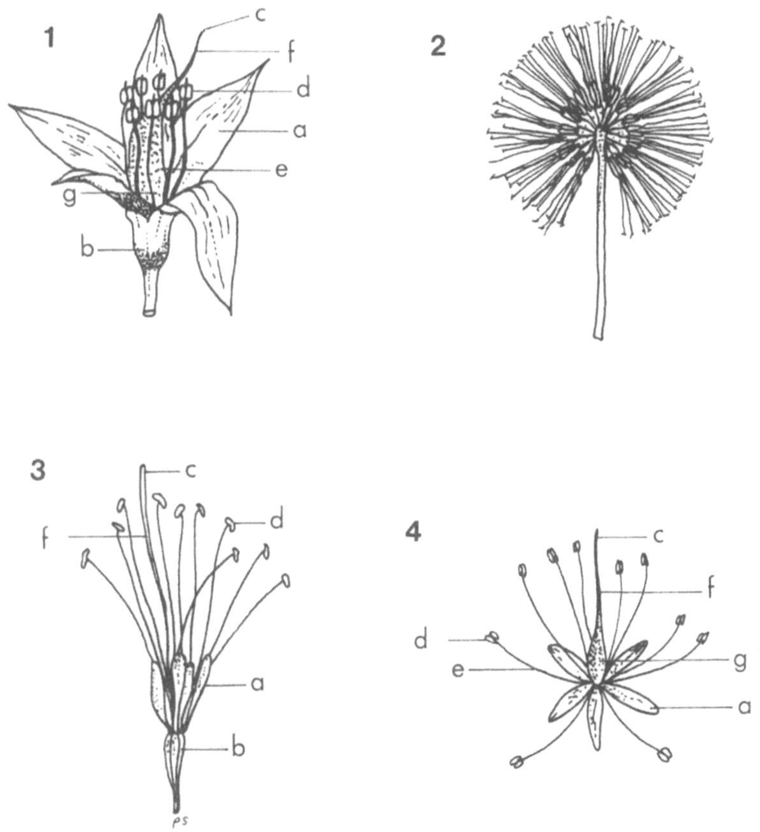

Figure A1.3 Subfamily *Mimosoideae*. (1) Floret of *Adenanthera pavonina*; (2) inflorescence (globose head) of *Leucaena leucocephala* in longitudinal section showing arrangement of florets on torus; (3) floret of *L. leucocephala* (side view); and (4) floret of *L. leucocephala* (top view). a, petal; b, sepal; c, stigma; d, anther; e, filament; f, style; and g, ovary.

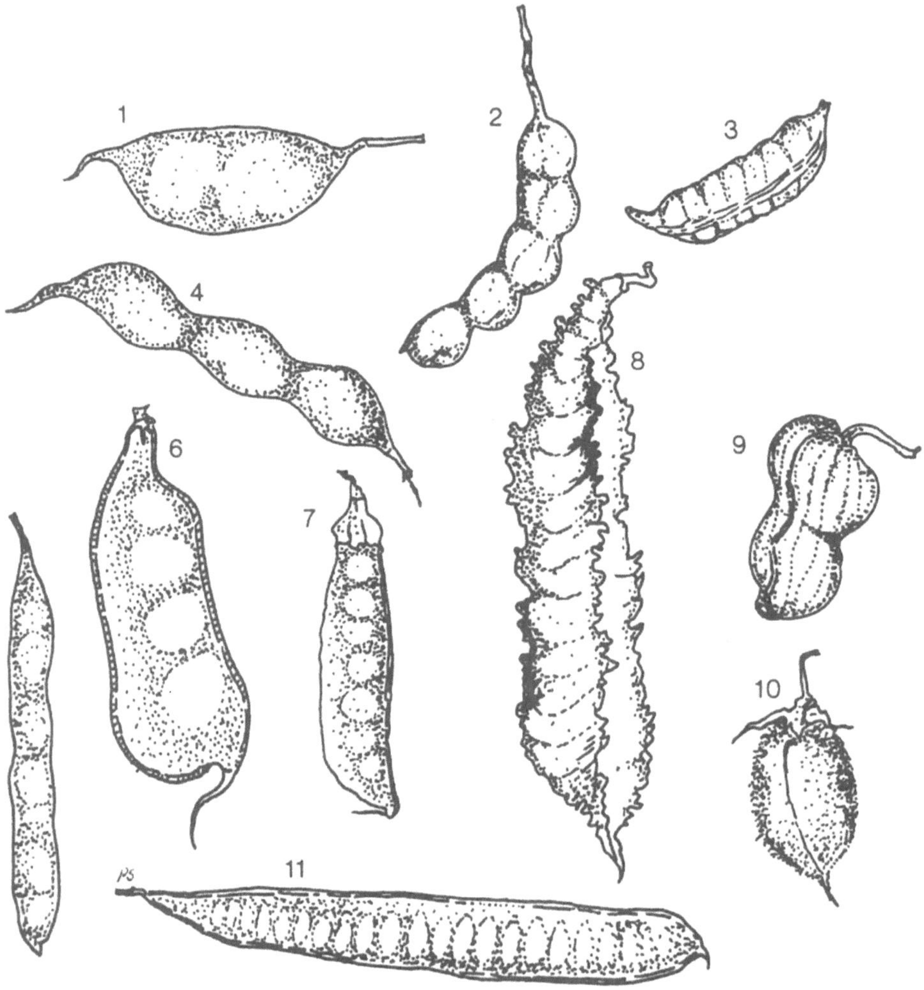

FIGURE A1.4 Legume pods. (1) *Strongylodon lucidus*; (2) *Tamarindus indica*; (3) *Acacia farnesiana*; (4) *Parkinsonia aculeata*; (5) *Prosopis pallida*; (6) *Lablab purpureus*; (7) *Pisum sativum*; (8) Psophocarpus *tetragonolobus*; (9) *Arachis hypogaea*; (10) *Cicer arietinum*; and (11) *Leucaena leucocephala*.

FIGURE A1.5 Leaves of legumes and associated structures. Leaf shapes: (1) oblong; (2) cuneate; (3) cordate; (4) linear; (5) lanceolate; (6) ovate; and (7) oval. Leaf arrangements: (8) bipinnate; (9) pinnate; (10) palmate; (11) simple; (12) trifoliate; (13) branch of *Pisum* showing 5-branched tendril (a) and stipule (b); (14) bipinnate leaf showing position of pulvinus (c); and (15) *Acacia* seedling showing simple phyllodes (d) and true compound leaves (e).

Leaf Shapes

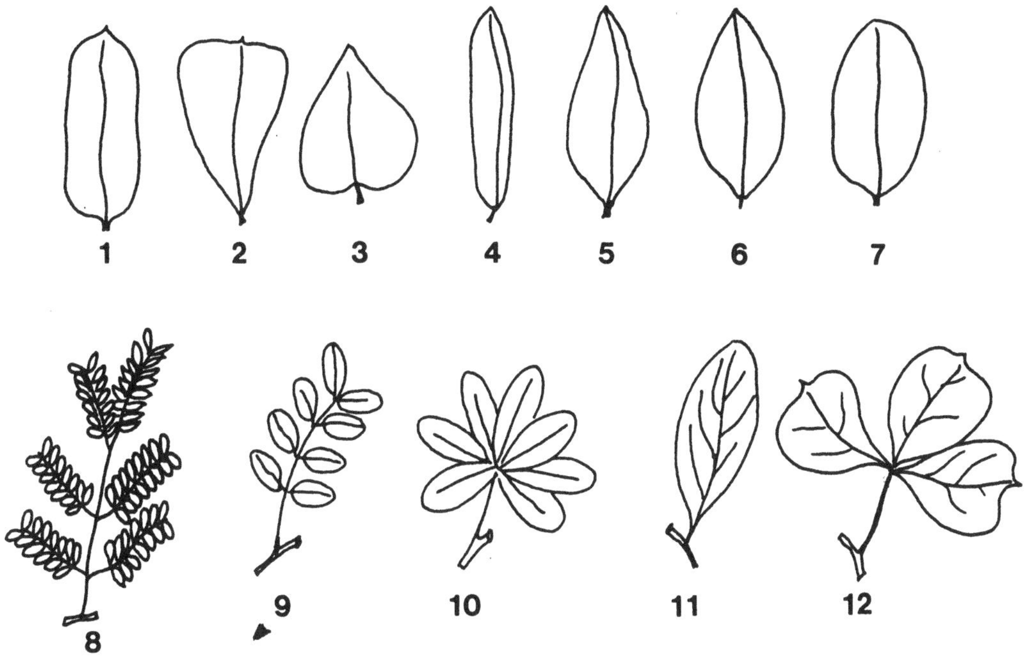

Leaf Arrangements

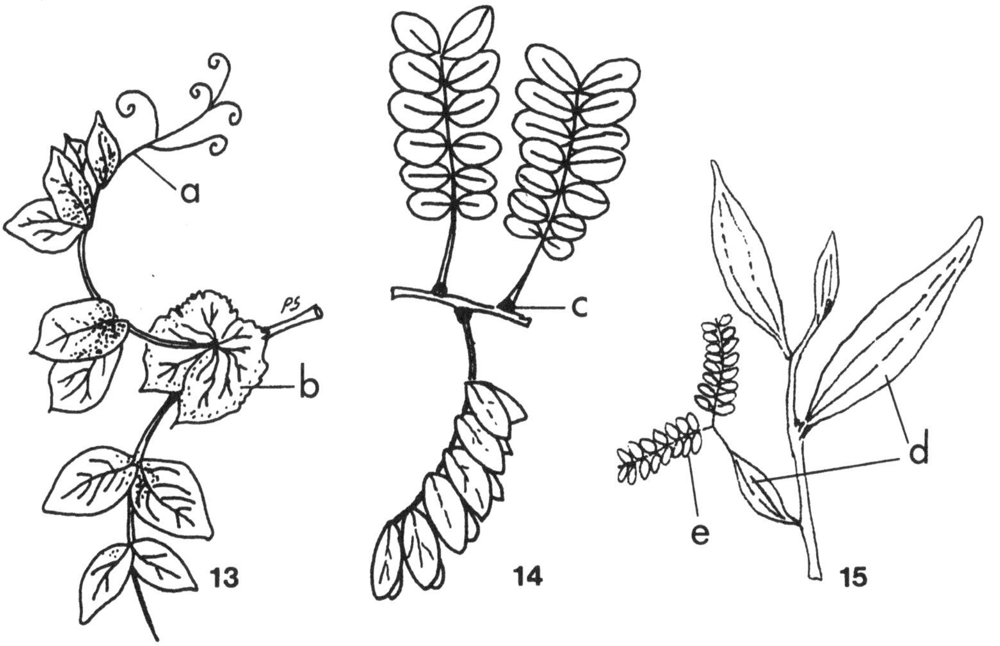

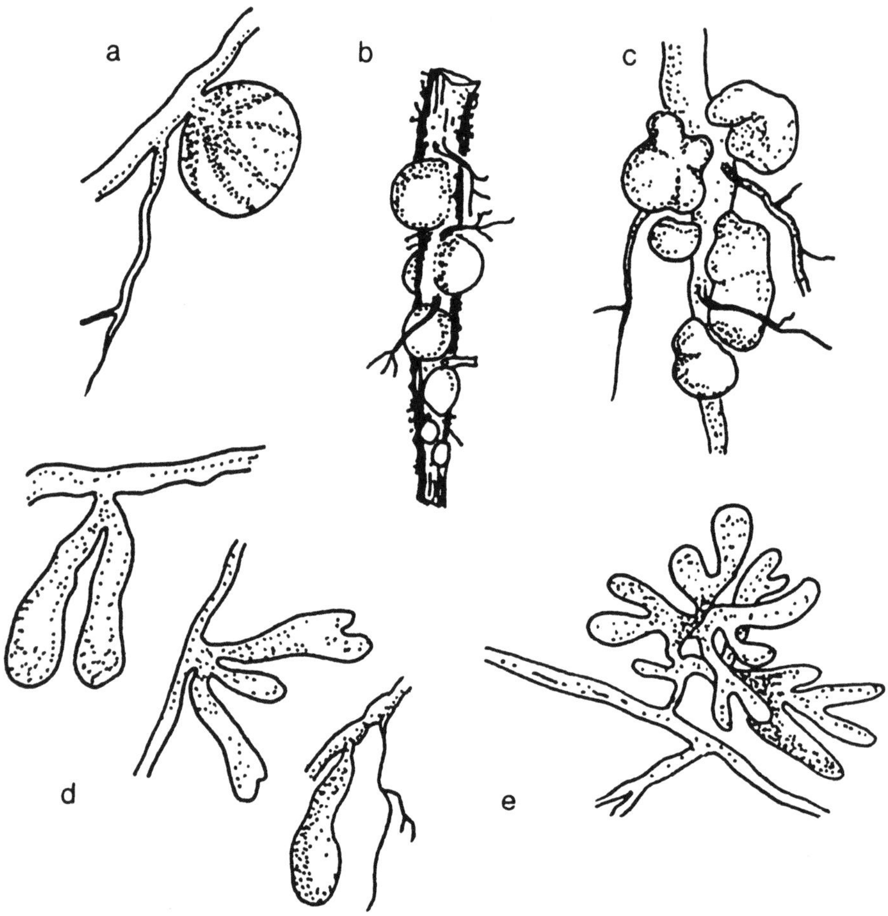

FIGURE A1.6 Some representative shapes of leguminous nodules. Spherical: (a) globose and streaked, e.g., *Glycine max*, *Calopogonium*, and *Vigna radiata*; (b) peanut (*Arachis hypogaea*); (c) semiglobose with smooth surface, e.g., *Vigna unguiculata* and *Psophocarpus*. Finger-like forms: (d) elongate and lobed, e.g., *Leucaena* and *Mimosa*. (e) Fan-shaped or coralloid, e.g., *Crotalaria* and *Calliandra*.

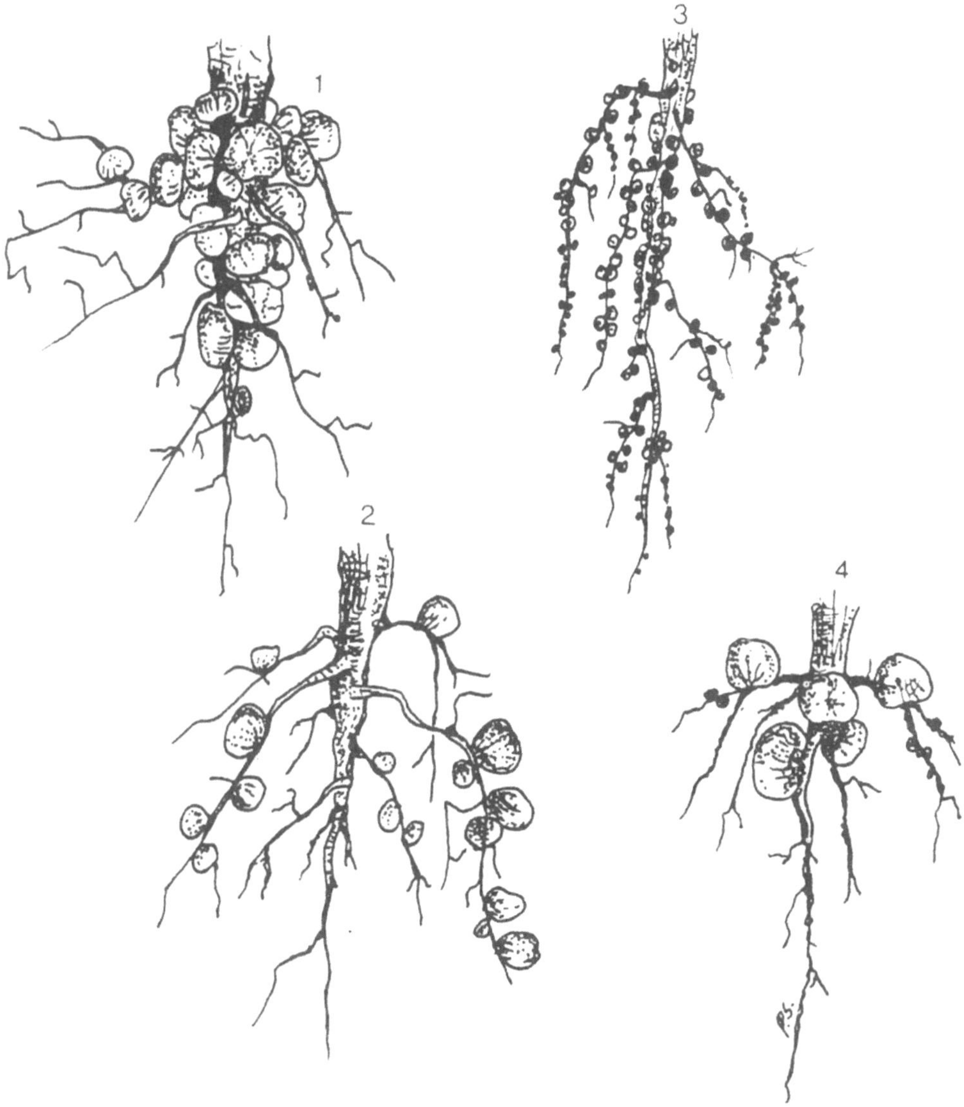

FIGURE A1.7 Some examples of nodule distribution on roots.

Appendix 2

The Nodule Preservation Vial

The apparatus diagrammed in Figure A2.1 is convenient for the preserving nodules collected during field trips. Nodules preserved this way can last 6–12 months, though rhizobial recovery during isolation may vary, depending on the legume species and storage temperature of the vials. Screw-cap vials (15 × 45 mm or 17 × 60 mm) are cheaper and a more convenient size than screw-cap test tubes.

PLANT SELECTION TO SAMPLE NODULES

Nodules should be collected from healthy, green plants. Such plants (if nodulated) may have large nodules with pink/red interiors, which may indicate effective fixation. Excavate plants carefully and remove adhering soil particles. Excise each nodule from the roots, leaving a small piece of root attached. Place the nodules (at least five) in the vial and cap tightly. For tree legumes, seedlings are the best source of nodules.

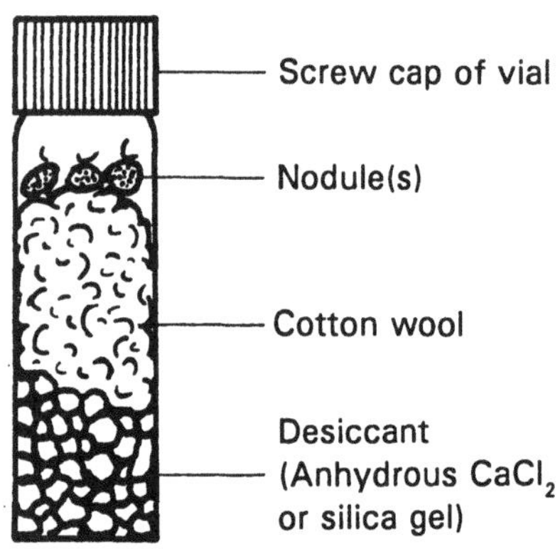

Figure A2.1 Nodule preservation vial.

APPENDIX 3

Bacterial Growth Media and Plant Nutrient Solutions

GROWTH MEDIA FOR RHIZOBIA

Arabinose-Gluconate Medium
(Kuykendall, 1987, modified by M. Sadowsky)

Constituents per liter:

Hepes-MES (HM) buffer[1]	10.0 ml
Salts[2] (six)	10.0 ml each
Arabinose	1.0 g
Na gluconate	1.0 g
Yeast extract	1.0 g
Agar (if required)	15.0 g

Preparation:

- Make stock solutions of each salt in distilled water.
- Add 10 ml of buffer and each of the stock salt solutions to 930 ml of distilled water; add 15 g of agar if a solid medium is required.
- Check pH and adjust to 6.6 if necessary.
- Autoclave at 121°C for 15 min.

---

[1]HM buffer, pH 6.6:

Hepes (Sigma NO. H-3375)	130 g
MES (Sigma NO. M-8250)	110 g

- Dissolve Hepes and MES in 1 liter of distilled or deionized water.
- Adjust pH to 6.6 with 10 N NaOH.

[2]Salt solutions:

1. $FeCl_3 \cdot 6H_2O$	0.67 g
2. $MgSO_4 \cdot 7H_2O$	18.00 g
3. $CaCl_2 \cdot 2H_2O$	1.30 g
4. Na_2SO_4	25.00 g
5. NaCl	32.00 g
6. Na_2HPO_4	12.50 g

- Dissolve each salt in 1 liter of distilled or deionized water. Store at 4°C.

Bergersen's Defined Medium
(Bergersen, 1961; modified after Sherwood, 1970)

Constituents per liter of distilled water:

K_2HPO_4	22.00 g
$MgSO_4 \cdot 7H_2O$	0.10 g
$FeCl_3$	0.02 g
$CaCl_2$	0.04 g
Glutamic acid–sodium salt	1.10 g
Mannitol	10.00 g
Thiamine[3] (1 ml of stock)	100.00 μg
Biotin[3] (0.1 ml of stock)	10.00 μg

Preparation:

- Under continuous stirring, add salts to 1 liter of distilled water (dissolve $FeCl_3$ in 1 N HCl; add the dissolved $FeCl_3$ and $CaCl_2$ last).
- Add mannitol, thiamine, and biotin.
- Adjust pH to 6.8.
- Autoclave at 121°C for 15 min.
- If a solid medium is required, add agar 20 g l-1 (Difco Bacteriological, Difco-Laboratories, Detroit, MI) to broth medium and dispense well before autoclaving.

Fermentor Broth (Burton, 1967)

Constituents per liter:

Mannitol	2.0 g
Sucrose	10.0 g
K_3PO_4	0.2 g
KH_2PO_4	0.4 g
$MgSO_4 \cdot 7H_2O)$	0.2 g
NaCl	0.06 g
$CaCO_3$	0.2 g
$CaSO_4 \cdot H_2O$	0.04 g
Yeast extract	0.5 g
$(NH_4)_2HPO_4$	0.1 g

[3]Stock solutions: Thiamine–0.01 g/100 ml of distilled water; biotin–0.01 g/100 ml of distilled water.

Micronutrient stock solution[4]	1.0 ml
Water	1000 ml

Preparation:

- Dissolve mannitol, sucrose, yeast extract, and salts in 1 liter of distilled water.
- Add 1 ml of micronutrient stock solution to fermentor broth.
- Autoclave at 121°C for 15 min.

Luria-Bertani (LB) Medium

Tryptone	10 g
Yeast extract	5 g
NaCl	5 g
Water	1000 ml

Adjust to pH 7.4 with 1 N NaOH. (Add 15 g per liter of agar before autoclaving to make LB agar.)

Mannitol Nitrate Medium (Vincent, 1970) (For Phage Typing)

Solution A:

Mannitol	10.0 g
$Na_2HPO_4 \cdot 12H_2O$	0.45 g
$N_2SO_4 \cdot 10H_20$	0.06 g
KNO_3	0.60 g
$FeCl_3 \cdot 6H_20$	0.01 g
Thiamine-HCl	100 μg
Biotin	0.5 μg
Agar	7.5 g
Distilled water	1.0 liter

[4]Micronutrient Stock Solution

Constituents:

H_3BO_3	2.78 g
$MnSO_4 \cdot 7H_2O$	1.54 g
$ZnSO_4 \cdot 7H_2O$	0.21 g
Na_2MoO_4	4.36 g
$FeCl_3 \cdot 6H_2O$	5.00 g
$CoSO_4 \cdot 6H_2O$	0.004 g
Lactic acid (88%)	580 ml
Distilled water	420 ml

Addition of 1.0 ml per liter of medium gives: B, 0.5 μg; Mn, 0.5 μg; Zn, 0.05 μg; Mo, 1.0 μg; Fe, 100 μg; and Co, 0.0005 μg per liter (or ppm).

Solution B:

$MgCl_2 \cdot 6H_20$	0.1 g
$CaCl_2 \cdot 6H_20$	0.1 g
Distilled water	100 ml

Sterilize solution A and B separately and add 0.5 ml of B to the Petri dish before adding 5 ml of the melted agar (A).

Peptone Glucose Agar with Bromcresol Purple (PGA-BCP)

Ingredients per liter:

Glucose	5 g
Peptone	10 g
Agar	15 g

Preparation:

- Dissolve glucose and peptone in 1 liter of distilled water.
- Add 10 ml of BCP stock solution[5] to achieve a BCP concentration of 100 μg ml per liter.
- Add agar and suspend evenly.
- Autoclave at 121°C for 15 min.

Terrific Broth (TB) (Tartof et al., 1987)

To 900 ml of deionized water, add, per liter

Bacto-tryptone	12 g
Bacto-yeast extract	24 g
Glycerol	4 ml

Shake until the solutes have dissolved and sterilize by autoclaving for 20 min. at 15 lb/in^2 on liquid cycle. Allow the solution to cool to 60°C or less, and then add 100 ml of a sterile solution of 0.17 M KH_2PO_4, 0.72 M K_2HPO_4. (This solution is made by dissolving 2.31 g of KH_2PO_4 and 12.54 g of K_2HPO_4 in 90 ml of deionized water. After the salts have dissolved, adjust the volume of the solution to 100 ml with deionized water and sterilize by autoclaving for 20 min. at 15 lb/in^2 on liquid cycle.)

Tryptone-Yeast (TY) Medium

Tryptone	5.0 g
Yeast extract	3.0 g
$CaCl_2 \cdot H_2O$	0.87 g
Deionized water	1000 ml

[5]Prepare BCP stock solution by dissolving 1 g of BCP in 100 ml of ethanol.

Adjust pH to 6.8–7.2 with 1 N NaOH. Autoclave for 15 min. A preciptate forms after autoclaving. (For TY agar, add 12 g of agar per liter before autoclaving.)

Yeast-Mannitol Broth (YMB) (Vincent, 1970)

Constituents:

Mannitol	10.0 g[6]
K_2HPO_4	0.5 g
$MgSO_4 \cdot 7H_2O$	0.2 g
NaCl	0.1 g
Yeast extract	0.5 g
Distilled water	1.0 liter

Preparations:
- Dissolve salts in 1 liter of distilled water.
- Add mannitol and yeast extract.
- Dissolve under continuous stirring.
- Adjust pH to 6.8 with 0.1 N NaOH.
- Autoclave at 121°C for 15 min.

Yeast-Mannitol Agar (YMA)

Constituents:

YMB	1 liter
Agar	15 g

Preparation:
- Prepare YMB.
- Add agar, shake to suspend evenly, and autoclave.
- After autoclaving, shake flask to ensure even mixing of melted agar with medium.

YMA with Antibiotics

YMA with Streptomycin (YMA str)

Preparation:
- To 1 liter of YMA, add 10 ml of str stock solution[7] to achieve a concentration of 40 μg str ml^{-1}.
- Add 20 ml of str stock solution if a concentration of 80 μg str ml^{-1} is desired.
- Autoclave at 121°C for 15 min.

---

[6]This amount has been used traditionally, however, more recent findings (H. Keyser, unpublished observations) show that 1 g per liter is sufficient for most rhizobia.

[7]str stock solution: Dissolve 400 mg of streptomycin sulfate (Sigma Chemical Company, St. Louis, Mo.) in 100 ml of water.

YMA with Spectinomycin (YMA spc)

Preparation:
- To 1 liter of YMA, add 10 ml of spc stock[8] solution to achieve a concentration of 250 μg spc ml^{-1}.
- Add 20 ml of the spc stock solution if a concentration of 500 μg ml^{-1} is desired.

Media with Incorporated Dyes

YMA with Congo Red Indicator (CR YMA)

Preparation:
- Add 10 ml of CR stock solution[9] to 1 liter of YMA to achieve a CR concentration of 25 μg ml^{-1}
- Autoclave at 121°C for 15 min.

YMA with Bromthymol Blue Indicator (BTB YMA)

Preparation:
- Add 5 ml of BTB stock solution[10] to 1 liter of YMA to achieve a concentration of 25 μg ml^{-1}.
- Autoclave at 121°C for 15 min.

YMA with Brilliant Green (BG YMA)

Preparation:
- Add 1 ml of BG stock solution[11] to 1 liter of YMA to achieve a concentration of 1.25 μg BG ml^{-1} YMA.
- Autoclave at 121°C for 15 min.

Preparation of Soybean Water

Grind 100 g of soybean (*Glycine max*) seeds to a coarse flour and place in 1000 ml of water. Boil slowly for 2 h, replacing the lost water regularly. Allow to cool and centrifuge at 6000 × g. Remove the supernatant, autoclave, and store. For rhizobia media, use 100 ml per liter. Nitrogen sources can also be prepared from other grain legume seeds in the same way.

---

[8]spc stock solution: Dissolve 2.5 g of spectinomycin dihydrochloride (Sigma Chemical Company) in 100 ml of distilled water.
[9]CR stock solution: Dissolve 250 mg of CR in 100 ml of water.
[10]BTB stock solution: Dissolve 0.5 g of BTB in 100 ml of ethanol.
[11]BG stock solution: 125 mg of BG in 100 ml of ethanol.

Yeast Water

Fresh, starch-free cakes of yeast are preferred in making yeast water. Suspend 100 g of yeast in 1000 ml of water and boil slowly or steam for 3–4 h, replacing the water lost regularly. Allow the cooled suspension to stand until yeast cells have settled to the bottom (usually 10–12 h). Siphon off the clear, straw-colored liquid; adjust the liquid to pH 6.6–6.8 with sodium hydroxide; bottle and autoclave for 30–40 min at 121°C. Following sterilization, the yeast water may be stored at room temperature.

Dried yeast may also be used in making yeast water. One kilogram of dry yeast is equivalent to about 2.5 kg of wet yeast. Suspend 40 g of dry yeast in 1 liter of water. Boil, decant, bottle, and sterilize in the same way as described for fresh yeast. One hundred milliliters of yeast water should contain about 75 mg of N.

Yeast extract powders prepared by spray drying aqueous autolyzed yeast preparations are available in many countries. When these are available, about 0.5 g per liter of the dried preparation is used to replace yeast water. Dry preparations are convenient and usually satisfactory. The media containing yeast may foam excessively when aerated vigorously in fermentor vessels. Adding a small amount of sterile white mineral oil or silicone emulsion can control foaming.

PLANT NUTRIENT SOLUTIONS

Fahraeus N-free Medium (Fåhraeus, 1957)

Constituents per liter of water:

$CaCl_2$	0.100 g
$MgSO_4 \cdot 7H_2O$	0.120 g
KH_2PO_4	0.100 g
$Na_2HPO_4 \cdot 2H_2O$	0.150 g
Ferric citrate	0.005 g
Trace element stock solution[12]	1.0 ml

Preparation:

- Mix all constituents; adjust pH to 6.8–7.0 with NaOH.
- Autoclave at 121°C for 20 min.

[12]Trace elements stock solution.

Constituents per liter of water:

H_3BO_3	2.86 g
$MnSO_4 \cdot 4H_2O$	2.03 g
$ZnSO_4 \cdot 7H_2O$	0.22 g
$CuSO_4 \cdot 5H_2O$	0.08 g
$Na_2MoO_4 \cdot 2H_2O$	0.14 g

- Add quarter-strength medium for topping up Leonard jars or pots.
- For nitrate controls, add 0.5 g KNO_3 per liter (0.05%), e.g., clover (*Trifolium* spp.), and alfalfa (*Medicago* spp.); or 1.0 g KNO_3 per liter (0.1%), e.g., soybean (*Glycine max*), cowpea (*Vigna unguiculata*), and mung beans (*Vigna radiata*).

Modified Jensen's N-free Medium (Roughley, 1984)

Constituents per liter of water:

$CaHPO_4$	1.0 g
K_2HPO_4	0.2 g
$MgSO_4 . 7H_2O$	0.2 g
NaCl	0.2 g
Trace elements stock solution[13]	1.0 ml
$FeCl_3$	0.1 g

Preparation:

Procedures are the same as for Fahreus medium.

N-free Nutrient Solution (Broughton and Dilworth, 1970)

Stock Solution	Chemical	g/liter
1	$CaCl_2 \cdot 2H_2O$	294.1
2	KH_2PO_4	136.1
3	$FeC_6H_5O_7 \cdot 3H_2O$	6.7
	$MgSO_4 \cdot 7H_2O$	123.3
	K_2SO_4	87.0
	$MnSO_4 \cdot H_2O$	0.338
4	H_3BO_3	0.247
	$ZnSO_4 \cdot 7H_2O$	0.288
	$CuSO_4 \cdot 5H_2O$	0.100
	$CoSO_4 \cdot 7H_2O$	0.056
	$Na_2MoO_2 \cdot 2H_2O$	0.048

Preparation:

- Prepare stock solutions; use warm water to get the feric-citrate into solution.
- Make 10 liters of full-strength plant culture solution as follows.
- To 5 liters of water, add 5 ml of each stock solution and mix.
- Dilute to 10 liters by adding another 5 liters of water.
- Adjust pH to 6.6–6.8 with 1 N NaOH

[13]Trace element stock solution as in Fahreus medium.

- For plus-N control treatment, KNO_3 (0.05%) is added, giving N concentration of 70 ppm.

Seedling Agar Slants

Any of the plant culture media can be used to prepare solid medium in growth tubes.

Procedure:

- Prepare liquid culture medium.
- Add 12 g of agar per liter of medium.
- Autoclave for 10 min. or microwave to melt agar; shake to mix while hot.
- Dispense appropriate volumes into growth tubes.
- Autoclave at 121°C for 15 min.
- Allow agar in tubes to solidify at an incline to present a 5–10-cm agar face for seedling growth.

APPENDIX 4

Reagents and Buffers

ANTIBODY SOLUTIONS

Primary Antibody Solution [for Enzyme-Linked Immunosorbent Assay (ELISA)]

Dilute species-specific primary antibody to the appropriate titer (e.g., 1:4000) by adding 25 μl to 100 ml of phosphate-buffered saline (PBS).

Primary Antibody Solution (for Immunoblot)

Dilute species-specific primary antibody to the appropriate titer (e.g., 1:4000) by adding 25 μl to 100 ml of Tris-buffered saline-Tween (TBST).

Second Antibody Solution (for ELISA and Immunoblot)

Dilute second antibody, either goat anti-rabbit immunoglobulin G (IgG) or sheep anti-rabbit IgG alkaline phosphatase conjugate 1–4000 by adding 25 μl to 100 ml of TBST. For ELISA, use PBS instead of TBST.

BUFFERS

Carbonate Buffer (pH 9.8)

$NaHCO_3$ (0.1 M)	8.4 g
$MgCl_2 \cdot 6H_2O$ (1.0 mM)	0.203 g

Dissolve in distilled water and adjust to pH 9.8 with NaOH. Adjust volume to 2 liter with distilled water.

Coating Buffer (pH 9.6) (Carbonate-Bicarbonate Buffer, 0.05 M)

Na_2CO_3	1.59 g
$NaHCO_3$	2.93 g
NaN_3	0.2 g

Dissolve in 1 liter of distilled water; store at 4°C for not more than 2 weeks.

Enzyme Substrate Buffer (for ELISA) (Diethanolamine Buffer, 10%)

Diethanolamine	97 ml
NaN_3	0.2 g
$MgCL_2 \cdot 6H_20$	100 mg

Dissolve in 800 ml of distilled water and adjust pH to 9.8 with HCl. Adjust volume to 1 liter with distilled water. Store at room temperature in an amber bottle.

Phosphate Buffer, 0.1 M (pH 8.0) [Used in the Conjugation of Fluorescein Isothiocyanate (FITC)]

Na_2HPO_4 (anhydrous)	14.2 g

Dissolve in 800 ml of distilled water. Adjust the pH to 8.0 by the dropwise addition of 1 N HCl. Dilute to 1000 ml with distilled water.

Phosphate Buffer, 0.15 M (pH 8.0)

Na_2HPO_4 (anhydrous)	21.3 g

Dissolve in about 800 ml of distilled water. Adjust the pH to 8.0 by the dropwise addition of 1 N HCl. Dilute to 1000 ml with distilled water. Check pH occasionally.

0.15 M Phosphate Buffer, 0.15 M (pH 9.0) (Used in the Conjugation of FITC)

Prepare the same as described previously, but without adding HCl.

PBS, 0.01 M (pH 7.1) [for Fluorescent Antibody (FA)]

NaCl	8.5 g
Na_2HPO_4 (anhydrous)	1.08 g
$NaH_2PO_4 \cdot 2H_20$	0.31 g
Merthiolate	0.1 g

Dissolve in 1 liter distilled water and store at 4°C

PBS, 0.015 M (pH 7.4) (for Immunoblot)

NaCl	8.0 g
KH_2PO_4	0.2 g
$Na_2HPO_4 \cdot 12H_20$	2.9 g
KCl	0.2 g
NaN_3	0.2 g

Dissolve in 1 liter of distilled water and store at 4°C.

PBS-Tween (PBST)

Tween 20	0.5 ml

Dissolve in 1 liter of PBS and store at 4°C.

TBS

Tris base	2.41 g (20 mM)
NaCl	29.24 g (500 mM)

Add Tris and NaCl and bring to 1 liter with distilled or deionized water. Adjust to pH 7.5 with HCl.

TBS Acidified (pH 2.8)

TBS adjusted to pH 2.8 with 1 N HCl

TBST (Tween 20 Wash Solution)

Tris-base	2.41 g (20 mM)
NaCl	29.24 g (500 mM)
Tween 20	0.5 ml (0.05%)

Add Tris-base, NaCl, and Tween 20, and bring to 1 liter with distilled or deionized water. Adjust to pH 7.5 with HCl.

REAGENTS

5-Bromo-4-chloro-3-indolyl-phosphate (BCIP)/Nitro Blue Tetrazolium (NBT) Alkaline Phosphatase Color Development Solution

1. Prepare 1 ml of 70% DMF (N, N-dimethyl-formamide) solution by mixing 0.7 ml of DMF with 0.3 ml of H_2O. Dissolve 30 mg of NBT (Nitro Blue Tetrazolium) in this 70% DMF solution. Label solution A.
2. In a second vial, dissolve 15 mg of BCIP (5-bromo-4-chloro-3-indolyl-phosphate) in 1 ml of DMF. Label solution B.
3. Just prior to starting color development, add solution A and solution B (previously described) to 100 ml of room-temperature carbonate buffer. Use immediately. The final concentrations should be 0.3 mg ml^{-1} for NBT and 0.15 mg ml^{-1} for BCIP.

Biuret Reagent

In 500 ml of distilled or deionized water, dissolve:

$CuSO_4 \cdot 5H_2O$	1.5 g
$NaKC_4H_4O_6 \cdot 4H_2O$	6.0 g

To this mixture, add 300 ml of CO_2-free 10% NaOH slowly under continuous stirring. Add CO_2-free H_2O to make this reagent up to 1 liter and store in a tightly screw-capped polyethylene or glass bottle at 40°C.

Blocking Solution (for ELISA)

Skim milk powder	30 g
PBS	1 liter

Dissolve milk powder in PBS and refrigerate until needed.

Blocking Solution (for Immunoblot)

Skim milk powder	20 g
TBS	1 liter

Dissolve milk powder in TBS and refrigerate until needed.

Enzyme Substrate (for ELISA)

p-nitrophenyl phosphate	5 mg
Enzyme substrate buffer	5 ml

Immediately before use, dissolve a tablet of p-nitrophenyl phosphate in enzyme substrate buffer at room temperature. [The tablets (5 mg) are stored a −20°C in the dark until use]. The substrate must be used the same day.

Gelatin-Rhodamine Isothiocyanate (RhITC) Conjugate

1. Prepare a 2% gelatin solution.
2. Add 1 N NaOH dropwise until pH reaches 10.0–11.0.
3. Autoclave for 10 min at 15 lb/in^2 and 121°C.
4. After cooling, add gelatin-rhodamine isothiocyanate (RhITC) dissolved in a minimum volume of acetone to provide 8 μg of dye per 1 mg of gelatin. Remove residues by filtration through a 45-μm membrane filter.
5. Allow conjugation to proceed overnight with gentle stirring.
6. The conjugate is separated from unreacted RhITC by gel filtration on Sephadex G-25, using PBS pH 7.1 (alternatively the preparation could be dialyzed against PBS pH 7.1 until no further color is detected in the dialysate).

7. Add Merthiolate to the conjugate (1:10,000) and distribute the conjugate in small volumes into screw-cap tubes and store at −20°C. Alternatively, the bulk of the conjugate could be freeze dried and stored in a desiccator. When needed, the desired amount of the dry sample should be reconstituted in distilled water.

Mounting Solutions (Kawamura, 1969)

Buffered glycerol or Elvanol is commonly employed. Fluorescence fades in a short time (about 30% overnight and then more gradually) in glycerol, but remains for a longer time in Elvanol. The fluorochrome of the rhodamine series dissolves in Elvanol, however, and therefore it cannot be used, except with FITC-labeled antiserum. The pH of the buffered glycerol is normally 7.0 to 7.5. However, we have used it at a pH of 8.5 with good results.

1) Buffered Glycerol Solution

0.5 M carbonate buffer (pH 9.5)	1 volume
Glycerine (reagent grade, free of autofluorescence)	9 volume

The two reagents are mixed thoroughly (with a magnetic stirrer). The final pH should be 8.5.

2) Elvanol (Elvanol-Buffered Glycerine Mixture)

Elvanol (polyvinyl alcohol, 51-05 grade)	1 volume
0.5 M carbonate buffer (pH 9.0)	4 volume

The two reagents are mixed with a magnetic stirrer for 16 h. One volume of reagent grade glycerine is mixed with two volumes of the previously mentioned mixture. The final mixture is stirred again with a magnetic stirrer for 16 h, centrifuged for 60 min at 1650 × g and the pH of the supernatant corrected to 8.5. The final product should be kept in an air-tight container. It is best when stored in tubes and kept in the dark. It will harden under the cover glass and then fix it firmly.

Nessler's Reagent

In 15 ml of distilled water, dissolve:

Mercuric chloride	1.0 g
Potassium bromide	5.0 g
Sodium hydroxide	2.5 g

Dilute to 100 ml and refrigerate. Allow to sit in the refrigerator for 5 days. Use the upper clear solution only, or filter.

STAINS

Carbol fuchsin Stain

Basic fuchsin	1 g
Ethanol	10 ml
5% Phenol solution 100 ml	

The fuchsin stain should be diluted 5–10 times with distilled water before use.

Gram Stain Solutions (Vincent, 1970)

Solution I: Crystal violet solution	
Crystal violet	10 g
Ammonium oxalate	4 g
Ethanol	100 ml
Distilled water	400 ml
Solution II: Iodine solution	
Iodine	1 g
Potassium iodide	2 g
Ethanol	25 ml
Distilled water	100 ml
Solution III: Alcohol	
Distilled water	5 ml
Ethanol	95 ml
Solution IV: Counterstain	
2.5% Safranin in ethanol	10 ml
Distilled water	100 ml

Molecular Biology Reagents and Buffers

DENATURATION SOLUTION

NaOH	20 g
NaCl	88 g
Water	1000 ml

Dissolve in 800 ml of distilled or deionized water and bring up the final volume to 1 liter.

DEPURINATION SOLUTION (0.2 M HCl)

6 N HCl	20 ml
Water	580 ml

Make up the solution by adding the acid to distilled or deionized water.

DENHARDT'S SOLUTION (50x)

Ficoll	0.5 g
Polyvinylpyrrolidone	0.5 g
Bovine serum albumin (BSA)	0.5 g

Dissolve in 50 ml of 2x sodium chloride/sodium citrate (SSC) to obtain a 1% (w/v) of the listed components. Aliquot and store in a freezer at −20°C.

ECKHARDT SOLUTION A

1. Dissolve 2 g of Ficoll (Type 400, MW 400,000) in 9 ml of 1x Tris-borate-EDTA (TBE). Autoclave 15 min to dissolve and sterilize.

2. Prepare 0.5% bromphenol blue (0.05 g of dye per 10 ml of 1× TBE). Autoclave for 15 min.
3. Add 10 mg lysozyme to cooled Ficoll solution. Shake well by hand or on a rotary shaker.
4. Add 1.0 ml of bromphenol dye solution to the lysozyme/Ficoll mixture.
5. Add 10 μl of heat-treated RNase to the Ficoll/lyzozyme/bromphenol mixture. (To prepare RNase, add 2 mg RNase per 1.0 ml of sodium acetate—acetic acid buffer. Heat in a water bath at 98–99°C for 15 min. Dispense 100 μl of treated RNase into sterile microfuge tubes and store in a freezer.)
6. Mix final solution containing lysozyme, Ficoll, bromphenol, and RNase. Dispense 600 μl in sterile microfuge tubes and store in a freezer.

ECKHARDT SOLUTION B

1x TBE	20 ml
Sodium dodecyl sulfate (SDS)	0.4 g
Ficoll	2.0 g

Autoclave for 15 min. Dispense 600 μl into sterile microfuge tubes. Store in a freezer.

ECKHARDT SOLUTION C

1x TBE	40 ml
SDS	0.8 g
Ficoll	2.0 g

Autoclave for 15 min. Dispense 1 ml in microfuge tubes. Store in a freezer.

ETHIDIUM BROMIDE (EtBr) STOCK SOLUTION

Use CAUTION when working with ethidium bromide (EtBr), which is a powerful mutagen. Wear gloves and use a particle mask.

Weigh 0.1 g of (EtBr) into 10 ml of deionized water in a 50-ml Erlenmeyer flask. Stir to dissolve (with a stirring bar) in a dark or diffused light environment. Wrap the flask in foil and store refrigerated. Prepare staining solution for gels by adding 100 μl of stock per 100 ml of 1x TBE.

76% ETHANOL/10 mM AMMONIUM ACETATE

100% Ethanol	380 ml
7.5 M ammonium acetate	0.667 ml
Water	119 ml

Mix reagents and store in a refrigerator.

5 M GUANIDINE ISOTHIOCYANATE/0.1 M EDTA

Guanidine isothiocyanate	59.08 g
Na_2 EDTA · $2H_2O$	3.7 g
Water	100 ml

Dissolve in distilled or deionized water. Adjust to pH 7.0.

HYBRIDIZATION SOLUTION

5x SSC
1x Denhardt's solution
20 mM sodium phosphate, pH 6.5
0.5% SDS
5% Dextran sulfate
0.2 mg ml^{-1} salmon sperm DNA (denatured)

The previously listed components are final concentrations based on the volume of hybridization solution needed for the blot. (Hybridization solution is needed at the rate of 50–100 μl per cm^2 blot.) Mix the solution and filter through a 0.2-μm membrane filter.

LOADING BUFFER (10x)

20% Ficoll	20 g Ficoll Type 400
0.1 M EDTA	3.7 g Na_2 EDTA · $2H_2O$
1% SDS	1.0 g SDS
0.2% dye	0.2 g bromphenol blue
Water	100 ml

Make up solution in deionized or distilled water. Aliquot in microfuge tubes after sterilization by autoclaving. Store in a freezer at −20°C.

NEUTRALIZATION SOLUTION

Tris-base	13.4 g
Tris-HCl	140.4 g
NaCl	88 g

Dissolve in 800 ml of distilled or deionized water and bring up the final volume to 1 liter.

PHENOL

Take a 500-g bottle of good quality phenol crystals. Add 300 ml of warm 25 mM NaCl to this bottle of phenol . (Prewarm the NaCl solution before adding.) Then add 10 g of Tris-base (not Tris-HCl) and 0.9 g of 8-hydroxyquinoline. The pH will be approximately 7.8. Dissolve the phenol completely overnight and store in a refrigerator.

PREHYBRIDIZATION SOLUTION

5x SSC
5x Denhardt's solution
25 mM sodium phosphate, pH 6.5
0.5% SDS
5% Dextran sulfate
0.5 mg ml^{-1} salmon sperm DNA (denatured)

The previously listed components are final concentrations based on the volume of prehybridization solution needed for the blot. (Prehybridization solution is needed at the rate of 50–100 μl per cm^2 blot.) Mix this solution and filter through a 0.2-μm membrane filter.

RESTRICTION ENZYME BUFFERS

High Salt Buffer (10x)

0.5 M Tris-HCl (pH 8.0)	7.88 Tris-HCl
1.0 M NaCl	5.85 g NaCl
0.1 M $MgCl_2$	0.95 g $MgCl_2$
Water	100 ml

Make up the solution in distilled or deionized water. Sterilize by autoclaving. Dispense 1.0-ml quantities in microfuge tubes. Store in a freezer at −20°C.

Medium Salt Buffer (10x)

0.5 M Tris-HCl (pH 8.0)	7.88 g Tris-HCl
0.5 M NaCl	2.9 g NaCl
0.1 M $MgCl_2$	0.95 $MgCl_2$
Water	100 ml

Make up the solutions in distilled or deionized water. Sterilize by autoclaving. Dispense 1.0-ml quantities in microfuge tubes. Store in a freezer at −20°C.

RNase (DNase-free)

Dissolve pancreatic RNase (RNase A) at a concentration of 10 mg ml^{-1} in 10 mM Tris-HCl (pH 7.5) and 15 mM NaCl. Heat to 100°C for 15 min and allow to cool slowly to room temperature. Dispense into aliquots and store at −20°C.

SALMON SPERM DNA (5 mg ml^{-1})

To be effective in saturating any nonspecific DNA binding sites during hybridization experiments, salmon sperm DNA must be sheared into small fragments. Dissolve 50 mg of DNA in 10 ml of sterile deionized water in a sterile test tube. (Use Type III DNA sodium salt, Sigma Chemical Co., St. Louis) Vortex vigorously to dissolve. Pass several times through an 18-gauge hypodermic needle to shear the DNA into small fragments. Denature in a boiling water bath for 15 min. Immediately transfer to ice to achieve quick cooling. Store 0.5-ml aliquots in microfuge tubes at −20°C.

SA-AP (STREPTAVIDIN-ALKALINE PHOSPHATASE)

1.0 mg ml^{-1} of SA-AP
3 M NaCl
1 mM $MgCl_2$
0.1 mM $ZnCl_2$
30 mM triethanolamine (pH 7.6)

0.1% SARKOSYL-TEN BUFFER SOLUTION

N-Lauroylsarcosine	0.05 g
TEN buffer	50 ml

Autoclave and store in a refrigerator.

SODIUM ACETATE/ACETIC ACID BUFFER

Solution I: Prepare 0.4 M sodium acetate by dissolving 1.36 g salt in 25 ml deionized water.

Solution II: Mix 2.29 ml of glacial acetic acid in 100 ml of deionized water.

Make buffer by mixing 18 ml of solution I and 82 ml of solution II. pH should be around 4:0. Store in refrigerator. Autoclave before use.

SOLUTION I

50 mM glucose	9.0 g $C_6H_{12}O_6$
25 mM Tris-HCl (pH 8.0)	4.0 Tris-HCl
10 mM EDTA (pH 8.0)	3.72 Na_2 EDTA·$2H_2O$
Water	1000 ml

Prepare solution and sterilize by autoclaving for 15 min at 10 lb/in^2 and store in a refrigerator at 4°C.

SOLUTION II

0.2 N NaOH (freshly diluted from a 10 N stock) 1% SDS.

SOLUTION III

5 M potassium acetate	60 ml
Glacial acetic acid	11.5 ml
Water	28.5 ml

(The resulting solution is 3 M with respect to potassium and 5 M with respect to acetate). Store solution in a refrigerator at 4°C.

20x SSC (SODIUM CHLORIDE SODIUM CITRATE)

NaCl	175.3 g
Na_3 citrate · $2H_2O$	88.2 g
Water	1000 ml

Dissolve in 800 ml of distilled or deionized water. Adjust to pH 7.0 with HCl. Bring up the final volume to 1 liter.

SODIUM CHLORIDE-TRIS-EDTA (STE) BUFFER

0.1 M NaCl	5.85 g NaCl
10 mM Tris-HCl (pH 8.0)	1.57 g Tris-HCl
1 mM EDTA (pH 8.0)	0.37 g Na_2 EDTA · $2H_2O$
Water	1000 ml

Dissolve in 800 ml of distilled or deionized water and adjust the pH to 8.0. Bring the volume to 1 liter. Autoclave to sterilize. Store in a refrigerator at 4°C.

10x TBE BUFFER

Tris-base	108 g
Na_2 EDTA · $2H_2O$	9.3 g
Boric acid (H_3BO_3)	55 g
Distilled water	1000 ml

Dissolve with stirring in 800 ml of distilled or deionized water. Adjust pH to 8.3. Adjust final volume to 1000 ml. Filter through a 0.22-μm filter. Autoclave and store. To obtain 1x TBE, dilute 1:10.

TRIS-EDTA (TE) BUFFER

10 mM Tris-HCl	1.21 g Tris-base
1 mM EDTA	0.37 g Na_2 EDTA · $2H_2O$

Dissolve in 900 ml of distilled or deionized water. Adjust pH to 8.0 with HCl and make up the final volume to 1 liter with distilled or deionized water. Autoclave to sterilize.

TE_{25} BUFFER

10 mM Tris-HCl	1.21 g Tris-base
25 mM EDTA	9.3 g Na_2 EDTA · $2H_2O$

Dissolve in 900 ml of distilled or deionized water. Adjust pH to 8.0 with HCl and make up the final volume to 1 liter with distilled or deionized water. Autoclave to sterilize.

TEN BUFFER

Tris-base	3.03 g
Na_2 EDTA · $2H_2O$	1.86 g
NaCl	1.46 g
Distilled or deionized water	500 ml

Adjust pH to 8.0. Autoclave and store.

TES BUFFER

10 mM Tris-HCl	1.21 g Tris-base
25 mM EDTA	9.3 g Na_2 EDTA · $2H_2O$
150 mM NaCl	8.8 g NaCl

Dissolve in 900 ml of distilled or deionized water. Adjust pH to 8.0 with HCl and make up the final volume to 1 liter with distilled or deionized water. Autoclave to sterilize.

APPENDIX 6

McFarland Nephelometer Barium-Sulfate Standards[1]

PREPARING THE STANDARDS

1. Prepare 1% aqueous barium chloride and 1% aqueous sulfuric acid solutions.
2. Add the amounts indicated in Table A6.1 to clean dry ampoules. Ampoules should have the same diameter as the test tube to be used in the subsequent density determinations.
3. Seal the ampoules and label them.

TURBIDITY ADJUSTMENT OF THE BACTERIAL SUSPENSION

For bacterial agglutinations, the cell suspension is usually adjusted to approximately 1×10^9 cells ml^{-1}. In the McFarland standards, tubes 3 and 4 will have approximately 9.0

TABLE A6.1 Preparing McFarland Nephelometer Barium-Sulfate Standards

Tube	Barium Chloride 1% (ml)	Sulfuric Acid 1% (ml)	Corresponding Approximate Density of Bacteria (million ml^{-1})
1	0.1	9.9	300
2	0.2	9.8	600
3	0.3	9.7	900
4	0.4	9.6	1200
5	0.5	9.5	1500
6	0.6	9.4	1800
7	0.7	9.3	2100
8	0.8	9.2	2400
9	0.9	9.1	2700
10	1.0	9.0	3000

[1]From E.H. Lenette, A. Balows, W.J. Hausler, and J.P. Truant. 1974. *Manual of Clinical Immunology*. American Society for Microbiology, Washington, DC.

$\times 10^8$ (1×10^9) and 1.2×10^9 cells ml^{-1}, respectively. The arbitrary selection of these two densities will yield satisfactory results for many systems.

With dust-free saline in a tube (blank) similar in diameter to the standards, set the nephelometer to a low nephelometric unitage. Read the corresponding unitage on tubes 3 or 4. With approximately 8 ml of saline in another clean tube, add the turbid washed suspension of rhizobial cells dropwise with a Pasteur pipette until a turbidity is reached that is slightly lower than the corresponding standard chosen. Place the tube in the nephelometer and adjust the turbidity to the required unitage by further additions of the turbid rhizobial suspension. If a nephelometer is not available, the turbidity is adjusted to fall between tubes 3 and 4 by visual comparison.

APPENDIX 7

Preparing Seedling-Agar Slants and NifTAL-Tubes for Culturing Small-Seeded Legumes

Small-seeded legumes can be cultured enclosed in tubes if these plants are to be used for authenticating rhizobia or for enumerating rhizobia by the plant-infection technique. One of the limitations of strain evaluation in enclosed tubes is that a tube environment restricts growth conditions and proper differentiation of the plant. An N-free nutrient solution is solidified with agar for slant preparation or without agar for NifTAL tubes.

SEEDLING-AGAR SLANTS

1. Tubes 250 × 25 mm (Figure A7.1) are required. Tubes are stoppered with cotton plugs that are sufficiently loose to allow good air exchange and simultaneously filter off contaminants.

2. A total of 1.62 liters of the N-free nutrient solution is needed for 54 tubes at the rate of 30 ml per tube. For convenience, divide the nutrient solution into manageable volumes in beakers or Erlenmeyer flasks prior to adding the agar powder. (Example: It is convenient to have 500 ml of the N-free nutrient solution in a 1-liter container because this will greatly facilitate stirring when the agar is being melted or dispensed). Add 1.5% (w/v) agar to the N-free nutrient solution (24.3 g of agar powder will be needed for 1.62 liters of N-free nutrient solution). Melt the agar either by steaming in an autoclave or by direct heating over a Bunsen flame. If direct heating is used, the mixture must be constantly stirred over gentle heat to prevent charring the agar on the bottom of the container.

3. Dispense the melted agar in 30-ml portions into the tubes and plug. To facilitate agar dispensing, a simple setup is illustrated in Figure A7.1 that is adequate for approximate volumes. Arrange tubes in suitable metal baskets and autoclave at 121°C for 30 min. To make slants, support the tubes at an angle as illustrated.

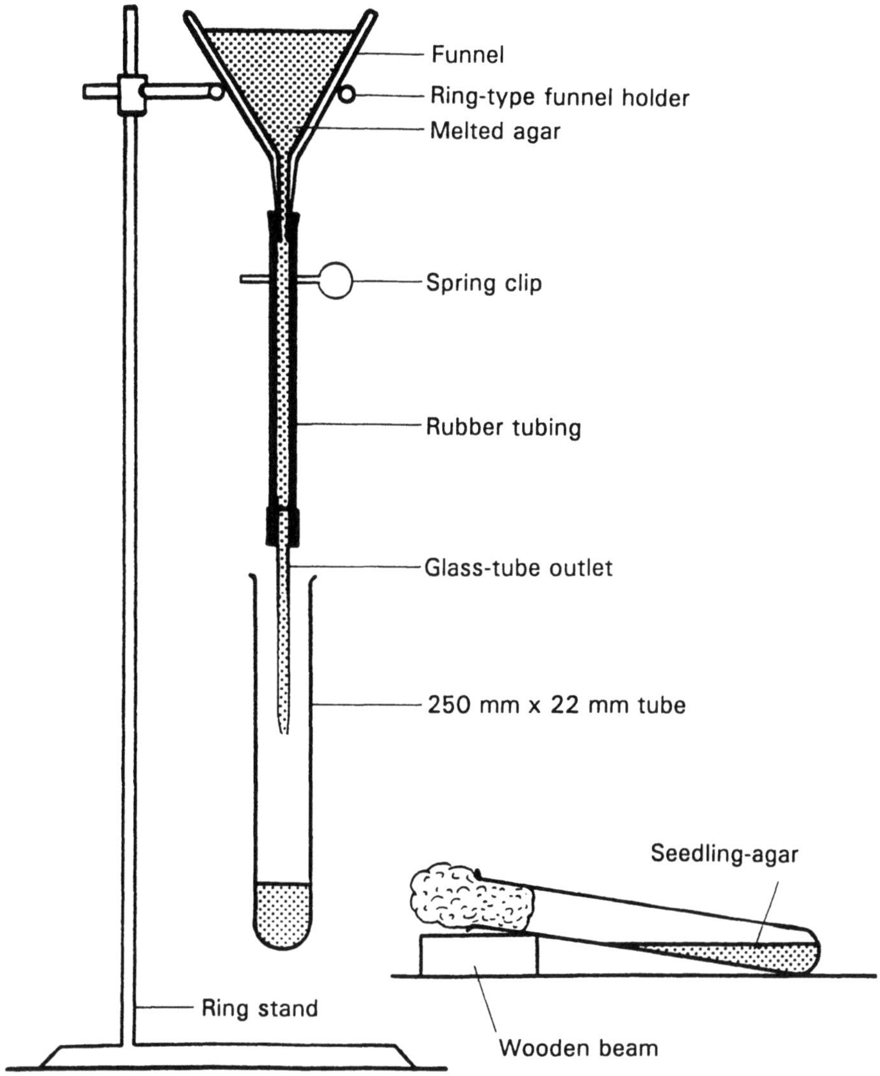

FIGURE A7.1 Simple setup for dispensing seedling agar into tubes and forming slants.

NifTAL TUBES

1. Tubes 250 × 25 mm (Figure A7.2) are required. Instead of using cotton plugs, the tubes are closed with autoclavable polypropylene caps of the tube-size specifications. The caps are first modified by cutting two narrow vents (20 mm long and 3 mm wide) on the cap as shown in Figure A7.2 During the growth of the seedling, the vents allow efficient gas exchange between the tubes internal environment and the outside of the tube. (Caps of the specified size without vents for 250 × 25 mm tubes

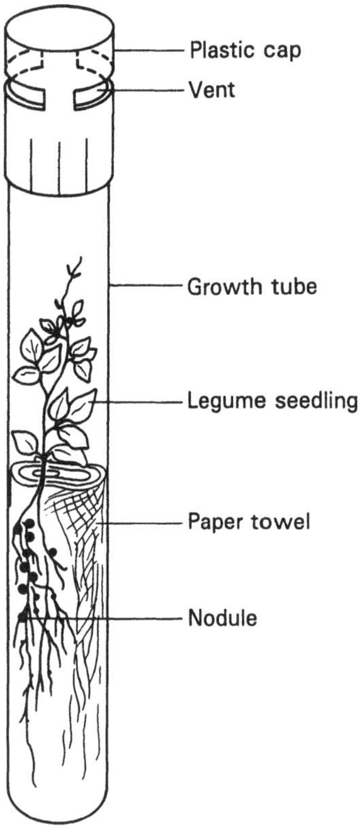

FIGURE A7.2 Modified enclosed tube (NifTAL tube) culture system for legume seedlings.

can be purchased from Bellco Glass, Inc., Vineland, NJ). Vents of the dimensions described are easily cut on the cap using a bench saw once the adjustments are made. To facilitate cutting the vent on the cap, the cap is held in place with a length of wooden dowel (25-mm diameter).

2. A rolled-up bleached multifold paper towel is placed inside the tube. This serves as a wick and as a solid support for the legume seedling. The paper towel is nontoxic and measures 9.25 × 9.5 in (235 × 238 mm) when fully spread out. The paper towel is folded back once to reduce the length and then rolled up before placement in the tube with the help of long forceps. (The type of paper towels described here are available from James River Corp., Norwalk, CT). Twenty-five milliliters of N-free nutrient solution are pipetted into the tube. This amount is sufficient to wet the paper towel and lasts for 10–15 days. When required, more moisture is provided by replenishing with sterile water or half-strength N-free nutrient solution.

3. The tube is completely closed during autoclaving. The pressure fins molded on the inside of the caps grip the tube and allow steam to enter the tube during sterilization. The tubes are allowed to cool before planting the germinated seeds.

4. Germinated seeds are secured in position by carefully inserting the radicle between the paper towel and the inner wall of the tube. The cap is then adjusted to fit in the venting position. Aluminum foil or black paper is used to wrap the tubes to prevent exposure of the developing roots to light. The roots proliferate on the surface of the towel and nodules are easily seen.

Appendix 8

Building a Rack for Growth Pouches

In an effort to keep growth pouches standing upright, researchers have improvised different types of racks. Gramophone record holders have frequently been used for this purpose. More suitable racks may be built from galvanized or stainless steel wire of at least 14 gauge and a wooden board as shown in Figure A8.1. The spacing between the wire frames should be 1.0–1.5 cm. Tools needed are: a drill with a bit of a slightly smaller diameter than the wire, wire cutter, small vise, and a hammer.

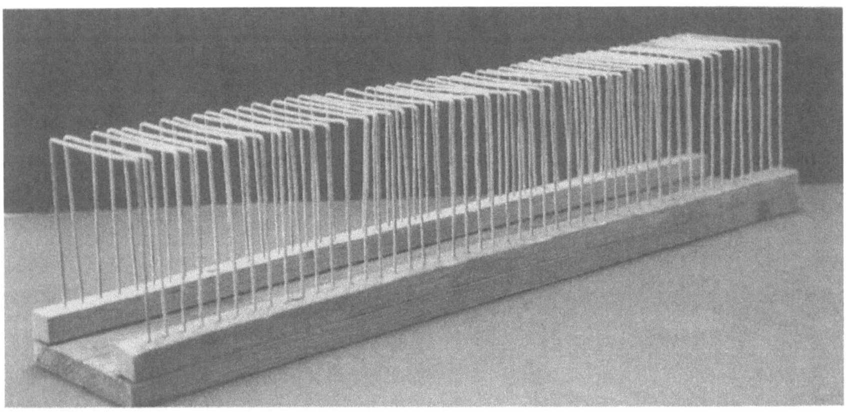

Figure A8.1 Improvised rack for growth pouches.

APPENDIX 9

Recommendations of Legumes and Growth Systems for Authentication

Authentication is the demonstration of the ability of a pure culture of a presumptive isolate of rhizobia to form true nodules on the roots of the homologous (parent) legume (or another species in the same cross-inoculation group) grown under axenic conditions. Choice of the legume for the authentication (Table A9.1) depends very much on the specificity of the host. Most temperate and tropical legumes nodulated by fast-growing, acid-producing rhizobia are usually specific and would require the parent host. In most instances, the host-dependent classification for rhizobia may serve as a useful guide for selecting the legume for use in authentication. If the legume from which the presumptive isolate is made is identified and its cross-inoculation group is known, but no seeds of the parent (homologous) host are available, the cross-inoculation group should be consulted to select an alternative (heterologous) host. However, this is sometimes difficult as with the pink *Bradyrhizobium* sp. from *Lotononis bainesii*, which requires only the parent host because there seems to be no substitute. Most known tropical legumes are nodulated by the slow-growing, alkali-producing rhizobia (*Bradyrhizobium*), in which case a "guinea-pig" legume like *Macroptilium atropurpureum* (siratro) can be confidently used for authentication. Over 90% of bradyrhizobia will nodulate siratro.

The choice of the growth system (Table A9.1) will depend on the seed size of the host selected for authentication, and the size of the plant. Some small-seeded species, e.g., *Vigna aconitifolia and Macrotyloma uniflorum*, produce plants of an unsuitable size for tubes, but manageable in growth pouches. If the size of the plant is known, most small-seeded species can be cultured in tubes or growth pouches in growth (environmental) chambers. It is important to bear in mind that there are legumes that will not nodulate easily in tubes or pouches, resulting in a false negative authentication. Chickpea (*Cicer arietinum*) and *Leucaena retusa* are notable examples. For these species, authentication must be performed in Leonard jars. The environment (growth chamber or greenhouse) where the authentication is done must be absolutely clean and adequately constructed to keep out insects and other contamination sources.

TABLE A9.1 Recommended Hosts and Growth Systems for Authentication of Presumptive Isolates of Rhizobia

Parent Host	Type of Rhizobia[1]	Host for Authentication	Growth System
Phaseolus vulgaris *P. coccineus*	f, ac	Parent hosts	Growth pouches or Leonard jars
P. acutifolius	s, al	Parent host	Growth pouches or Leonard jars
Medicago spp. *Melilotus* spp. *Trigonella* sp.	f, ac	Parent hosts or *Medicago sativa*	Tubes or growth pouches
Trifolium spp.	f, ac	Parent hosts	Tubes or growth pouches
Pisum spp. *Vicia* spp. *Lens culinaris*	f, ac	Parent hosts or *Vicia faba*	Tubes, growth pouches, or Leonard jars
Glycine max	f, ac and s, al	Parent host	Growth pouch or Leonard jars
Lupinus spp. *Ornithopus* sp.	s, al	Parent hosts	Growth pouches or Leonard jars
Sesbania spp.	f, ac	Parent hosts	Growth pouches or Leonard jars
Leucaena leucocephala *L. diversifolia*	f, ac	Parent hosts or *L. leucocephala*	Tubes or growth pouches
L. retusa	f, ac	Parent host or *L. leucocephala*	Leonard jars
Lotononis bainesii	s, al	Parent host	Tubes or growth pouches
Cicer arietinum	f, ac or neutral	Parent host	Leonard jars
Calliandra spp.	f, ac and s, al	Parent host	Tubes, growth pouches, or Leonard jars
Phaseolus vulgaris P. coccineus	f, ac	Parent hosts	Growth Pouches or Leonard jars
Acacia senegal	f, ac	Parent host	Growth pouch or Leonard jar

Table A9.1 (continued)

Parent Host	Type of Rhizobia[1]	Host for Authentication	Growth System
Acacia auriculaeformis	s, al	Parent hosts or siratro (*Macroptilium atropurpureum*)	Tubes, growth pouches or Leonard jars
A. mearnsii			
A. albida			
Arachis hypogaea			
A. glabarata			
Alysicarpus vaginalis			
Cajanus cajan			
Calopogonium mucunoides			
Canavalia spp.			
Stylosanthes spp.			
Aeschenomene spp.			
Macrotyloma spp.			
Glycine wightii (syn. *Neonotonia wightii*)			
Voandzeia subterranea			
Desmodium spp.			
Centrosema spp.			
Crotalaria spp.			
Clitoria spp.			
Lablab purpureus			
Cyamopsis tetragonoloba			
Psophocarpus tetragonolobus			
Vigna spp.			
Phaseolus lunatus			
Zornia spp.			
Pacyrrhizus spp.			
Sphenostylis macrocarpa			
Macroptilium spp.			

[1]f, ac and s, al indicate fast-growing, acid-producing and slow-growing, alkali-producing, respectively.

APPENDIX 10

Seed Surface Sterilization and Germination

Surface sterilizing legume seeds is dependent on the purpose and nature of the experiment. Authentication, strain selection, and the enumeration of rhizobia by the plant-infection technique require legumes to be raised from surface-sterilized seeds to ensure strict microbiological control. Sterilants frequently used for surface sterilizing seeds are solutions of sodium hypochlorite (2.5% commercial bleach), acidified mercuric chloride (0.2%), hydrogen peroxide (3%), or concentrated sulfuric acid. Only hard-coated seeds are treated with concentrated sulfuric acid, which scarifies (softens) the seed coat besides effective surface sterilization. Selected seeds must be of good viability (more than 70%), clean, and damage free. Treated seeds (pesticides, fungicides, or insecticides) must be rinsed quickly in water, then dried on paper towels.

METHOD a

Sterilization with mercuric chloride, sodium hypochlorite, or hydrogen peroxide solutions.

1. Place seeds in an Erlenmeyer flask (wide-mouthed and previously sterilized by autoclaving). Cover the mouth of the flask with half of a sterile Petri dish. The space the seeds take up should be about 25% of the volume of the flask since too many seeds will affect the efficiency of the sterilization. The Petri dish cover should be kept in place throughout the operation.

2. Rinse the seeds in 95% alcohol for 10 s to remove waxy material and trapped air. Drain off the alcohol.

3. Add mercuric chloride, sodium hypochlorite, or hydrogen peroxide solutions in sufficient volumes to immerse the seeds completely. Swirl the contents gently to bring the seeds and sterilant into contact. After 3–5 min, drain off the sterilant.

4. Rinse with at least six changes of sterile water. Observe aseptic procedures throughout the rinsing. After the sixth rinse, pour in sufficient water to submerge the seeds, then leave in the refrigerator for 4 h so the seeds imbibe. (Some seeds, e.g., the

California black-eye variety of *Vigna unguiculata* should not be allowed to imbibe in water because the cotyledons fall apart.)

5. After 4 h, rinse the seeds with two or more changes of water and plate the seeds in 0.75% (w/v) water agar in Petri dishes. (Seeds can easily be scooped out of the flask with long spoons to transfer the seeds onto the agar.) Evenly spread the seeds on the agar and avoid overcrowding. About 20–100 seeds are recommended per plate, depending on the size. Large Petri dishes are needed to plate species with large seeds (e.g., *Canavalia* spp. and *Vicia faba*). Large-seeded species are more conveniently germinated in sterile (autoclaved) vermiculite. The vermiculite is moistened and sterilized 1 day in advance. Obtain a 5–10-cm layer of horticultural grade vermiculite in a shallow autoclavable polypropylene tray. Moisten the vermiculite by alternate additions of water and mix. Cover the tray with aluminum foil and sterilize by autoclaving for 15 min. Allow the vermiculite to cool overnight. Remove the foil in a laminar flow hood or other clean environment. Make furrows in the vermiculite with a sterile spatula. Sow the seeds in furrows and cover with vermiculite. Replace the aluminum foil cover.
6. Incubate at 25–30°C. Invert the plates for small-seeded species (clover, medic, siratro, etc.) with seed diameters of 3 mm and less. Inverting the plates allows the development of straight radicles from the seeds.

METHOD b

Sterilization with concentrated sulfuric acid.

1. Place seeds in a sterile Erlenmeyer flask and cover with half a sterile Petri dish as in method a.
2. Add just enough acid to coat the seeds. Allow sterilization and scarification to proceed for 10 min. Drain off excess acid.
3. Add sterile water in sufficient volume to dissipate the heat generated by the exothermic reaction. Rinse and pour out the water. The first rinse should be done quickly to avoid heat-killing the seeds. Continue rinsing the seeds with another five changes of water.
4. Leave the seeds (with some water) overnight in the refrigerator to imbibe. Rinse with two changes of sterile water.
5. Plate the sterilized seeds on water agar and incubate at 25–30°C or germinate in sterile vermiculite as described in method a.

Methods of seed sterilization for the various leguminous species are shown in Table A10.1.

TABLE A10.1 Methods of Seed Surface Sterilization and Germination

Legume Species (Common Name)	Sterilization: Method[1] (a or b)	Sterilization: Sterilant	Recommended Germination Medium[2]
Arachis hypogaea (peanut, groundnut)	a	Peroxide/bleach	v
Glycine max (soybean)	a	Peroxide/bleach	v
Cicer arietinum (chickpea)	a	Peroxide/bleach	v
Lens culinaris (lentil)	a	Peroxide/bleach	wa
Lupinus spp. (lupines)	a	Bleach/peroxide	v/wa
Vigna unguiculata (cowpea)	a	Bleach/peroxide	v
Canavalia sp. (jackbean)	a	Bleach/peroxide	v
Phaseolus lunatus (lima bean)	a	Peroxide/bleach	v
P. acutifolius (tepary bean)	a	Peroxide	wa/v
Voandzeia subterranea (bambara groundnut)	a	Bleach/peroxide	v
P. vulgaris (bean)	a	Bleach/peroxide	v
P. coccineus (scarlet runner bean)	a	Bleach/peroxide	v
Vigna mungo (green gram)	a	Peroxide/bleach	wa
V. radiata (urd bean)	a	Peroxide/bleach	wa
V. angularis (adzuki bean)	a	Peroxide/bleach	wa
V. umbellata (rice bean)	a	Peroxide/bleach	wa
V. aconitifolia (mat or moth bean)	a	Peroxide/bleach	wa
Pisum spp. (pea)	a	Peroxide/bleach	v
Centrosema pubescens (centro)	b	Acid	wa
Clitoria ternatea (butterfly pea)	b	Acid	wa (invert plates)
Cajanus cajan (pigeon pea)	b	Acid	v/wa
Sesbania sp.	b	Acid	wa
Medicago spp. (medics)	a	Peroxide	wa (invert plates)
Trifolium spp. (clovers)	a	Peroxide	wa (invert plates)
Glycine wightii	a	Acid	wa (invert plates)
Pachyrrhizus spp. (yam bean)	b	Acid	v
Psophocarpus tetragonolobus (winged bean)	b	Acid	v

TABLE A10.1 (continued)

Legume Species (Common Name)	Sterilization Method[1] (a or b)	Sterilization Sterilant	Recommended Germination Medium[2]
Lotononis bainesii (lotononis)	a	Peroxide	wa (invert plates)
Desmodium spp.	b	Acid	wa (invert plates)
Lotus spp.	a	Peroxide	wa (invert plates)
Stylosanthes spp.	b	Acid	wa (invert plates)
Leucaena spp.	b	Acid	wa
Macroptilium atropurpureum (siratro)	a	Acid	wa (invert plates)
Calopogonium mucunoides (calopo)	b	Acid	wa (invert plates)
Pueraria phaseoloides (tropical kudzu)	b	Acid	wa (invert plates)
Acacia spp.	b	Acid	wa (invert plates)

[1]a refers to seed surface sterilization using sodium hypochlorite (bleach) or hydrogen peroxide (peroxide); b refers to seed surface sterilization and scarification using concentrated H_2SO_4. The sterilants are indicated in order of preference though both can be used in surface sterilization.
[2]v refers to vermiculite; wa refers to water agar.

Appendix 11

Preparing Leonard Jars

The modified Leonard jar assembly (Figure A11.1) consists of a 700-ml capacity beer bottle with the lower portion cut off. This bottle is inverted into a heavy glass jar (reservoir), 1-liter minimum capacity. The mouth of the bottle should be 2–3 cm above the base of the reservoir. The growth medium (sand or vermiculite) in the bottle is irrigated by a centrally positioned cotton wick running the length of the bottle and extending out of the mouth and into the reservoir containing the nutrient solution. Various types of wick material have been used with Leonard jars, e.g., braided cotton lantern wicks,

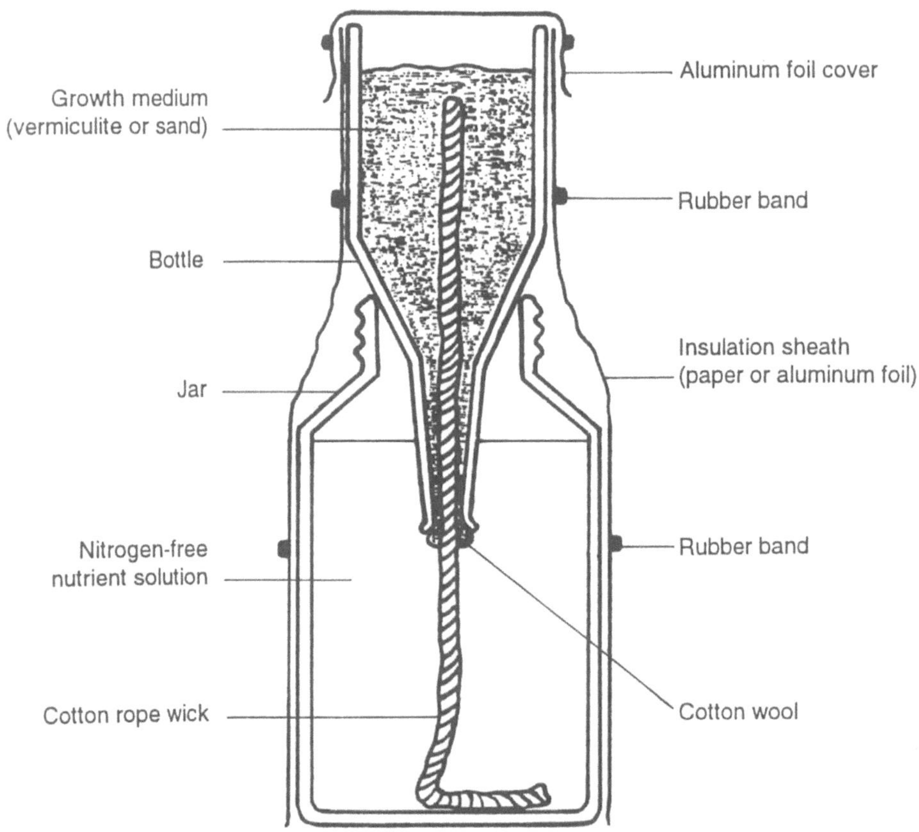

Figure A11.1 The Leonard jar.

cotton rope, strands from cotton mop heads, coiled cotton wool, and braided or twisted nylon rope. New wick materials should be tested for their ability to conduct water and their compatibility with plants. Generally, a 12-mm cotton rope is adequate and easy to obtain.

Place approximately 50 cm of wick material into the bottle, with about 10 cm extending out of the mouth. A small amount of absorbent cotton stuffed into the neck of the bottle will aid in securing the position of the wick, and prevent the growth medium from settling in the reservoir. Wick material of cotton rope should be boiled in water and squeezed dry prior to use. This removes air trapped in the wick and improves water conductivity. While holding the wick in a central position, fill the bottle with growth medium (well-washed river sand or horticultural grade vermiculite). Pack the medium to minimize air spaces. Sand is easier to pack when dry. For vermiculite, it is more convenient to pack when wet. The vermiculite should be soaked overnight and the water drained off prior to packing into the bottles.

Position the bottle in the reservoir. The bottle should fit firmly on the rim of the reservoir. Moisten the growth medium in the bottle by adding 150–200 ml of the N-free nutrient solution. Allow the nutrient solution to saturate the medium and the excess to drain into the reservoir. Fill the reservoir with 800 ml of the nutrient solution; use 1600 ml if the reservoir has a 2-liter capacity. Wrap the bottle and jar assembly with white or brown moisture-proof paper and secure with rubber bands at critical points along the jar. Tape may also be used. Aluminum foil wrapping may be used if it is inexpensive and available. Cap the open end of the bottle with either aluminum foil or wrapping paper. Hold the assembly by the reservoir when moving it. Sterilize the complete assembly and nutrient solution by autoclaving for 1.5–2.0 h at 121°C and 15 lb/in^2. For convenience, cool the assembly in the autoclave overnight.

Appendix 12

Injecting and Bleeding Rabbits

Use healthy, 6–12-month-old rabbits for antiserum production. Label each animal with an ear tattoo or tag. Maintain a record for each rabbit and record their individual treatments. During ear (intravenous) injections, intramuscular injections, and trial bleedings, the rabbit must be restrained (immobilized). The recommended method is to roll the rabbit in a large towel, tightly securing the fore and rear limbs. For intraperitoneal injections, the animal may be strapped to a rack or held on its side by another person. During cardiac puncture, a bleeding rack is used to hold the rabbit on its back (Figure A12.1). Another approach is to sedate the rabbit with an injected tranquilizer such as Rompun [(Xylazine), Haver-Lockhart Bayvet Division, Cutter Laboratories, Inc. Shawnee, KS]. Avoid using ether or chloroform.

The following schedules have been used successfully for antisera development in rabbits.

FIGURE A12.1 A bleeding rack.

SCHEDULE 1[1]

Day	Procedure
1	Inject 0.5 ml intravenously (IV)
2	Inject 1.0 ml IV
3	Inject 1.5 ml IV
7	Inject 1.5 ml IV
8	Inject 2.0 ml IV
9	Inject 2.0 ml IV and 2.0 ml subcutaneously (SC)
16	Test bleeding and titer determination
18	Cardiac bleed (30–50 ml)
25	Inject 2.0 ml SC
32	Cardiac bleed (30–50 ml)
39	Inject 2.0 ml SC
46	Cardiac bleed (30–50 ml)

SCHEDULE 2[2]

Day	Procedure
1	Inject 1 ml of mixture of equal parts culture suspension and Freund's complete adjuvant intramuscularly (IM)
28	1 ml IV (antigen alone)
30	Bleed from ear 10–20 ml
32	Bleed from ear 10–20 ml
34	Bleed from ear 10–20 ml

SCHEDULE 3[3]

Day	Procedure
1	Inject 1 ml SC of emulsion of equal parts of antigen suspension and Freund's complete adjuvant
14	1 ml IV (antigen suspension alone)
28	Test bleeding and titer determination
30	Cardiac bleed (30–50 ml)
37	Inject 1.0 ml IV
44	Cardiac bleed (30–50 ml)

[1]Schmidt et al., 1968.

[2]Dudman, 1964.

[3]P. Somasegaran, unpublished observations.

An IM injection is used to start the immunization schedule (Chapter 8). Immobilize the rabbit by rolling it tightly into a large towel. Free one of the rear legs, and use alcohol to swab a small area of the skin covering the thigh muscle. Insert the needle about 1.5 cm into the muscle and inject. A large needle (20 gauge) is recommended to introduce the emulsion quickly and reduce the animal's discomfort. SC booster injections are usually given to maintain the antibody titer. Inject the antigen under the skin in the shoulder area. Use a 3–5-ml syringe fitted with a 22-gauge needle. IV injections are given into the marginal ear vein of one ear. Expose the vein by shaving a small section of the ear with a razor blade. Swab the shaved area with alcohol (70%) and inject the antigen with a 1–2-ml syringe fitted with a narrow (25 gauge) needle. If the schedule calls for several consecutive injections, make the first injection at the distal end of the ear. Progress toward the base of the ear with each successive injection.

For test bleeding, extract blood from the ear not used for injections. Shave a small area along the marginal ear vein and swab the area with alcohol (70%). To prevent blood from spreading into the fur, apply petrolatum around the area to be nicked. Use a scalpel with a small pointed blade (no. 11) and make a small nick in the vein. Collect 1–2 ml of blood in a test tube. Stop the bleeding by applying light pressure to the injury with the thumb and forefinger. If additional bleedings are necessary, progressively nick the ear closer to its base. Alternatively, blood may be drawn from the marginal ear vein with a 1–2-ml syringe equipped with a 26-gauge needle.

There are various methods of extracting larger volumes of blood from rabbits. Among those frequently practiced are cutting the jugular vein, ear bleeding with the help of a vacuum, and cardiac puncture. Cardiac puncture (Figure A12.2) is recommended here because it is fast and efficient. The rabbit is tied to the inclining bleeding rack. The area above the sternum is shaved and swabbed with 70% alcohol. The blood is extracted with a large syringe (50 ml) fitted with an 18-gauge needle and emptied into a sterile screw-capped tube. About 50 ml of blood can be taken from a 10–12 lb (approximately 4.5–5.5 kg) rabbit without endangering the animal's life.

A bleeding rack may be built by nailing two wooden rails to a board (Figure A12.1) and elevating one side with a wooden support to provide an incline of approximately 12°. The distance between the rails should be 4–6 cm, depending on the neck size of the rabbits used. The rabbit's head is held by the rails at the upper end, while the legs are tied to a cleat at the lower end.

The Bellco (Bellco Biotechnology, Vineland, NJ) rabbit bleeding apparatus (Figure A12.3) is another convenient means of obtaining large quantities of blood from a rabbit. Bellco's instructions provide the following information.

Equipment required: vacuum pump (or line), a sharp razor blade, receptacle (culture tube or flask with appropriate size rubber stopper), and a short piece of heavy rubber or plastic hose for attachment to the vacuum line.

The ear of the animal is disinfected, a single slit is made through the marginal ear vein, and the ear is inserted into the large opening of the apparatus. The vacuum line

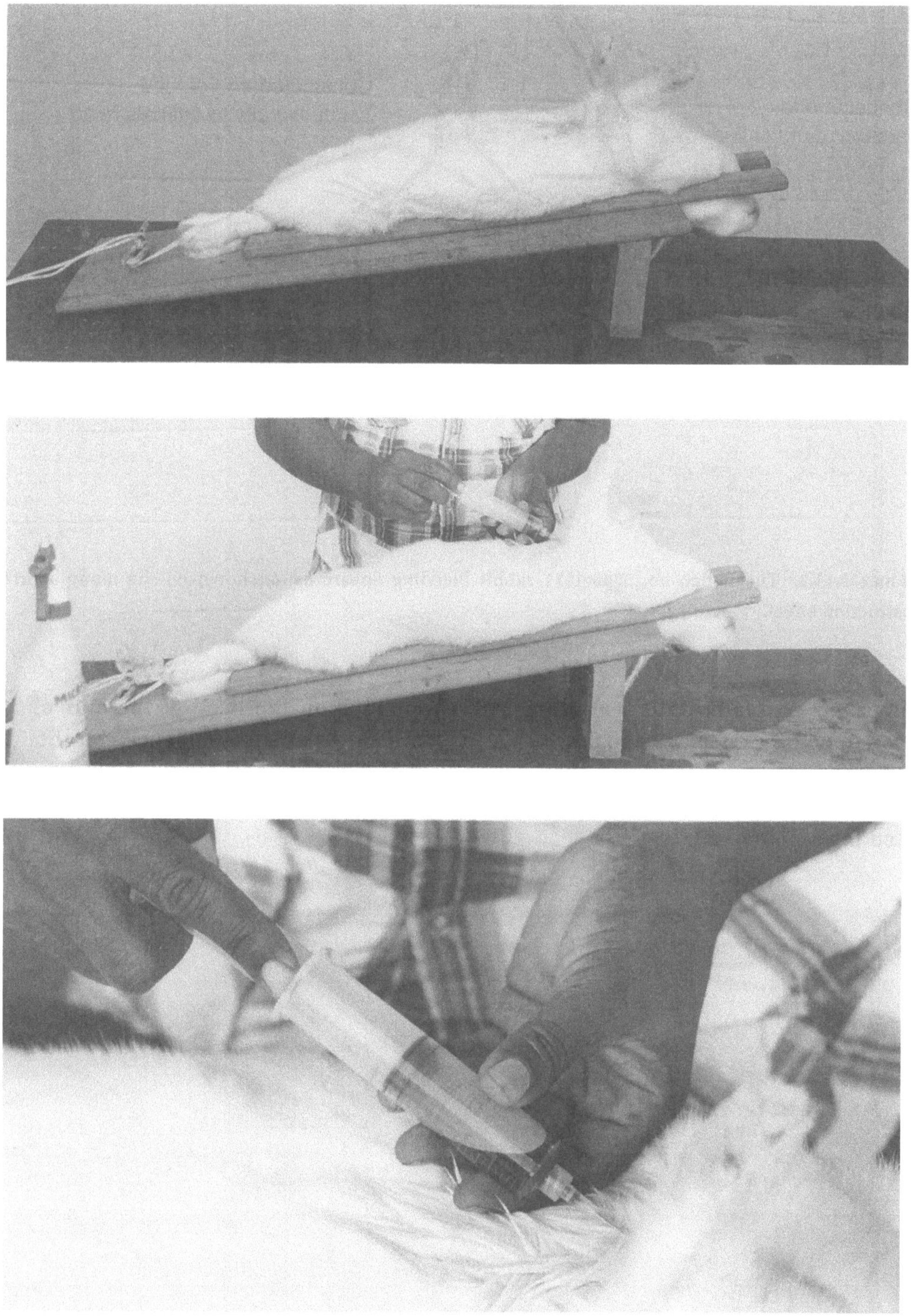

FIGURE A12.2 Collecting blood from a rabbit by cardiac puncture. (a) Rabbit is secured to the bleeding rack; (b) drawing the blood; and (c) a close-up of the draw.

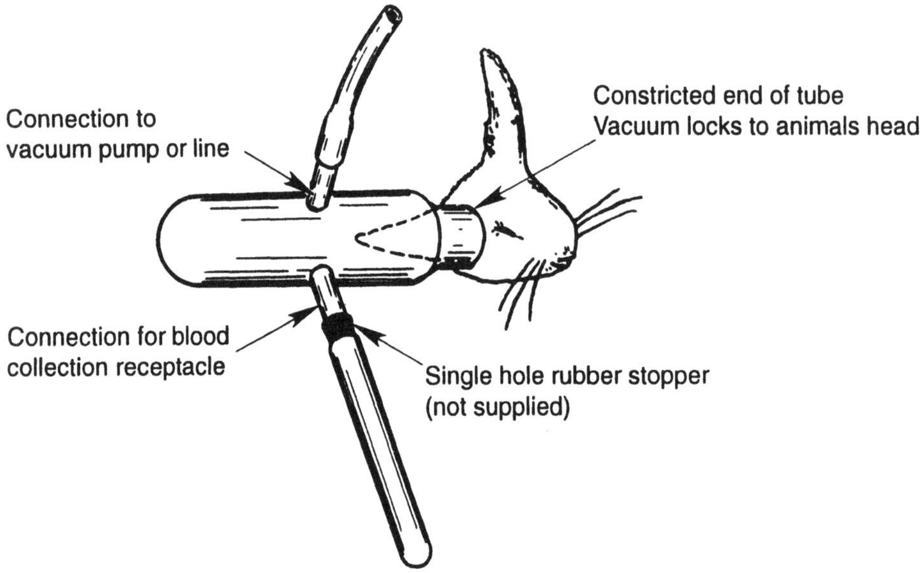

FIGURE A12.3 The Bellco no. 5640-1111 rabbit bleeding apparatus as shown on the manufacturer's instruction sheet.

is opened gradually until a vacuum lock is obtained on the head of the animal. Immediately the blood begins to flow in a steady stream. As much as 50 ml can be obtained in 1 min. without any sign of trauma to the animal. The entire rabbit bleeding apparatus is autoclaved, the ear of the rabbit is treated with a disinfectant, and only one tube is used for each animal.

APPENDIX 13

The Indirect Fluorescent Antibody Technique

The indirect fluorescent antibody (FA) technique uses antibodies (antisera) of rabbit and goat (or sheep). The specific antiserum for the rhizobial strain is produced as described in Chapter 8, but the antiserum is not conjugated with fluorescein isothiocyanate (FITC). Purified gamma globulins from a rabbit (not immunized previously with rhizobia) are injected into a goat as antigen to produce antibodies against the rabbit gamma globulin. The antibodies from the goat, commonly referred to as GARGG (goat-anti-rabbit gamma globulin), are then conjugated with FITC.

In the identification procedure, rhizobial cells are smeared on a slide and heat fixed. This smear is reacted with the unconjugated rabbit antiserum specific for the rhizobial strain. After reaction, unreacted rabbit antiserum is washed off. This is followed by staining with the GARGG (or SARGG from sheep).

While GARGG is available commercially (e.g., from Difco Laboratories, Detroit, MI), some investigators prefer to produce their own. NifTAL (Paia, HI) has produced GARGG successfully using a 1% solution of purified rabbit gamma globulin as antigen. The following injection schedule has proven successful (intramuscular, IM; intravenous, IV; intraperitoneal, IP; and subcutaneous, SC).

INJECTION SCHEDULE[1]

Day	Procedure
1	1:1 emulsion of antigen: Freund's complete adjuvant—20 ml IM (10 ml into each thigh muscle)
14	1:1 emulsion of antigen: Freund's incomplete adjuvant—4 ml IM (2 ml into each thigh muscle). Antigen—2 ml SC (1 ml above each shoulder) Antigen—2 ml IP Antigen—2 ml IV (optional)

[1]Hoben, unpublished observations.

28	1:1 emulsion of antigen: Freund's incomplete adjuvant—4 ml IM (2 ml into each thigh muscle) Antigen—2 ml SC (shoulders) Antigen—2 ml SC (1 ml into each hip region)
33	Trial bleeding
34	Blood collection
40	Blood collection
54	First booster injection (same as day 28). Booster injections can be made on 28-day cycles.

Injection and blood collections may be continued beyond day 34. The blood may be collected 6 and 12 days after each set of booster injections. The booster injections follow the same protocol as day 28. Complete adjuvant should only be used in the beginning of the immunization. Incomplete adjuvant should be given on subsequent injection days.

One person is required to hold the animal down, while another gives the injections. A tranquilizer, such as Rompun (made by Bayer Leverkusen, Leverkusen, Germany) is recommended to subdue the animal during blood collection. When the tranquilizer is injected intramuscularly according to the manufacturers instructions, the animal will fall asleep within 5–15 min., and awaken after 2 h.

The goat is bled from the jugular vein as follows. Shave the appropriate area on the neck and locate the vein by touch. Press the thumb of your left hand onto the vein. This will block the blood flow and enlarge the vein just above your thumb. Swab this area with 70% ethanol and insert a sterile 20-gauge needle (holder-needle assembly for use with Vacutainer glass tubes) into the jugular vein. Place the Vacutainer glass tube into the holder and collect the blood. Keep exchanging Vacutainer glass tubes until the desired amount is collected. A 50-lb (approximately 23 kg) goat can safely deliver 300 ml in one bleeding.

The blood is handled as described in Chapter 8 and the resulting antiserum is checked for quality by immunodiffusion (Chapter 11) as follows.

Dilute the goat (sheep) antiserum in twofold steps from 1:2 to 1:32. Using the hexagonal immunodiffusion pattern, place the different dilutions into the outer wells and the antigen (1% rabbit globulin solution) into the center well. If sufficient antibodies are present in the serum, strong precipitin bands will be produced at dilutions of 1:4 or higher. Antisera of acceptable quality are then conjugated with FITC (Chapter 13).

The indirect FA technique eliminates the need for conjugating rabbit antisera. It is considered more sensitive than the direct FA technique. The indirect method can be used with any rhizobial antisera produced in a rabbit, even those with low titer, which are not suitable for conjugation. It differs from the direct method mainly by including the additional reaction step, while most of the procedures detailed in Chapter 13 for the direct technique remain the same. Since nonspecific fluorescence may occasionally occur with the indirect method, a control smear treated only with conjugated GARGG should be included.

The staining is done as follows:

1. Make a thin smear and heat fix.
2. Cover the smear with 1:10 diluted rabbit antiserum and incubate for 20 min.
3. Briefly wash off the excess antiserum with phosphate-buffered saline (PBS).
4. Cover the smear with diluted FITC conjugate of goat anti-rabbit globulin and incubate for 20 min. (A suitable dilution of the conjugated GARGG to be used is determined by its staining titer.)
5. Wash off excess FITC conjugate and place in PBS for 20 min.
6. Complete the washing process by placing the slide in distilled water for 10 min.
7. Air dry, mount, and observe under the UV microscope.

APPENDIX 14

Additional Information on the Plant Infection Count

ASSESSING THE QUALITY OF MOST-PROBABLE-NUMBER (MPN) RESULTS[1]

Acceptability of Results, the Range of Transition (ROT)

The range of transition (ROT) is the number of dilution steps between entirely positive and entirely negative dilutions. This is a direct measure of the experimental compliance with the principle assumptions underlying the most-probable-number (MPN) procedure, namely, that a single cell is capable of producing a root nodule and that the cells follow a Poisson distribution.

In Table A14.1, the results of a six-step tenfold dilution series of four replicates yield experimental results of 4-4-3-1-0-0. The number of dilution steps from the first

TABLE A14.1 Calculating the ROT from Dilution with Four Replicate Tubes per Dilution Level ($A = 10$, $n = 4$).

	Nodulation					
	Replications				No. of Nodulated	
Dilution	I	II	III	IV	Units	ROT
10^{-1}	+	+	+	+	4	
10^{-2}	+	+	+	+	4	
10^{-3}	+	+	−	+	3	+[1]
10^{-4}	−	+	−	−	1	+[1]
10^{-5}	−	−	−	−	0	
10^{-6}	−	−	−	−	0	

[1]The ROT for these results is 2.

[1]From J.E. Bennet, P. Woomer, and R.S. Yost, 1990. *User's Manual for MPNES Most-Probable-Number Enumeration System*, University of Hawaii, NifTAL Project, Paia, HI.

not entirely positive to the last not entirely negative dilution yields the ROT, 2 in this case. To test the compliance probability of a dilution series, the ROT is compared with a tabular value applicable to the dilution series/replicate combination (Table A14.2). When the column for tenfold dilution is located on the table for four replicates, a ROT value of 2 yields a probability of 0.271; the dilution series is acceptable. An experimental code of 4-3-3-0-1-0 developed under similar experimental conditions has a ROT of 4 and a probability of 0.0036 (shown as 0.004 in Table A14.2). We are certain to 0.9964 that the results of this dilution series does not comply with underlying assumptions and the results are discarded.

The probabilities of the ROT for many dilution series/replicate combinations are presented in Table A14.2. Stevens (1958) suggests that this test of technique not be applied to individual series until a bulk of results has been examined, and that this technique be used to discover and remove procedural deficiencies. Following this, researchers may adopt a rule of rejecting results at $p = 0.01$. Researchers should be aware that statistical methods allow for tests of experimental technique and, when possible, collect and test data using the ROT.

Table A14.2 A Test of Technique of Dilution Series Results. The Expected Frequency of Equaling or Exceeding the ROT[1]

	Probability of technique		
	Dilution Ratio		
Range	2	4	10
	Two replicates per dilution level		
1	0.930	0.717	0.525
2	0.820	0.373	0.114
3	0.625	0.123	0.013
4	0.415	0.034	0.001[2]
5	0.246	0.009[2]	0.0001
6	0.136	0.002	0.00001
	Four replicates per dilution level		
1	NA	0.955	0.838
2	NA	0.682	0.271
3	NA	0.294	0.035
4	NA	0.088	0.004[2]
5	NA	0.023	0.0004
6	NA	0.006[2]	0.00004

[1]Adapted from J.E. Bennet, P.L. Woomer, and R.S. Yost, 1990. *User's Manual* for *MPNES Most-Probable-Number Enumeration System*, University of Hawaii NifTAL Project, Paia, HI.

[2]Marked areas indicate failure of technique ($p \leq 0.01$). NA, frequency distribution not available.

EXAMPLES FOR THE CALCULATION OF THE MPN

Example: Determine the number of *Bradyrhizobium japonicum* cells contained in 1 g of a 100-g bag of inoculant made from nonsterile peat.

1. Dilute the 100 g of inoculant in 900 ml of water.
2. Make a tenfold dilution series (Table A14.3)
3. Set up plants in quadruplicates as described in Chapter 6 and inoculate each plant with 1 ml of the dilutions.
4. Record nodulation (+ or −).
5. Beside each dilution, write the number of nodulated (+) units.
6. Add the total of the nodulated units, assuming the results shown in Table A14.3.
7. Note that the number of replications, $n = 4$; dilution steps, $s = 10$; number of nodulated units, $(+) = 27$; and lowest dilution in the series, $d = 10^{-1}$.
8. Use MPN tables. Table A14.4 is calculated for twofold dilutions, Table A14.5 for fourfold dilutions, and, Table A14.6 for tenfold dilutions. Locate 27 (for 27+ units) in column $n = 4$ of Table A14.6.

TABLE A14.3 Example of Data Obtained for the MPN Enumeration of *B. japonicum* in an Inoculant Prepared with Nonsterile Peat

	Nodulation					
	Replications					
Dilution	I	II	III	IV	No. of Nodulated Units	ROT
10^{-1}	+	+	+	+	4	
10^{-2}	+	+	+	+	4	
10^{-3}	+	+	+	+	4	
10^{-4}	+	+	+	+	4	
10^{-5}	+	+	+	+	4	
10^{-6}	+	+	+	+	4	
10^{-7}	−	−	+	−	2	+[1]
10^{-8}	+	−	−	−	1	+[1]
10^{-9}	−	−	−	−	0	
10^{-10}	−	−	−	−	0	
				Total	27	

[1]Rate of transition is 2.

TABLE A14.4 Number (m) of Rhizobia Estimated by the Plant Infection Count (After Vincent, 1970). A. Twofold dilutions ($A = 2$)[1]

Positive Tubes		Dilution Steps (s)			
n = 4	n = 2	s = 10			
40	20	>520			
39					
38	19	520			
37		370			
36	18	290			
35		220			
34	17	180			
33		140	s = 8		
32	16	120	>130		
31		95			
30	15	78	130		
29		65	93		
28	14	54	72		
27		45	55		
26	13	37	45		
25		31	35	s = 6	
24	12	26	29	>33	
23		21	24		
22	11	18	19	33	
21		15	16	23	
20	10	13	13	18	
19		11	11	14	
18	9	8.9	9.3	11	
17		7.4	7.7	8.9	s = 4
16	8	6.3	6.4	7.4	>8.3
15		5.2	5.4	6.0	
14	7	4.4	4.6	4.9	8.3
13		3.7	3.8	4.1	5.9
12	6	3.2	3.2	3.4	4.6
11		2.6	2.6	2.7	3.4
10	5	2.2	2.2	2.3	2.8
9		1.8	1.9	1.9	2.2
8	4	1.5	1.5	1.6	1.8
7		1.2	1.3	1.3	1.4
6	3	1.0	1.0	1.0	1.1
5		0.79	0.79	0.81	0.97
4	2	0.60	0.60	0.62	0.66
3		0.42	0.43	0.43	0.46
2	1	0.27	0.27	0.27	0.29
1		<0.2	<0.2	<0.2	<0.2
0	0				
Approximate range		2000	500	120	30
Factor, 95% Fiducial limits		n = 2 2.7			
(×, ÷)		n = 4 2.0			

[1]Calculated from Table VIII_2 of Fisher and Yates (1963).

TABLE A14.5 Number (m) of Rhizobia Estimated by the Plant Infection Count (After Vincent, 1970). B. Fourfold Dilutions ($A = 4$)[1]

Positive Tubes		Dilution Steps (s)			
$n = 4$	$n = 2$	$s = 10$			
40	20	$>2.0 \times 10^5$			
39					
38	19	2.0×10^5			
37		1.2			
36	18	8.1×10^4			
35		5.5			
34	17	3.8			
33		2.6	$s = 8$		
32	16	1.8	$>1.3 \times 10^4$		
31		1.3			
30	15	9.1×10^3	1.3×10^4		
29		6.3	7.9×10^3		
28	14	4.5	5.1		
27		3.5	3.5		
26	13	2.2	2.4		
25		1.6	1.7	$s = 6$	
24	12	1.1	1.1	$>7.9 \times 10^2$	
23		8.0×10^2	8.0×10^2		
22	11	5.6	5.6	7.9×10^2	
21		4.0	4.0	5.0	
20	10	2.8	2.8	3.2	
19		2.0	2.0	2.2	
18	9	1.4	1.4	1.5	
17		1.0	1.0	1.0	$s = 4$
16	8	7.1×10^1	7.1×10^1	7.2×10^1	$>5.0 \times 10^1$
15		5.0	5.0	5.1	
14	7	3.5	3.5	3.5	5.0×10^1
13		2.5	2.5	2.5	3.2
12	6	1.8	1.8	1.8	2.0
11		1.3	1.3	1.3	1.4
10	5	8.9×10^0	8.9×10^0	8.9×10^0	9.6×10^0
9		6.3	6.3	6.3	6.6
8	4	4.5	4.5	4.5	4.6
7		3.2	3.2	3.2	3.2
6	3	2.2	2.2	2.2	2.2
5		1.6	1.6	1.6	1.6
4	2	1.1	1.1	1.1	1.1
3		7.2×10^{-1}	7.2×10^{-1}	7.2×10^{-1}	7.2×10^{-1}
2	1	4.4	4.4	4.4	4.4
1					
0	0	$<4.4 \times 10^{-1}$	$<4.4 \times 10^{-1}$	$<4.4 \times 10^{-1}$	$<4.4 \times 10^{-1}$
Approximate range		5×10^5	3×10^4	2×10^3	1×10^2
Factor for 95% Fiducial limits		$n = 2$ 4.0			
(×, ÷)		$n = 4$ 2.7			

[1]Calculated from Table VIII_2 of Fisher and Yates (1963).

TABLE A14.6 Number (m) of Rhizobia Estimated by the Plant Infection Count (After Vincent, 1970). C. Tenfold Dilutions ($A = 10$)[1]

Positive Tubes		Dilution Steps (s)			
n = 2	n = 2	s = 10			
40	20	>7 × 10^8			
39					
38	19	6.9			
37		3.4			
36	18	1.8			
35		1.0			
34	17	5.9 × 10^7			
33		3.1	s = 8		
32	16	1.7	>7 × 10^6		
31		1.0			
30	15	5.8 × 10^6	6.9		
29		3.1	3.4		
28	14	1.7	1.8		
27		1.0	1.0		
26	13	5.8 × 10^5	5.9 × 10^5		
25		3.1	3.1	s = 6	
24	12	1.7	1.7	>7 × 10^4	
23		1.0	1.0		
22	11	5.8 × 10^4	5.8 × 10^4	6.9	
21		3.1	3.1	3.4	
20	10	1.7	1.7	1.8	
19		1.0	1.0	1.0	
18	9	5.8 × 10^3	5.8 × 10^3	5.9 × 10^3	
17		3.1	3.1	3.1	s = 4
16	8	1.7	1.7	1.7	>7 × 10^2
15		1.0	1.0	1.0	
14	7	5.8 × 10^2	5.8 × 10^2	5.8 × 10^2	6.9
13		3.1	3.1	3.1	3.4
12	6	1.7	1.7	1.7	1.8
11		1.0	1.0	1.0	1.0
10	5	5.8 × 10^1	5.8 × 10^1	5.8 × 10^1	5.9 × 10^1
9		3.1	3.1	3.1	3.1
8	4	1.7	1.7	1.7	1.7
7		1.0	1.0	1.0	1.0
6	3	5.8 × 1	5.8 × 1	5.8 × 1	5.8 × 1
5		3.1	3.1	3.1	3.1
4	2	1.7	1.7	1.7	1.7
3		1.0	1.0	1.0	1.0
2	1	0.6	0.6	0.6	0.6
1		<0.6	<0.6	<0.6	<0.6
0	0				
Approximate range		10^9	10^7	10^5	10^3
Factor, 95% Fiducial limits[2]		n = 2 6.6			
(×, ÷)		n = 4 3.8			

[1]Calculated from Table VIII_2 of Fisher and Yates (1963).

[2]Cochran; *Biometrics* (1950) 6: 105.

9. Find the most likely number (m) in column $s = 10$ corresponding to 27 in the $n = 4$ column. The most likely number is $m = 1.0 \times 10^6$.

10. Multiply the most likely number with the reciprocal of the lowest dilution used in the series ($d = 10^1$).

11. Divide the product of m and d by the aliquot used for inoculation ($v = 1$ ml). Result: The MPN per gram of inoculant was:

$$X = \frac{m \times d}{v} = \frac{1.0 \times 10^6 \times 10}{1} = 1.0 \times 10^7$$

The ROT value was 2 and the probability, as indicated in Table A14.2, was 0.271. The results of this MPN count are acceptable.

Determining the MPN of Rhizobia in Soil

The MPN count is often used to determine the number of rhizobia present in soil. Whereas a tenfold dilution series with two or four replicates is sufficient for most peat inoculants, which usually have a relatively high number of rhizobia ($>10^8$ cells g^{-1}), a fourfold or even twofold dilution series with replications in quadruplicate is usually chosen for soil. The smaller dilution steps provide a more precise estimate when less than 10,000 cells of rhizobia per gram of soil are expected. The first sample of the series, however, is frequently diluted tenfold or 100-fold.

Example: A tropical soil was sampled to determine the size of the native "cowpea-type" rhizobial population. One hundred grams of the soil were diluted in 900 ml of sterile water. A dilution series with four replications was prepared ranging from 4^{-1} to 4^{-8}. Aliquots of 2 ml were used for the inoculations of siratro (*Macroptilium atropurpureum*) seedlings grown in growth pouches. The experiment was terminated after 3 weeks and the results are shown in Table A14.7.

The following parameters were extracted from the data presented in Table A14.7:

$n = 7$
$s = 8$
$d = 4$

Total + units = 22

Use Table A14.5 for fourfold dilutions in this appendix. In column $n = 4$, locate 22 for 22 + units to determine the value of m.

$m = 5.6 \times 10^2$
d = (Reciprocal of the lowest dilution) = 4^1
$v = 2$ ml

$$X = \frac{5.6 \times 10^2 \times 4}{2} = 1.12 \times 10^3$$

TABLE A14.7 Example of Data Obtained for the MPN Enumeration of a Population of Native Bradyrhizobia in a Field Soil Using Siratro As the Trap Host

	Nodulation					
	Replications				No. of Nodulated	
Dilution	I	II	III	IV	Units	ROT
4^{-1}	+	+	+	+	4	
4^{-2}	+	+	+	+	4	
4^{-3}	+	+	+	+	4	
4^{-4}	+	+	+	+	4	
4^{-5}	−	+	+	+	3	+[1]
4^{-6}	+	+	−	−	2	+[1]
4^{-7}	+	−	−	−	1	+[1]
4^{-8}	−	−	−	−	0	
				Total	22	

[1]Rate of transition is 3.

Because the soil was diluted 1:10 before the actual dilution series was started, the final result must be multiplied by 10. Therefore, the MPN of rhizobia in the soil that nodulated siratro was 1.12×10^4 cells g^{-1}.

LIMITATIONS IN THE USE OF PLANT INFECTION COUNTS

Ideally, each plant infection count would yield results in which the nodulation of the plants, in growth units, proceeded in an orderly transition from the entirely positive to the partially positive to the entirely negative. Unfortunately, this is not always the case. Irregular results, as shown in the following examples, are not uncommon.

Skips at Lower Dilution Levels

In lower dilutions, especially in soil, many other microorganisms are present that may interfere with the infection of the plants by rhizobia, resulting in negative infections at the beginning of the series. For example, a tenfold dilution series with four replicates ($A = 10$, $n = 4$) may yield the following results:

10^{-1}	10^{-2}	10^{-3}	10^{-4}	10^{-5}	10^{-6}	10^{-7}	10^{-8}
0	0	4	4	2	1	0	0

The skip tubes (10^{-1} and 10^{-2}) should be considered positives in such a case and the series should be evaluated for:

10^{-1}	10^{-2}	10^{-3}	10^{-4}	10^{-5}	10^{-6}	10^{-7}	10^{-8}
4	4	4	4	2	1	0	0

MPN = 1.0×10^5

Isolated Positives at Higher Dilutions

In the following tenfold dilution series with four replications ($A = 10$, $n = 4$), the nodulated units yield results that express an orderly transition from entirely positive at the lowest dilution to entirely negative results at the highest dilution. Note the interruption of this normal transition by a positive (10^{-6}) near the end of the series.

10^{-1}	10^{-2}	10^{-3}	10^{-4}	10^{-5}	10^{-6}	10^{-7}	10^{-8}
4	4	2	0	0	1	0	0

MPN = 100

The isolated positive at the higher dilution levels may result from contamination. The experiment should be discarded if uninoculated controls show nodulation. If no contamination can be found in any of the controls, the isolated positive may be ignored and the series should be evaluated as follows:

10^{-1}	10^{-2}	10^{-3}	10^{-4}	10^{-5}	10^{-6}	10^{-7}	10^{-8}
4	4	2	0	0	0	0	0

MPN = 58

Alternatively, both series may be evaluated and the average taken from the results. The difference will be slight.

Incomplete Dilution Series

Occasionally, a dilution series yields positive and partially positive results, even at the highest dilutions. This is illustrated by the tenfold dilution series with four replications ($A = 10$, $n = 4$).

10^{-1}	10^{-2}	10^{-3}	10^{-4}	10^{-5}	10^{-6}	10^{-7}	10^{-8}
4	4	4	4	4	3	2	2

In this case, the researcher has underestimated the number of rhizobia expected in the sample. A dilution series that does not lead to "extinction" is incomplete and cannot be evaluated.

Nonsensical Results

Sometimes, dilution series yield results that cannot be interpreted. For example:

10^{-1}	10^{-2}	10^{-3}	10^{-4}	10^{-5}	10^{-6}	10^{-7}	10^{-8}
1	4	2	4	0	3	2	2

The previous results may be attributed to several failures including mislabeling, contamination, microbial antagonism, and random error. The results of this dilution series should be discarded.

For statistical analysis of MPNs, it is necessary to have a sufficient number of replications (i.e., several MPNs) of the sample being enumerated. Data from replications may be recorded as shown in Figure A14.1. A blank table (Figure A14.2) is provided, and may be copied if required. The data in Figure A14.1 were evaluated by the most-probable-number enumeration system (MPNES) computer program, which can accommodate designs (e.g., when $n = 3$ or $n = 5$) that are not presented in most MPN tables. The MPNES program may be obtained from the NifTAL Center, (Paia, HI).

Experiment title: Quality check on bean inoculant

Test legume: *Phaseolus vulgaris* (small black-seeded variety)

Date planted: April 04, 1992 Date inoculated: April 10, 1992

Time of nodule score; 2 week ( ), 3 week (X), 4 week ( ), other ______

Inoculant type: Peat (nonsterile)

Dilution factor A = 2 ( ) 4 ( ) 5 ( ) 10 (X); Dilution steps s = 10 ( ) 8 ( ) 6 (X) 4 ( )

Volume inoculated (ml) = 1.0 Weight or volume of sample (g) = 10

Type of growth unit: Growth pouch ( ) Agar slant ( )

NifTAL tube (X) Leonard jar ( )

Replications of growth unit (R): 3 MPN replication: 3

Dilution Steps	MPN-1	MPN-2	MPN-3	MPN-4
10^{-5}	3	3	3	
10^{-6}	3	3	3	
10^{-7}	3	3	3	
10^{-8}	3	3	3	
10^{-9}	2	2	3	
10^{-10}	0	0	0	
Σ Nodulated Units	14	14	15	
MPN	8.0×10^{8}	8.0×10^{8}	1.2×10^{9}	
ROT	1	1	0	

Notes/remarks: Inoculant was about 8 weeks old. Storage temperature was 4°C. MPN computed using MPNES program.

Investigator: Padma Date: May 15, 1992

FIGURE A14.1 Completed MPN data record sheet.

Experiment title: ______________________________

Test legume: ______________________________

Date planted: ______________ Date inoculated: ______________

Time of nodule score; 2 week ( ), 3 week ( ), 4 week ( ), other ______________

Inoculant type: ______________________________

Dilution factor A = 2 ( ) 4 ( ) 5 ( ) 10 ( ); Dilution steps s = 10 ( ) 8 ( ) 6 ( ) 4 ( )

Volume inoculated (ml) = ________ Weight or volume of sample (g) = ________

Type of growth unit: Growth pouch ( ) Agar slant ( )

NifTAL tube ( ) Leonard jar ( )

Replications of growth unit (R): ________ MPN replication: ________

Dilution Steps	MPN-1	MPN-2	MPN-3	MPN-4
Σ Nodulated Units				
MPN				
ROT				

Notes/remarks: ______________________________

____

____

Investigator: ______________________________ Date: ______________

Figure A14.2 MPN data record sheet.

APPENDIX 15

The Acetylene Reduction Assay for Measuring Nitrogenase Activity

The nitrogenase enzyme complex is responsible for biological N_2 fixation in the root nodules of legumes. Nitrogenase is synthesized by bacteroids in the nodules and also the reduction of molecular N_2 to NH_3 takes place within the cytoplasm of the bacteroids. The enzyme complex consists of two distinct protein components with Fe atoms common to both and Mo present in only one of the two components. Both the Fe-protein and Mo-Fe-protein are essential for nitrogenase activity. During the reduction process, molecular N_2 is converted to NH_3 through a series of steps involving enzyme(s) and ATP. Though molecular N_2 is the natural substrate for nitrogenase, other triple bonded "nitrogen-analogues" like acetylene (HC≡CH), cyanide (H-C≡N), nitrous oxide (N≡N-O), and methyl isocyanide (CH_3-N≡C) can also undergo reduction mediated by the nitrogenase complex. Because of its lack of toxicity and easy availability, acetylene is frequently used to assay for nitrogenase activity. In the assay procedure, root nodules of legumes are exposed to 5–25% of acetylene-in-air mixture and incubated at 25–30°C. The ethylene (C_2H_4) produced by the reduction of the acetylene is measured by gas chromatography. Although acetylene reduction is a sensitive method for assaying nitrogenase activity, the reduction information may not be translated into N_2 fixed because of frequent theoretical disagreement in the stoichiometry of the two reduction processes.

THE GAS CHROMATOGRAPH

A gas chromatograph with a hydrogen flame-ionization-detector (FID) is usually used in the assay. A stainless steel column, 3 m long and 1 mm in diameter, is filled with molecular sieve material, usually Porapak media (produced by Waters Associates Inc., Farmingham, MA). Porapak is a porous polymer composed of ethylvinylbenzene cross-linked with divinylbenzene to form a uniform structure of distinct pore size. It is available in bead form with different mesh sizes. Porapak N of 100–200 mesh size allows good separation of C_2H_2 and C_2H_4 using N_2 as the carrier gas.

A column temperature of 50–70°C with a carrier gas flow rate of 50 ml min^{-1} is used for routine work. Air and hydrogen gas are adjusted to flow at the rates of 300 and 90 ml min^{-1}, respectively. Different gases have different retention times in the column, therefore the gas chromatograph recorder will trace out the peaks in order of emergence.

A gas mixture containing CH_4, C_2H_2, and C_2H_4 will have the trace pattern illustrated in Figure A15.1. It is important for the operator to become familiar with the acetylene and ethylene peaks traced out on the gas chromatograph recorder chart.

SOURCE OF ACETYLENE

Acetylene is available commercially in cylinders. Very pure acetylene (99%) is available in small cylinders for analytical work. For laboratory work, small amounts can be con-

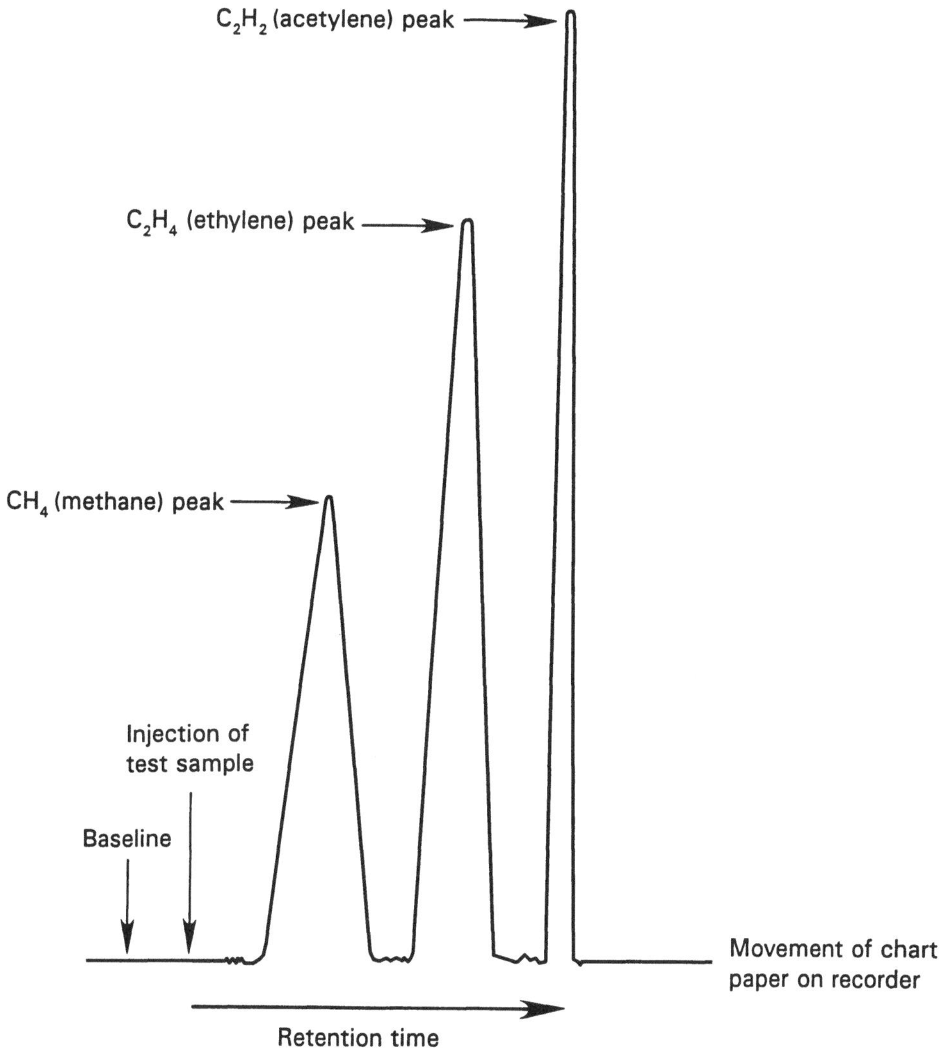

FIGURE A15.1 Trace pattern from an injection of a gas mixture containing CH_4 C_2H_2, and C_2H_4 showing the sequence of emergence of the different peaks. (Adapted from Postgate, 1971.)

veniently prepared by reacting calcium carbide with water. About 15 ml of water is used for each gram of calcium carbide. A simple apparatus for generating acetylene is shown in Figure A15.2. Acetylene generated this way contains very minute quantities of phosphine, ethylene, and methane.

CALIBRATION OF THE GAS CHROMATOGRAPH

In the calibration process, exact amounts of ethylene have to be injected into the gas chromatograph and the peak heights measured. The concentration of ethylene giving a particular peak height is computed. A calibration curve is obtained by plotting peak height (y axis) against ethylene concentration (x axis). The calibration curve should be linear and pass through the origin.

1. Obtain a small volume of the 99% pure ethylene for the calibration.
2. Dilute the pure ethylene as follows: Determine the true volume of a 1000-ml volumetric flask. Fill the flask completely with water to the mouth. Without trapping

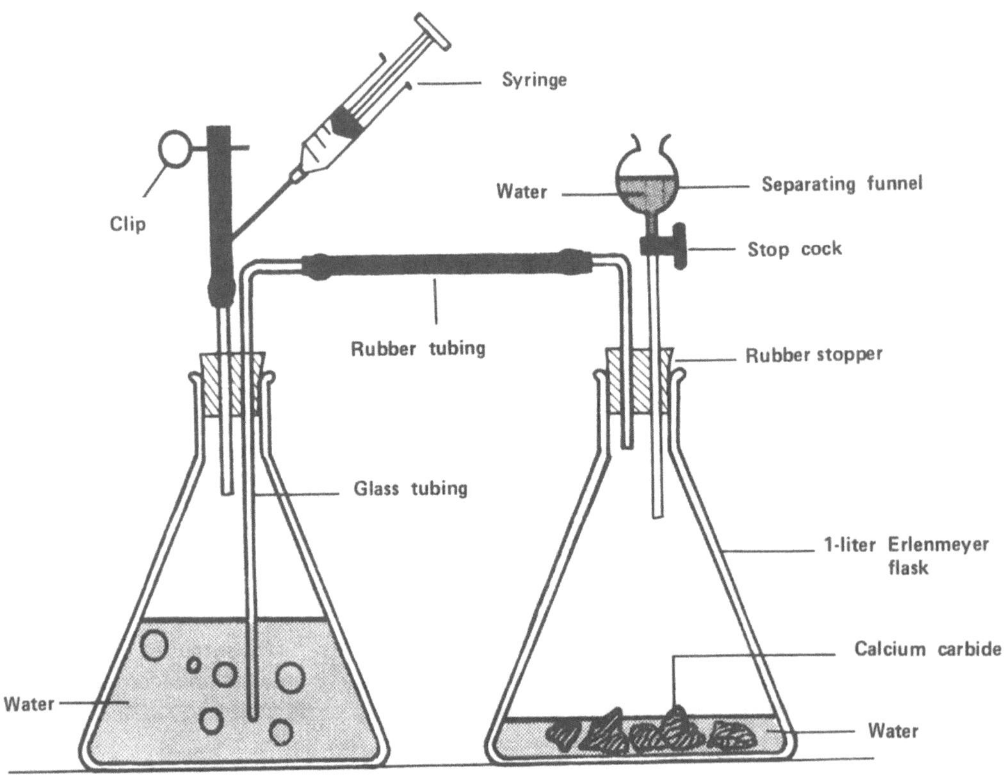

FIGURE A15.2 A simple apparatus for generating small amounts of acetylene (C_2H_2), in the laboratory.

any air, carefully place a Suba-seal (W. Freeman & Co., Led., Barnsely, Yorkshire, UK) or a long, sleeved rubber stopper for serum bottles (Wheaton Scientific, Millville, NJ) to contain the water. Invert the flask to detect air trapped during the placement of the seal or stopper. Repeat filling and sealing if too much air is trapped. Pour the water into a measuring cylinder and record the volume.

3. Flush out the flask with N_2 and seal again. Remove one ml of the air from the sealed flask with a syringe. Then inject 1 ml of pure ethylene into the sealed flask and allow to stand for 10 min at room temperature to equilibrate.

4. With a 1-ml plastic syringe, pierce the rubber seal, remove 1 ml of the diluted ethylene from the flask, and inject it into the gas chromatograph. Measure the height of the ethylene peak from the trace. Inject two more 1-ml samples to check for reproducibility of the peaks. Note the column temperature of the gas chromatograph.

CALCULATIONS OF THE CALIBRATION

Suppose the diluted ethylene (now referred to as the standard) was equilibrated at 23°C and 756 mm of Hg pressure. Convert these values to normal temperature pressure (NTP) using the gas law relationship:

$$\frac{P_1V_1}{T_1} = \frac{P_2V_2}{T_2}$$

$P_1 = 760$ mm Hg: $V_1 =$ unknown, $T_1 = 273°K$
$P_2 = 756$ mm Hg: $V_2 = 1$ ml, $T_2 = (273° + 23°)K$

Therefore, volume of ethylene, V_1 at NTP

$$= 1 \times \frac{756}{760} \times \frac{273}{296}$$
$$= V_1 = 0.9174 \text{ ml}$$

According to molar volume, 1 mole of C_2H_4 at NTP will occupy 22.4 liters (22.4×10^3 ml).

Therefore, 0.9174 ml $C_2H_4 = \frac{0.9174}{22.4 \times 10^3}$ mole

$$= 0.041 \times 10^{-3} \text{ mole}$$

$$= (0.041 \times 10^{-3}) \times 10^9 \text{ nmole}$$

$$= 4.1 \times 10^4 \text{ nmole}$$

The accurate volume of the completely fixed volumetric flask (1 liter) was 1038 ml. Only 1 ml of the pure ethylene was diluted in the atmosphere of the flask.

$$\text{Therefore, 1 ml of the diluted ethylene} = \frac{4.1 \times 10^4 \text{ nmole}}{1038 \text{ ml}}$$

$$= 39.499 \text{ nmole}$$

When 1.0 ml of the diluted sample was injected into the gas chromatograph, an ethylene peak height of 75 divisions (on the recorder chart paper) at $\times 64$ attenuation was produced. Assume that one division on the recorder chart paper is equal to one (FID) unit. Therefore, (75×64) FID units = 39.499 nmole. Therefore, one FID unit at $\times 1$ attenuation $= \dfrac{39.499}{75 \times 64} = 0.0082$ nmole

$$1 \text{ FID unit} = 0.0082 \text{ nmole} = 8.2 \times 10^{-3} \text{ nmole}$$

From the standard ethylene preparation, inject in duplicate 0.2, 0.4, 0.6, 0.8, and 1.0 ml of the gas. Measure the peak heights corresponding to these volumes and nanomoles of the ethylene. Plot the calibration curve (i.e., peak height versus nanomoles of ethylene).

ASSAYING FOR NITROGENASE ACTIVITY WITH NODULATED ROOTS

To bring acetylene and nitrogenase into contact, the nodules must be contained in a suitable airtight vessel into which acetylene can be introduced. After a specified incubation period, samples are withdrawn and analyzed for ethylene produced with a gas chromatograph. Calibrate the gas chromatograph with the pure ethylene standard. This should be done very much in advance of bringing in incubated nodule samples for gas analysis.

Prepare incubation vessels from 1-liter Nalgene PVC (polyvinyl chloride) wide-mouthed bottles (or equivalent) for incubation of the nodulated roots. Carefully drill a 15-mm hole in the center of the cap of each bottle and fit a rubber septum (serum bottle flange-type stopper) of appropriate size to give a leak-proof fit. If metal caps are used, caps should have rubber liners to prevent leaks. Carefully excavate whole plants from the field or from Leonard jars. Cut off the tops at the point of the scar left by the cotyledons. Place the tops into labeled bags to be dried for dry weight determination. Remove as much of the soil or growth medium adhering to the roots as possible before placing it into the incubation vessel. Retrieve and include any nodule(s) that becomes detached during the excavating or cleaning operations.

Do not wash the roots to clean them because wetting decreases the nitrogenase activity significantly. A wet nodule probably traps the acetylene on the surface of the nodule by slight solution in water, thus making less acetylene available to the nitrogenase in the nodule. If the root system becomes wet, the nodules should be dried by blotting prior to being placed in the bottle. Nodulated roots from solution culture experiments should be treated similarly.

With a 50-ml plastic syringe (Beckton-Dickinson, Rutherford, NJ) fitted with an 18G needle and 1.5 inches (approximately 4 cm) long, remove 5 or 10% of the atmosphere in the incubation vessel. Replace this with a corresponding volume of acetylene. Record the time when incubation is initiated. Allow the incubation to proceed for 30–45 min, shaking the bottles periodically in between to permit good contact between the nodules and the acetylene. At the end of the incubation, shake the bottle, withdraw a 1-ml gas sample through the septum, and inject into the gas chromatograph. Duplicate the injections and note the attenuation. Other details can be indicated against each trace on the recorder chart paper.

Remove the nodulated roots from the incubation vessels after gas samples have been removed for analysis. Wash the roots and pick the nodules. Obtain the fresh weight of nodules after blotting dry, and finally, oven dry the nodules at 70°C. Calculate the nitrogenase activity from the information provided by the gas chromatograph as shown in the following example.

Example: Nodulated roots of two soybean (*Glycine max*) plants were placed in a 1000-ml incubation vessel (PVC wide-mouthed bottle). After the cap of the incubation vessel was secured tightly, 50 ml (5%) of the air was withdrawn from the incubation vessel (via the rubber septum in the cap) with a 50-ml plastic syringe and replaced with 50 ml of C_2H_2. After a 30-min incubation, 1 ml of the gas sample was withdrawn with a 1-ml syringe and injected into the gas chromatograph. A peak height of 40 divisions and ×32 attenuation was produced. Calculate the C_2H_4 produced by the nodules. (Use values of the standard calculated previously from the calibration of the gas chromatograph.)

Calculations:
Peak height = 40 divisions; attenuation = ×32
Incubation time = 30 min; volume injected = 1 ml
Total FID units = $40 \times 32 = 1280$
From the calibration, 1 FID unit = 8.2×10^{-3} nmole C_2H_4
Therefore, 1280 FID units = $(8.2 \times 10^{-3}) \times (1280)$ nmole C_2H_4

Since the volume of the incubation vessel was 1000 ml, then the total volume of C_2H_4 produced = $(8.2 \times 10^{-3}) \times (1280) \times (1000)$ nmoles

$$= 10496 \text{ nmoles}$$

$$= \frac{10496}{1000} = 10.496\ \mu\text{moles}$$

Two soybean roots produced 10.496 μmoles of C_2H_4 in 30 min.

Therefore, μmoles C_2H_4 per plant per hour $= \dfrac{10.496}{2} \times \dfrac{60}{30} = 10.496.$

The general formula for calculating nitrogenase activity is:

$$\text{Nitrogenase activity} = \frac{\text{Ethylene produced}}{\text{Time(h)} \times \text{no. of plants}}$$

Plot nitrogenase activity on the y axis and nodule (fresh or dry) weight on the x axis. Plot a similar graph, but with dry weight of plant tops on the x axis. Process both sets of plots statistically and obtain the coefficient of correlation (Appendix 18) for each of the two plots.

APPENDIX 16

Methods for Determining Lime Requirements of Acid Soils[1]

LIME REQUIREMENT OF ACID SOILS

Many procedures have been developed for measuring the lime requirement of soils, defined as the amount of lime needed to bring the pH value from its present value to any given pH value. Two methods are described here. The first method is the most reliable, but requires more time and equipment, and involves a direct titration with calcium hydroxide [Ca $(OH)_2$]. The second method, developed by Shoemaker (1959), depends on the depression in pH of a buffer solution when soil is added. It is rapid and involves a greater error, but can be used in the lime requirement estimation of large numbers of samples in relatively little time.

Calcium Hydroxide Titration

Reagents

Calcium hydroxide solution. Add 1 g of CaO or 1.5 g of $Ca(OH)_2$ per liter of CO_2-free water used. Mix and let stand protected from air until the excess has settled. Siphon off the solution. Store in a bottle protected from the CO_2 in the air.

Procedure

Place 10 g of acid soil in each of seven 100-ml beakers and add 0, 5, 15, 20, 30, 40, and 50 ml of $Ca(OH)_2$ solution to beakers 1–7, respectively. Add sufficient water to make each sample to a soil-to-water ratio of 1:5. Let stand for 3 days and determine the pH value of the soil-water suspension. Plot the pH against the milliequivalents (meq) of Ca added per 100 g of soil and determine the amount of lime needed to bring the pH to the desired level. One milliequivalent of Ca per 100 g is equal to 100 lb (45.35 kg) of lime per acre, assuming the lime is mixed with 2,000,000 lb (approximately 910,000 kg) of soil.

---

[1]Reproduced with permission from Chapman and Pratt, 1961.

Remarks

This method can be used if only a few samples are to be analyzed. If, however, there are large numbers, the space and time limitations become too great and the faster method described in the next section can be used. Three days are required for the reaction of $Ca(OH)_2$ with acid soil to come to an approximate equilibrium. Actually, about 97% of the reaction is complete in this time and the true equilibrium is attained after many days.

Buffer Method

Reagents

Buffer solution. Dissolve 1.8 g of p-nitrophenol, 2.5 ml of triethanolamine, 3.0 g of potassium chromate, 2.0 g of $Ca(OAc)_2 \cdot 2H_2O$, and 40.0 g of $CaCl_2 \cdot 2H_2O$ in approximately 800 ml of distilled water. Adjust the pH to 7.5 using hydrochloric acid or sodium hydroxide solutions, and dilute to 1 liter. Best results are obtained if 10–20 liters are prepared at one time. If protected from CO_2, this reagent will remain stable for 6 months or more. When titrated with hydrochloric acid, 50 ml of buffer should require 2.6–2.7 meq to bring the pH to 3.5 and the titration curve should be a straight line between pH 7.5 and 3.5.

Procedure

Weigh 10.0 g of soil and transfer to a 125-ml Erlenmeyer flask. Add 20 ml of buffer solution and shake for 10 min. Transfer to a 50-ml beaker and use a pH meter to determine the pH value. The lime requirement is proportional to the depression in pH of the buffer. The lime requirement can be determined from the data in Table A16.1, or the data in Table A16.1 can be plotted and the lime requirement obtained by reading from the pH versus lime requirement line.

If the pH of the soil-buffer suspension is greater than approximately 6.5, as is found with some highly acid, sandy soils, repeat the procedure using 50 g of soil and 20 ml of buffer, then divide the obtained lime requirement by 5. This modification gives better accuracy for poorly buffered soils of low lime requirement. The answer is obtained in terms of tons of pure $CaCO_3$ per 2,000,000 lb (approximately 910,000 kg) of soil to bring the pH to 6.5. Appropriate corrections must be made for variations in depth of mixing of lime or in bulk density of soils. A 6.5-in (16.5 cm) depth of soil over an acre in area will have 2,000,000 lb (approximately 910,000 kg) of dry soil if the bulk density is 1.35.

TABLE A16.1 Lime Requirement Scale for Buffer Method

Soil Buffer pH	Lime Requirement (tons $CaCO_3$)[1]	Soil Buffer pH	Lime Requirement (tons $CaCO_3$)
6.7	1.6	5.7	7.6
6.6	2.2	5.6	8.2
6.5	2.8	5.5	8.9
6.4	3.4	5.4	9.5
6.3	4.0	5.3	10.1
6.2	4.5	5.2	11.0
6.1	5.2	5.1	11.7
6.0	5.8	5.0	12.4
5.9	6.4	4.9	13.2
5.8	7.0	4.8	14.0

[1]Tons of pure $CaCO_3$ per 2,000,000 lb (approximately 910,000 kg) of soil or per acre if it is mixed with 6.5 in of soil having a bulk density of 1.35.

Analysis of Variance for a Rhizobial Strain Selection Experiment

The data in Table A17.1 presents the dry weight (grams) of plant tops from a strain selection experiment for soybean (*Glycine max* var. Jupiter). The experiment was a

TABLE A17.1 Data from a Strain Selection Experiment for Soybean

	Dry Weight of Plant Tops (g)				
	Blocks				
Treatments	B1	B2	B3	Treatment Total (T)	Treatment Means (x)
TAL 102	9.66	10.60	10.83	31.09	10.36
TAL 379	9.36	9.00	10.49	28.85	9.62
TAL 206	8.41	9.44	10.19	28.04	9.35
TAL 435	8.61	9.23	8.22	26.06	8.69
TAL 411	9.20	8.19	8.46	25.85	8.62
Allen 527	8.11	8.82	8.62	25.55	8.52
TAL 211	8.83	6.32	9.14	24.29	8.10
TAL 487	6.27	8.67	8.35	23.29	7.76
CB 1795	6.79	8.17	5.70	20.66	6.89
TAL 650	6.95	5.83	6.83	19.61	6.54
TAL 649	6.55	4.82	8.10	19.47	6.49
TAL 860	6.00	4.83	6.54	17.37	5.79
TAL 183	6.11	3.46	5.51	15.08	5.03
TAL 378	5.39	4.46	5.07	14.92	4.97
Control[1]	1.53	1.30	1.80	4.63	1.54
Control[2]	8.41	7.83	5.83	20.07	6.36
	114.18	110.97	119.68	344.83	116.36

[1]Uninoculated.

[2]70 ppm N.

randomized complete block design, with three blocks and 16 treatments (14 inoculated + 2 controls). Each treatment was replicated once within each block. Each treatment plot was a Leonard jar unit with two soybean plants. The plant tops were harvested at 32 days and oven dried at 70°C. The strains of *Bradyrhizobium japonicum* have been ranked according to dry weight.

SUMMARY OF CALCULATIONS FOR THE ANALYSIS OF VARIANCE FOR THE STRAIN SELECTION EXPERIMENT

No. of treatments = k = 16
No. of blocks = b = 3
No. of replicates per treatment per block = n = 1

Calculate the grand total (GT) by adding up all the treatment totals:

$$\begin{aligned} GT &= T_1 + T_2\, T_3 \ldots + T_k \\ &= 31.09 + 28.85 + 28.04 \ldots + 20.07 \\ &= \underline{344.83} \end{aligned}$$

Calculate the grand mean (X) by adding up all the treatment means:

$$\begin{aligned} X &= x_1 + x_2 + x_3 \ldots x_k \\ &= 10.36 + 9.62 + 9.35 \ldots + 6.36 \\ &= 116.13 \end{aligned}$$

Calculate the correction factor (CF):

$$\begin{aligned} CF &= \frac{(GT)^2}{bkn} = \frac{(344.83)^2}{3 \times 16 \times 1} \\ &= 2477.2444 \end{aligned}$$

Calculate the total sum of squares (SS):

$$\begin{aligned} SS &= \Sigma x^2 - CF \\ &= 9.66^2 + 10.60^2 + 10.83^2 \ldots + 5.83^2 - 2477.244 \\ &= 247.8507 \end{aligned}$$

Calculate the treatment sum of squares (SST):

$$SST = \frac{\Sigma T^2}{bn} - CF$$

$$= \frac{31.09^2 + 28.85^2 \ldots + 4.63^2 + 20.07^2}{3 \times 1} - 2477.2444$$

$$= 217.4785$$

Calculate the block sum of squares (SSB):

$$SSB = \frac{\Sigma B^2}{kn} - CF$$

$$= \frac{114.18^2 + 110.97^2 + 119.68^2}{16} - 2477.2444$$

$$= 2.4253$$

Calculate the error sum of squares (SSE):

$$SSE = SS - (SST + SSB)$$

$$= 247.8507 - (217.4785 + 2.4253)$$

$$= 27.9469$$

Prepare the analysis of variance according to Table A17.2. Using the previous formulations, substitute with actual figures from the calculations and prepare Table A17.3.

USE OF THE *F* DISTRIBUTION

The statistic F is a ratio of two variances and these variances are the mean squares. To identify the F distribution, the degrees of freedom (df) of each variance needs to be specified. The degrees of freedom of two variances may be represented as df_1 and df_2, where df_1 is the number of degrees of freedom in the numerator and df_2 is the number of degrees of freedom in the denominator.

From the calculations in Table A17.3 of analysis of variance for "treatments," $F(df_1, df_2) = F(15,30)$. From an F distribution table, the critical value for $F(15,30)$ with $p = 0.05$ is 2.01. Enter this tabular value into Table A17.3.

Similarly for "Blocks," the critical value for $F(2,30)$ with $p = 0.05$ is 3.32. Enter this tabular value into Table A17.3. Since the calculated F ratio for treatments is greater than the tabular value of F at the 5% level, the results indicate significant differences between the strains of *B. japonicum* in their N_2-fixing effectiveness. The calculated F ratio for blocks is less than the tabular value indicating the "blocking" of the experiment

Table A17.2 General Table for Analysis of Variance

Sources of Variation	Sum of Squares	df	Mean Squares	*F* Ratio
Treatments	SST	$k - 1$	$\frac{\text{SST}}{(k-1)}$	$\frac{\text{SST}}{\text{SSE}} \times \frac{(bkn - k - b + 1)}{(k-1)}$
Blocks	SSB	$b - 1$	$\frac{\text{SSB}}{(b-1)}$	$\frac{\text{SSB}}{\text{SSE}} \times \frac{(bkn - k - b + 1)}{(b-1)}$
Error	SSE	$bkn - k - b + 1$	$\frac{\text{SSE}}{(bkn - k - b + 1)}$	
Total	SS	$bkn - 1$		

TABLE A17.3 Analysis of Variance for Table A17.1

Sources of Variation	Sum of Squares	df	Mean Squares	*F* Ratio (Calculated)	*F* Ratio (Tabular 5%)
Treatments	217.4785	16 − 1 = 15	$\frac{217.4785}{15} = 14.4986$	$\frac{14.4986}{0.9316} = 15.5$	2.01
Blocks	2.4253	3 − 1 = 2	$\frac{2.4253}{2} = 1.2126$	$\frac{1.2126}{0.9316} = 1.30$	3.32
Error	27.9469	48 − 16 − 3 + 1 = 30	$\frac{27.9469}{30} = 0.9316$		
Total	247.8507	48 − 1 = 47			

did not create any significant disuniformity in the aeration, light, or other environmental factors in the greenhouse.

Calculate the least significant different (LSD):

$$LSD_{0.05} = t_{0.05}\sqrt{\frac{2s^2}{n}}$$

where $t_{0.05}$ = The tabular value of t for degrees of freedom for error at the 5% probability level

s^2 = Mean square for error

n = Number of replications

$$LSD_{0.05} = 2.042\sqrt{\frac{2 \times (0.9316)}{3}}$$

$$= 2.042 \times 0.79$$

$$= 1.60 \text{ g}$$

The LSD is used to compare values of two adjacent means. A pair of means that differ by more than the LSD is considered significantly different at the probability level of t employed. If comparison between means not adjacent to each other in a ranked array are made, the Duncan's multiple range test should be used. However, this test requires the computation of the Bayes LSD whose value may differ from the LSD as calculated previously. The calculation of the Bayes LSD is not presented here but its use is illustrated in Figure A17.1. Means not joined by the same line differ at $p = 0.05$ as given by Duncan's new multiple range test.

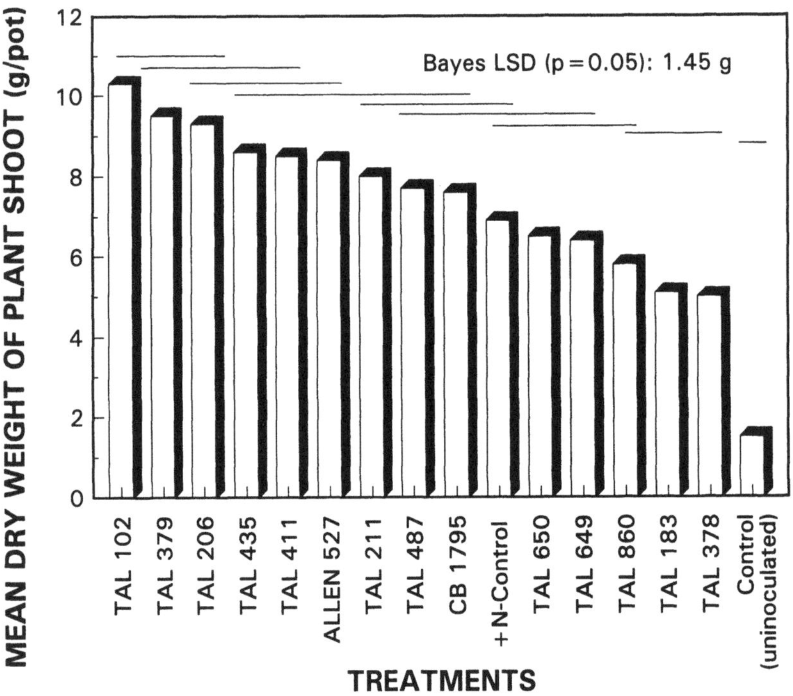

FIGURE A17.1 Effect of various strains of *B. japonicum* on the dry weight of shoots of soybean (*Glycine max* var. Jupiter).

APPENDIX 18

Computing the Coefficient of Correlation (r) to Show the Relationship between Shoot and Nodule Weights in a Rhizobial Strain Selection Experiment

The data in Table A18.1 present the dry weights (g) of the plant tops and nodules from a rhizobial strain selection experiment for cowpea (*Vigna unguiculata*). The experiment was a randomized complete block design with three blocks and 11 treatments (9 inoculated and 2 controls). Each treatment was replicated once within each block. Each treatment plot was a Leonard jar with two cowpea plants. The plants were harvested at 30 days, and the tops and nodules were oven dried at 70°C for 2 days. The plus-N control will be omitted from the correlation analysis.

TABLE A18.1 Dry Weights of Tops and Nodules from a Rhizobial Strain Selection Experiment

		Dry Weight(s)	
Rhizobia	Host of Isolation	Plant Tops	Nodules
TAL 173	*Vigna unguiculata*	2.29	0.21
TAL 651	*Calopogonium mucunoides*	2.08	0.23
TAL 209	*Vigna unguiculata*	1.71	0.29
TAL 309	*Macrotyloma africanum*	1.67	0.26
TAL 1147	*Desmodium intortum*	1.67	0.22
TAL 22	*Phaseolus lunatus*	1.63	0.18
TAL 310	*Dolichos biflorus*	1.30	0.19
TAL 647	*Pueraria phaseoloides*	0.97	0.15
TAL 379	*Glycine max*	0.75	0.13
Control (uninoculated)	—	0.14	0.00
Control (+70 ppm N)	—	4.03	0.00

Construct a new table containing the following data sets of plant tops (x) and nodules (y), as in Table A18.2. The number of pairs (n), excluding the plus-N control, is 10.

CALCULATIONS

Calculate the mean of x = M_x:

$$M_x = \frac{\Sigma x}{n} = \frac{14.21}{10} = 1.421$$

Calculate $M^2_x = (1.421)^2 = 2.019$

Compute the standard deviation (SD) for x:

$$SD_x = \sqrt{\frac{\Sigma x^2}{n} - M_x^2} = = \sqrt{\frac{23.94}{10} - 2.019} = 0.616$$

Similarly, compute the SD_y after determining M_y and M_y^2:

$$M_y = \frac{\Sigma y}{n} = \frac{1.86}{10} = 0.186$$

$$M_y^2 = (0.186)^2 = 0.0345$$

$$SD_y = \sqrt{\frac{\Sigma y^2}{n} - M_y^2} = \sqrt{\frac{0.4054}{10} - 0.0345}$$

$$SD_y = 0.077$$

TABLE A18.2 Data Sets for Use in the Computational Formula for r

x	x^2	y	y^2	xy
2.29	5.244	0.21	0.044	0.4809
2.08	4.326	0.23	0.053	0.4784
1.71	2.924	0.29	0.084	0.4959
1.67	2.788	0.26	0.068	0.4342
1.67	2.788	0.22	0.048	0.3674
1.63	2.657	0.18	0.032	0.2934
1.30	1.690	0.19	0.036	0.2470
0.97	0.941	0.15	0.023	0.1455
0.75	0.563	0.14	0.017	0.0975
0.14	0.019	0.00	0.000	0.0000
$\Sigma x = 14.21$	$\Sigma x^2 = 23.94$	$\Sigma y = 1.86$	$\Sigma y^2 = 0.4054$	$\Sigma xy = 3.0402$

The correlation coefficient:

$$r = \frac{\frac{\Sigma xy}{n} - (M_x \cdot M_y)}{SD_x \cdot SD_y}$$

From the table:

$$\frac{\Sigma xy}{n} = \frac{3.0402}{10} = 0.3040$$

Therefore

$$r = \frac{0.3040 - (1.421 \times 0.186)}{(0.616 \times 0.077)} = \frac{0.3040 - 0.2643}{0.0474}$$

$$r = \frac{0.0397}{0.0474} = 0.8375\ (0.84^{**})$$

**Denotes significance of r at the 1% level.

To test the significance of r at the 5% ($p = 0.05$) and 1% ($p = 0.01$) significance levels, consult a table giving the values of the correlation coefficient. The significant value of r depends on the degrees of freedom (df) as with the F test. Since the data used in the correlation is paired, the df $= n - 2$.

Whenever r is equal to or greater then the appropriate significant value, regardless of whether r is positive or negative, we can conclude that r is significant at the level of probability being used. From the table of correlation coefficients, for df $n = 8$, r is 0.632 at $p = 0.05$ and 0.765 at $p = 0.01$. Since r calculated from the data ($r = 0.84$) is greater than the tabulated value at both levels of significance, we conclude that r is highly significant. From the viewpoint of the top and nodule dry weights, the highly significant value of r indicates that there is a linear relationship between the dry weight of the plant tops and the nodule dry weight under the growth conditions of the experiment.

Once a relationship between the dry weights of tops and nodules has been established, the data can be represented graphically. The best straight line (the regression line) can then be drawn through the points after obtaining the regression equation describing the line. This line is easily transferred to the graph by drawing a line through any pair of points on it, preferably chosen as far apart as possible.

The equation for the regression line for a predicted value of y is given by:

$$y = \left[\frac{r \cdot SD_y}{SD_x}\right] x - \left[\frac{r \cdot SD_y}{SD_x}\right] M_x + M_y$$

Since $r = 0.84$, $SD_y = 0.077$, $SD_x = 0.616$, $M_x = 1.421$, and $M_y = 0.186$

$$y = \left[\frac{0.84 \times 0.077}{0.616}\right] x - \left[\frac{0.84 \times 0.077}{0.616}\right] 1.421 + 0.186$$

$$y = 0.105x + 0.037$$

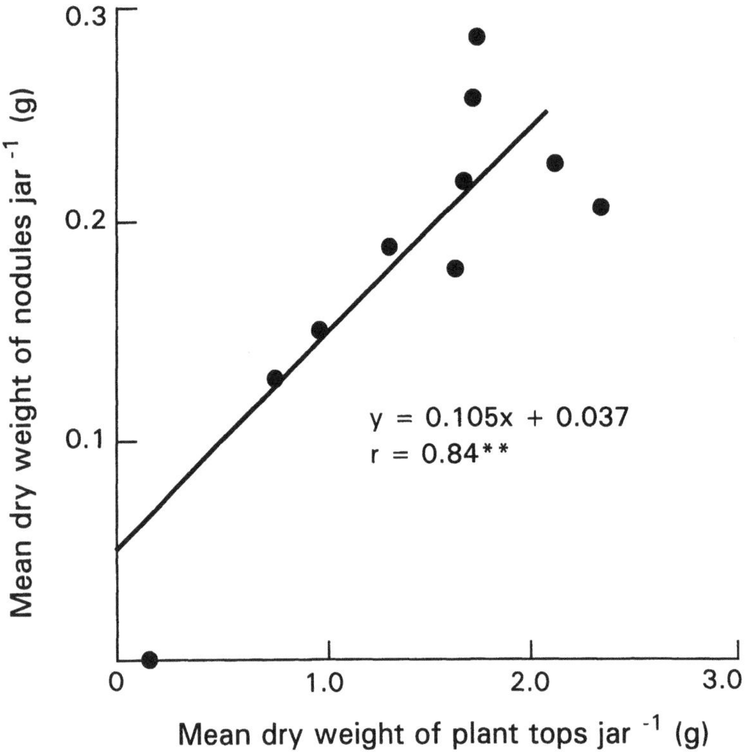

FIGURE A18.1 Relationship between dry weights of nodules and plant tops in cowpea.

By substituting into the regression equation, when $x = 2$, then $y = 0.246$, and when $x = 0.5$, then $y = 0.089$. These two points determine the regression line shown in Figure A18.1.

APPENDIX 19

Replicators and Microtiter Plates[1]

Replicators are used for the simultaneous application or inoculation of bacteria and viruses on different culture media. They are designed to be used in combination with 96-well microtiter plates. Two models allow either full-plate or half-plate inoculation. The inoculation pins are made of corrosion-resistant stainless steel (Figure A19.1).

Microtiter plates are made of rigid polystyrene and may be used for various types of microtiter work. Each of the 96 wells has a working capacity from 125–300 μl, depending on the type of application. Well bottoms are U-shaped or flat. U-bottomed plates are used for replication of cultures in studies such as intrinsic antibiotic resistance, carbohydrate utilization, and immunoblot. The flat-bottomed plates are used for the enzyme-linked immunosorbent assay (ELISA).

For sample replication, use the following procedure:

- Dip the prongs into a shallow dish of alcohol.
- Burn off the alcohol.
- Set the replicator on its side in a transfer chamber.

FIGURE A19.1 Replicators and microtiter plates.

[1]Available through West Coast Scientific, 1287 66th Street, Emeryville, CA 94608.

- Allow to cool.
- Fill the wells of a sterile microtiter plate with the cultures to be replicated.
- Carefully place the replicator over the microtiter plate until the pins are perfectly aligned with the wells.
- Slowly lower the replicator, allowing all the pins to sink into the wells.
- Allow the pins to rest in the bottoms of the wells for one second.
- Carefully raise the replicator out of the well and set briefly on the solid medium that is to be inoculated.
- Repeat this procedure as many times as required by the experiment.

If the pins of the replicator come in contact with a nonsterile object between replications, they must be resterilized by dipping into alcohol and flaming. Resterilizing is also required between replicating different sets of media. The 48-pin replicator may be used for replicating cultures onto solid agar media contained in standard size, round Petri dishes. Use the 96-pin replicator with media contained in oversized or square Petri dishes.

APPENDIX 20

Seed Inoculation Procedures[1]

THE DUSTING METHOD

Frequently, farmers do not use any stickers. They simply mix dry powdered inoculants such as the peat-based type with dry seed and no liquid. This dry dusting is the poorest inoculation method because dry inoculant does not adhere well to seed, and most of it will blow away during planting.

THE SLURRY METHOD

In slurry method of inoculant application, a slurry is prepared from inoculant and water or other sticker solution. Just before planting, premeasured amounts of sticker solution and inoculum are thoroughly mixed to make a smooth pourable suspension. The slurry is then poured into a suitable container with a preweighed amount of seed, and stirred continuously until the seeds are uniformly coated. The container should have a volume twice the volume of the seeds. A cement mixer is recommended for large amounts.

The rate of sticker solution to inoculant is dependent on the type of seed used. Smaller seeds require more sticker solution per seed weight in the slurry than legumes with larger seed because of the larger surface area to be coated. The slurry should be added to the seeds in small amounts because too much sticker solution will cause the seeds to clump together or swell. The amount of sticker solution and inoculant needed to inoculate 1 kg of various kinds of seeds are given in Table A20.1.

THE TWO-STEP METHOD

In the two-step method, the sticker and the powdered inoculant are applied to the seed separately. In the first step, the seeds are uniformly coated with the sticker solution. In the second step, the powdered inoculant is added to the sticky seeds. This method is especially useful when a large number of rhizobia must be applied to the seed. Approximately 10 times as many rhizobia can be bound to the seed in this procedure as compared with the slurry method.

---

[1]Adapted from *Applied BNF Technology—A Practical Guide for Extension Specialists.*

TABLE A20.1 Amounts of Peat Inoculant and Sticker Solution Required for the Slurry Inoculation of Seeds of Various Sizes and Resulting Numbers of Rhizobia per Seed

Legume	Seeds (no./kg)[1]	Inoculant (g kg^{-1} of seed)	Sticker (ml kg^{-1} of seed)	Rhizobia/ Seed[2]
Trifolium repens (white clover)	2000000	5	25	2.5×10^3
Medicago sativa (alfalfa)	500000	5	22	1.0×10^4
Desmodium intortum (intortum)	440000	5	22	1.1×10^4
Stylosanthes hamata (stylo)	400000	5	22	1.2×10^4
Coronilla varia (crown vetch)	250000	5	22	2.0×10^4
Macroptilium atropurpureum (siratro)	67000	5	20	7.5×10^4
Vigna radiata (green gram)	25000	5	20	2.0×10^5
Cajanus cajan (pigeon pea)	17000	5	20	3.0×10^5
Vigna unguiculata (cowpea)	10000	5	15	5.0×10^5
Glycine max (soybean)	5000	5	15	1.0×10^6
Phaseolus vulgaris (common bean)	2400	5	10	2.1×10^6
Cicer arietinum (chickpea)	2000	5	10	2.5×10^6
Arachis hypogaea (peanut)	2000	5	10	2.5×10^6
Vicia faba (broad bean)	1250	5	7	4.2×10^6

Note: Legumes are arranged in ascending order of seed size.

[1]Approximate values.

[2]Based on inoculant of 1×10^9 cells per g.

It is important to use the proper amount of sticker because too much sticker will make the seeds clump together and too little will cause uneven coating with the inoculant. When large amounts of seeds are coated, the sticker may be added in small amounts until the seeds are evenly wet. The amount of inoculant to be added is not as critical and can be adjusted. For example, in the case of soybean, 10–50 g of inoculant per kg of seeds may be applied. The recommendations given in Table A20.2 may be used as guidelines.

An easy coating procedure using a plastic bag is described in Chapter 29 and illustrated in Figure A20.1. The plastic bag method of seed coating can be done in batches

TABLE A20.2 Amount of Sticker Recommended for the Inoculation of Seed of Various Sizes by the Two-step Method

Legume	Seeds (no./kg)[1]	Stickers (ml kg^{-1} of seed): 20% Sugar Solution	M-E Cellulose	Gum Arabic	Rhizobia/ Seed[2]
Trifolium repens (white clover)	2000000	50	38	25	5.0×10^3
Medicago sativa (alfalfa)	500000	44	33	22	2.0×10^4
Desmodium intortum (Intortum)	440000	44	33	22	2.2×10^4
Stylosanthes hamata (stylo)	400000	44	33	22	2.4×10^4
Coronilla varia (crown vetch)	250000	44	33	22	4.0×10^4
Macroptilium atropurpureum (Siratro)	67000	42	32	21	1.5×10^5
Vigna radiata (green gram)	25000	40	30	20	4.0×10^5
Cajanus cajan (pigeon pea)	17000	40	30	20	5.9×10^5
Vigna unguiculata (cowpea)	10000	30	23	15	1.0×10^6
Glycine max (soybean)	5000	20	15	10	2.0×10^6
Phaseolus vulgaris (common bean)	2400	20	15	10	4.2×10^6
Cicer arietinum (chickpea)	2000	20	15	10	5.0×10^6
Arachis hypogaea (peanut)	2000	20	15	10	5.0×10^6
Vicia faba (broad bean)	1250	14	11	7	8.3×10^6

Note: Legumes are arranged in ascending order of seed size.

[1]Approximate values.

[2]Based on inoculant of 1×10^9 cells per g used at the rate of 10 g kg^{-1} seed.

up to 5 kg. However, above 3 kg the inflated bag should be rolled on the ground rather than shaken. Place the preweighed batch of seeds into a plastic bag.

Larger amounts of seed may be inoculated in a tumbler-type mixer such as a cement mixer. When using a cement mixer, check that the seeds are evenly coated with sticker

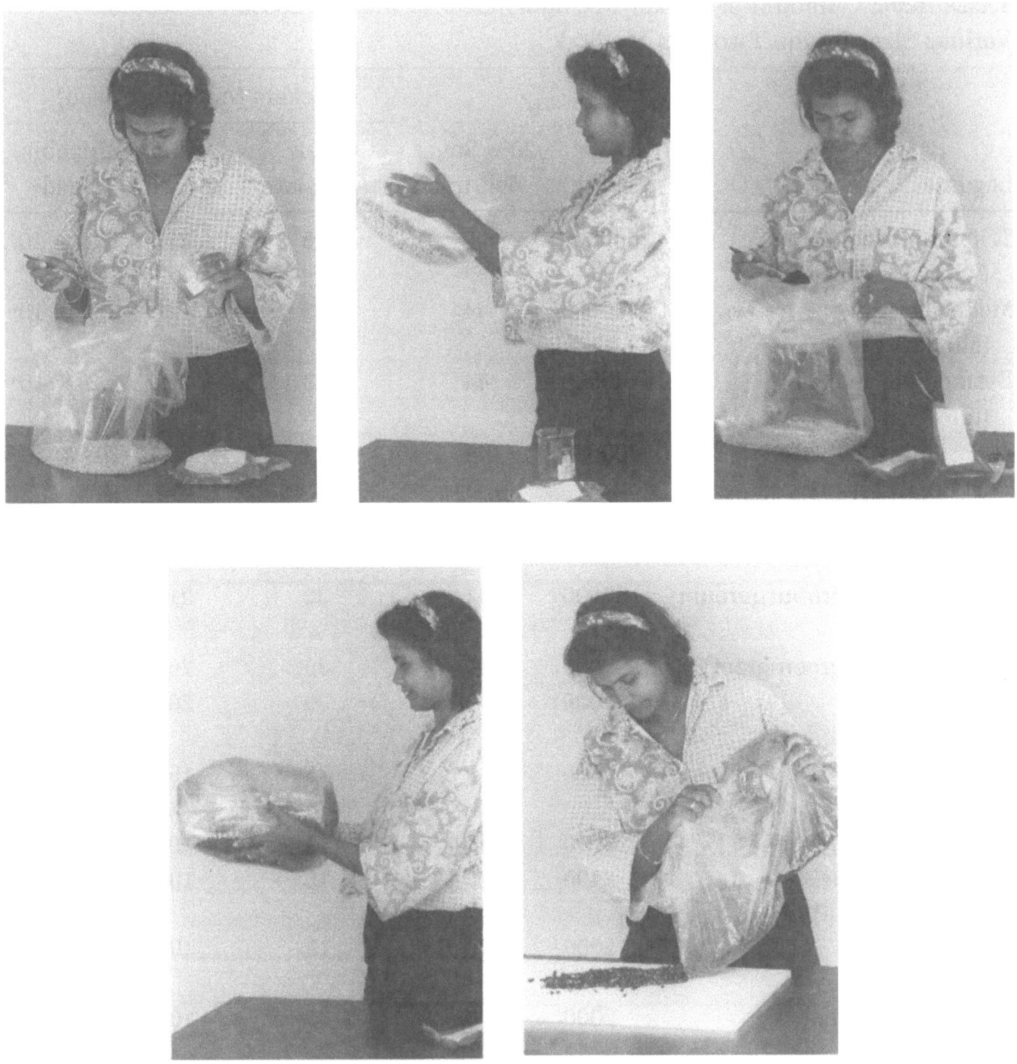

FIGURE A20.1 (a) Add sticker to seeds; (b) coat seeds by shaking in an inflated plastic bag; (c) add inoculant; (d) coat sticky seeds with inoculant by shaking in an inflated plastic bag; and (e) spread inoculated seed for air drying.

before adding the inoculant. If the seeds lump together or the seeds and adhesive form a coat on the wall of the mixer, you should stop the machine, break up the lumps, and scrape the mixer walls. Immediately after coating, spread the seeds on a clean surface and allow them to dry in the shade before sowing.

SEED PELLETING

It is sometimes an advantage to coat inoculated seeds with a protective layer of powdered lime or phosphate. First, the inoculant is applied as a slurry or by the two-step method using a sticker solution as an adhesive. The lime or phosphate should be ground to a very fine powder and sifted through a screen to remove lumps. The lime or phosphate powder is added immediately after inoculation while the seeds are still wet. They are quickly mixed with the pelleting material until they are thoroughly coated. The pelleted seeds will appear dry, but they should be spread on a canvas in a cool, shaded place to allow the pellet to solidify before sowing. Recommendations for inoculant and lime application are shown in Table A20.3.

There may be several reasons to pellet seeds:

- **When adverse weather conditions prevent immediate sowing of inoculated seeds.** Pelleting can prolong the survival of rhizobia on the seed until sowing.
- **When the soil is hot and dry.** Pelleting can help to preserve moisture in drought conditions. When seeds must be sown into dry soil or hot soil conditions, the protective pellet may help to preserve the rhizobia and the seed until there are suitable germination conditions. This protection is especially important when seeds are broadcast.
- **When insects are a problem.** In some areas, pelleting is used to protect the seeds from insects, especially seed-gathering ants.
- **When soils are very acid.** Lime pelleting can be beneficial in highly acid soils, or to protect the rhizobia from acid fertilizers applied to the soil.

TABLE A20.3 Amount of Sticker Solution and Powdered Limestone Required to Pellet Legume Seeds of Various Sizes After Coating with the Slurry Method

Legume	Seeds (no./kg)[1]	Sticker (ml kg^{-1} of seed)[2]	Limestone (g kg^{-1} of seed)	Rhizobia/Seed[3]
Trifolium repens (white clover)	2000000	5	25	5.0×10^3
Medicago sativa (alfalfa)	500000	5	22	2.0×10^4
Desmodium intortum (intortum)	440000	5	22	2.2×10^4
Stylosanthes hamata (stylo)	400000	5	22	2.2×10^4
Coronilla varia (crown vetch)	250000	5	22	4.0×10^4
Macroptilium atropurpureum (siratro)	67000	5	20	1.5×10^5
Vigna radiata (green gram)	25000	5	20	4.0×10^5
Cajanus cajan (pigeon pea)	17000	5	20	5.9×10^5
Vigna unguiculata (cowpea)	10000	5	15	1.0×10^6
Glycine max (soybean)	5000	5	15	2.0×10^6
Phaseolus vulgaris (common bean)	2400	5	10	4.2×10^6
Cicer arietinum (chickpea)	2000	5	10	5.0×10^6
Arachis hypogaea (peanut)	2000	5	10	5.0×10^6
Vicia faba (broad bean)	1250	5	7	8.2×10^6

[1]Approximate values.
[2]Sticker solution should be sugar (10%), methyl ethyl cellulose (4%), or gum arabic (40%).
[3]Inoculant of 1×10^9 rhizobia per g of inoculant is used at the rate of 10 g kg^{-1} of seed.

Appendix 21

Determining Field Capacity of Field Soil

Water content of soil at field capacity may be defined as the amount of water held in the soil profile after it has drained for 24 h. At this time, all the macropore spaces will be empty. The field capacity of soil is calculated so that the soil moisture of potted soil can be maintained at conditions optimum for plant growth. Differences in the water status of the potted soil increase experimental errors. Precise measurements of soil moisture tension at field capacity (usually 0.1×10^5 to 0.3×10^5 Pa tension) requires special equipment. An approximation of soil moisture at field capacity may be made by determining the moisture content of soil that has been wetted and drained in a column. This may be done as follows.

Fill a 1-liter (or larger) transparent cylinder with a sample of the air-dried (moist) soil used in the pots. The cylinder should be transparent to allow observations, and should have a sufficiently large diameter so soil in the middle of the column can later be removed. A hole in the bottom of the cylinder will allow air to escape when water is added to the soil column, but the hole is not necessary. Tamp soil in the cylinder to a similar consistency as that in pots. Cover the surface of the soil with a filter paper or paper towel.

Pour water into the cylinder in small increments until the migrating wet front indicates that one-half of the soil in the cylinder is wet. The migrating wet front can be observed through the wall of the cylinder. Allow the column to equilibrate in the laboratory for 24 h. The wet front will continue to move down in the soil column (Figure A21.1).

Record the weight of a weighing dish (tare weight). On the cylinder, mark off the middle 5 cm of the wet column (Figure A21.1). With a spatula or spoon, remove and discard soil above the top mark. Remove a 20–30-g soil sample from the marked area, and place it in the weighing dish. Immediately weigh and record the weight of the soil and the dish (WW), then oven dry the soil at 110°C, to constant weight. Weigh and

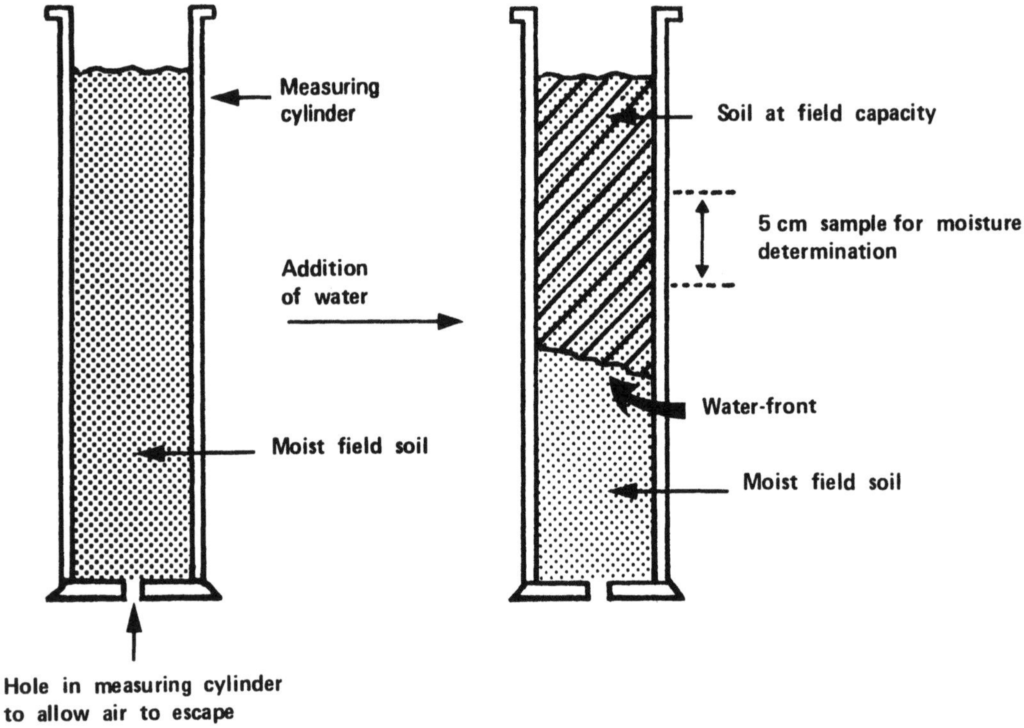

FIGURE A21.1 Determining field capacity of field soil.

record the weight of the dry soil and weighing dish (DW). The moisture fraction of the soil is calculated on the dry-weight basis as follows:

$$MF = \frac{WW - DW}{DW - TW}$$

where Tare weight (TW) = Weight of weighing dish
Wet weight (WW) = Weight of soil plus tare weight
Dry weight (DW) = Weight of oven-dried soil plus tare weight
MF = Moisture fraction

The oven-dry weight equivalent of the potted soil is also calculated because fertilizer amendments are based on this value. To determine the oven-dry weight equivalent, first determine the moisture fraction of the air-dried potting soil. Obtain a 20–30-g sample of air-dried soil from the pots and dry at 110°C to constant weight. Use the formula listed previously to calculate the moisture fraction of the air-dried soil.

The dry-weight equivalent is calculated as follows:

$$\text{DWE} = \frac{\text{AW of pot}}{(1 + \text{MF of AW})}$$

where DWE = Oven-dry weight equivalent
AW = Weight of air-dried soil

To supply potted plants with sufficient water, we must know how much water to add to bring the potted soil to field capacity. The example given here is based on the following assumptions:

1. The moisture fraction of the potted soil at field capacity is 0.35.
2. The moisture fraction of the air-dried soil is 0.15.
3. The weight of the air-dried soil in the pot is 2.4 kg.

The oven-dry weight equivalent (DWE) of the potted soil is:

$$\text{DWE} = \frac{\text{AW of pot}}{1 + \text{MF}} = \frac{2400}{(1 + 0.15)} = 2087$$

If the oven-dry weight equivalent is 2087 g, the weight of the water in the soil at field capacity is:

$$2087 \times 0.35 = 730 \text{ g}$$

and, the weight of potted soil at field capacity is:

$$2087 + 730 \text{ g} = 2817 \text{ g}$$

Since the potted soil already has a moisture fraction of 0.15, then 417 g of water per pot must be added to bring the potted soil to field capacity (2817 g − 2400 g = 417 g of water).

APPENDIX 22

The Simple Transfer Chamber

A simple transfer chamber for aseptic work may be constructed from the materials listed in this appendix, and according to the plans in Figures A22.1, A22.2, and A22.3. In this transfer chamber design, specific attention is given to the placement and position of the Bunsen burner because this is critical to producing a sterile environment suitable for aseptic work.

The Bunsen burner is inserted into the base of the chamber through a hole, allowing approximately 1 in. (2.5 cm) of the tip of the burner to protrude into the chamber. In this position, the gas supply line and the air-intake ports of the burner are left on the outside of the chamber. When the burner is lit, the flame eventually warms the air inside the chamber, resulting in a unidirectional warm-air current. This warm-air current exits through the open front, preventing entry of contaminants.

When using this chamber, the following instructions should be followed:

1. Open the hinged door and wipe the interior thoroughly with an antiseptic such as 70% ethanol. Allow the ethanol to dry.

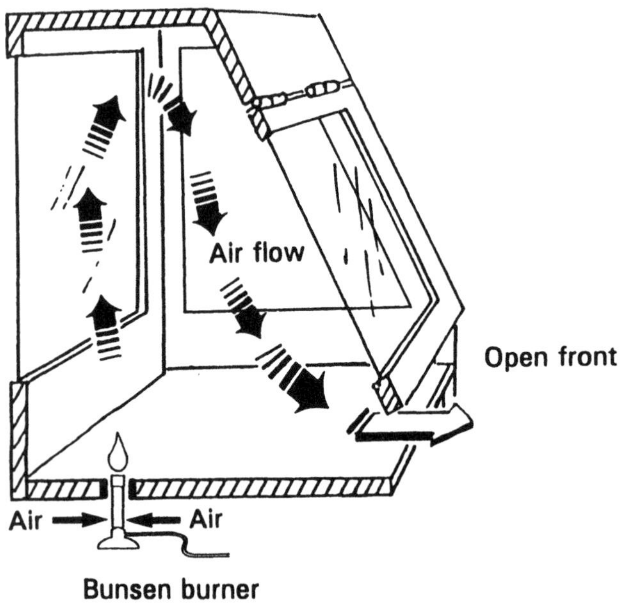

FIGURE A22.1 Cross-section of a chamber illustrating the working principle.

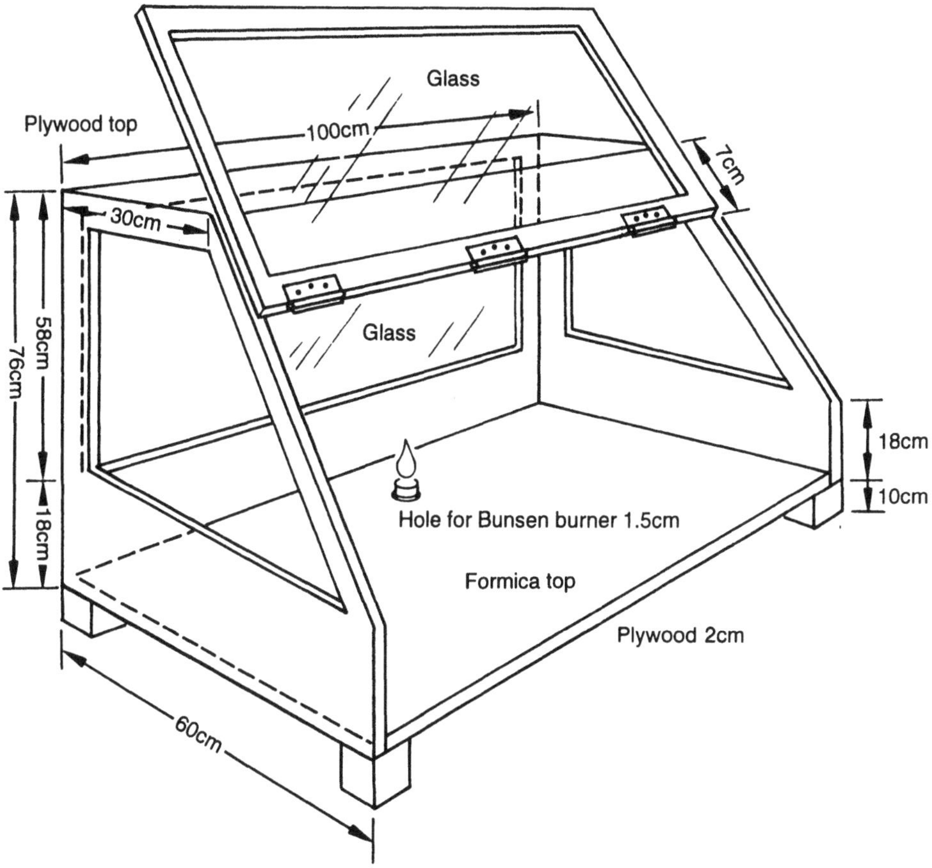

FIGURE A22.2 Simple transfer chamber.

2. Turn on the gas and light the burner. The flame should be blue and adjusted to no more than 6 cm in height.
3. Close the hinged door and wait 20 min. before using the chamber.

When you are through working in the chamber, turn off the flame and **disconnect the gasline** on the outside of the chamber. This is an important safeguard to prevent the possibility of gas leaking into and filling the chamber, which could result in an explosion the next time the burner is lit. Such an explosion is not only theoretically possible, but has happened when proper precautions were not observed. With correct practice and precautions, this transfer chamber can produce good results.

The components of the simple transfer chamber are the following:

1. Back: Made of plywood, hardwood, and glass (0.2–0.5-cm thickness).
2. Bottom: Made of plywood with Formica surface, includes 1.5–2-cm diameter hole for the Bunsen burner.

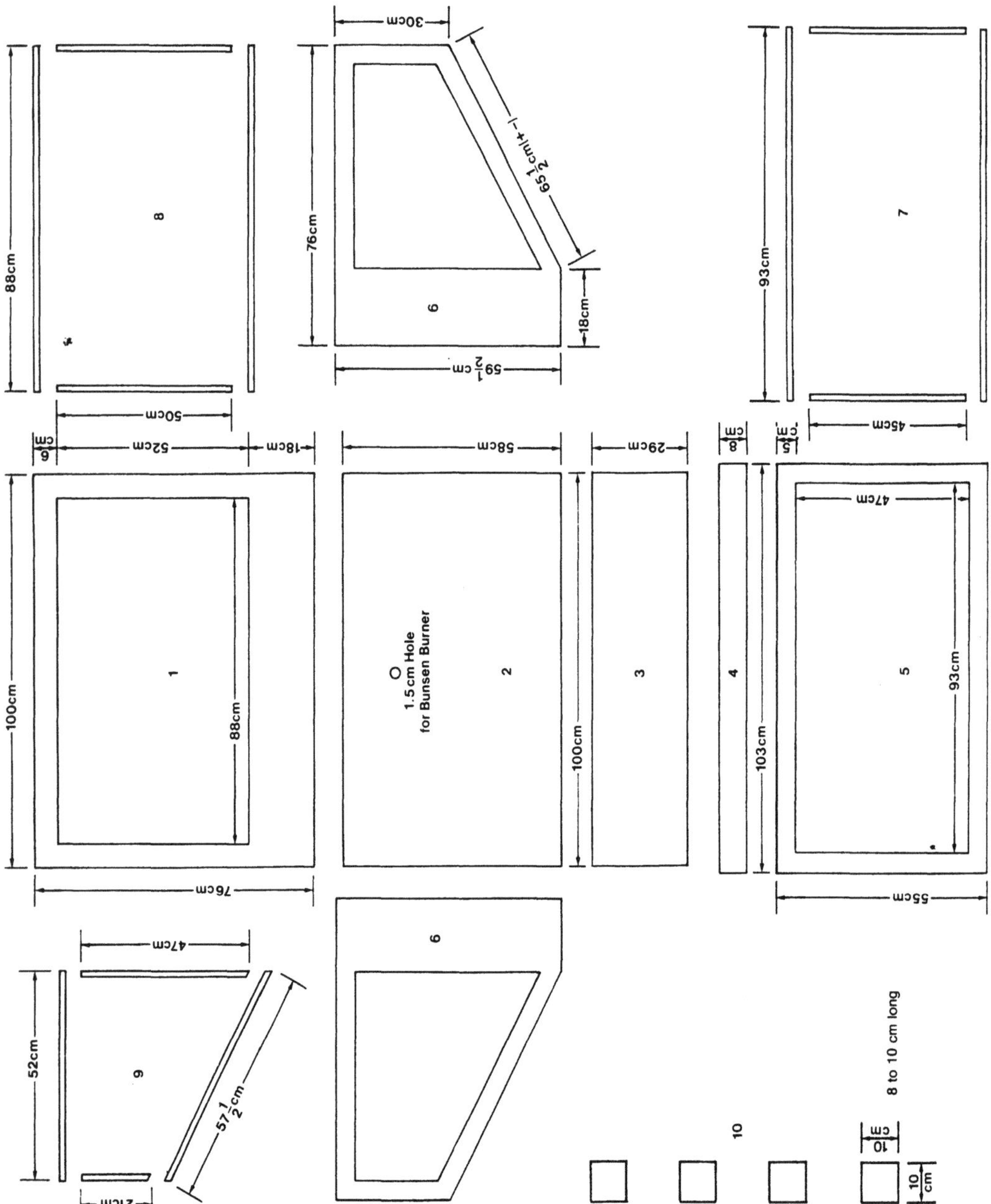

FIGURE A22.3 Components of a simple transfer chamber.

3. Top: Made of plywood.
4. Reinforcement: Made of hardwood or plywood; serves as an anchor for the door.
5. Door: Made of plate glass with hardwood frame; is attached to the reinforcement plate via hinges.
6. Two sides: Made of plywood and glass.
7. Eight wooden moldings: To hold glass for window and door.
8. Eight wooden moldings: To hold glass for the back window.
9. Sixteen wooden moldings: To hold glass for the side windows.
10. Four wooden legs: 10 cm high.

The plywood used should be 2-cm thick with a smooth finish on both sides. The chamber should be painted with an oil-based epoxy paint, leaving a hard, smooth coat.

APPENDIX 23

Freeze Drying Cultures of Rhizobia

Freeze drying or lyophilizing is a method of stabilizing materials of biological origin. This is one of the preferred methods for the long-term storage of microorganism cultures. Cultures of rhizobia remain viable for many years when freeze dried and vacuum sealed in glass ampoules.

SUMMARY OF THE PROCESS

Freeze drying allows moisture to be removed from the material without concurrent biological changes. This is done by removing moisture under a vacuum. To prevent frothing as air is withdrawn when the initial vacuum is applied, the culture is either prefrozen or subjected to centrifugation. In the former case, the ice will change directly from the solid to the vapor stage. In the latter, the temperature of the suspension falls as the water vapor is removed until it freezes and further drying occurs by sublimation. During freeze drying, the ice does not evaporate simultaneously from all parts of the material, but continuously from the outer boundary until only a dry cake is left, resembling the original sample in size and shape.

Freeze drying equipment may come with a variety of accessories. The essential components of freeze drying apparatus are: a vacuum chamber to hold the material to be freeze dried or a manifold to which ampoules or a vacuum vessel can be attached, a water trap, and a vacuum pump. A vacuum gauge is usually connected to the system between the water trap and the vacuum pump. The water vapor, which evolves during freeze drying, is captured in the water trap, and thus is prevented from entering the pump. Water traps may be chambers to which drying agents have been added, such as phosphorus pentoxide, or they may be refrigerated condensers with compressors capable of cooling temperatures below −40°C.

Evaporation may be hastened by heating the materials to be freeze dried. The action of the vacuum will keep the material frozen as long as it contains water. The rate and efficiency of the flow of water vapor from the material to the condenser chamber is directly related to the vapor pressure differential; that is, the vapor pressure of the frozen material minus the vapor pressure of the condenser chamber. Since vapor pressure and material temperature are inversely related, it is desirable to have a condenser temperature of about −40°–−50°C, and a material temperature as high as possible without causing a meltback of the material. For cultures in ampoules, room temperature is usually sufficient.

Freeze drying is carried out in two stages. During the primary stage, 90–95% of the moisture is removed. After the secondary stage, approximately 1% of the moisture remains. The retention of a small amount of moisture is essential for the survival of bacteria. This is achieved by suspending the cells in a medium that will not permit complete moisture removal. At NifTAL (Paia, HI), a mixture of peptone (5%) and sucrose (10%) is used. Since a high fatality rate occurs even under these conditions, highly concentrated cell suspensions are used.

It is often convenient to use one machine for the first stage and another machine for the second stage of freeze drying because the setup for each stage is different. Ampoules are constricted with an ampoule constrictor after the first stage of drying. This permits easier sealing under vacuum after the second stage has been completed. The ampoules are tested with a high-frequency tester to ensure successful sealing, then stored in the dark in a metal drawer cabinet at room temperature.

PRACTICE OF FREEZE DRYING

The practice of freeze drying may vary from laboratory to laboratory. The following methods described are performed at NifTAL.

Preparing Cotton Plugs

Cotton plugs are used to plug ampoules. No. 0 dental cotton balls (Richmond Dental Cotton Co., Charlotte, NC) may be used for this purpose. Prior to use they are placed into 100-ml beakers, covered with aluminum foil, and sterilized by autoclaving. This is followed by a 1-h drying period at 80°C in a dry-air oven.

Preparing Labels

Paper and ink must be compatible (nontoxic) with the rhizobia and resistant to moisture. Whatman no. 1 filter paper, purchased in large sheets, and ordinary typewriter ink of vegetable base are suitable. A computer equipped with a printer is used for typing the labels with the strain identification number and date. Only one identification number is printed at one time. The labels are cut manually to measure 4 × 25 mm. A margin of 10 mm is left on one side. This empty margin will later be touching the bottom of the ampoule, thus preventing the written part from being submerged and rendered unreadable when the cell suspension has been added.

Preparing Ampoules

Freeze drying ampoules of 0.5-ml capacity, inner diameter of 6 mm, and 100-mm length, are purchased from Edward's High Vacuum (W. Sessex, UK). They are checked for defects, such as cracks and pinholes, then soaked in 10% HCl overnight. They are then rinsed in tap water at least six times or until the pH of the last washing is neutral, indicating complete removal of the acid. This is followed by three rinses with deionized water and oven drying. Labels are added to the ampoules with forceps. The ampoules are then placed into a 250-mm beaker, covered with aluminum foil, and autoclaved. The now sterile ampoules are oven dried at 80°C for 1–2 h.

Preparing Freeze Drying Medium

A solution is made in distilled water containing 5% peptone and 10% sucrose. The peptone/sucrose solution is dispensed in 2-ml portions into snap-top culture tubes and sterilized by autoclaving. Including 10% sucrose or another sugar in the freeze drying medium will automatically cause it to retain 1% moisture after dehydration. This will improve the viability of the suspended organism. Total desiccation would result in death of all bacteria.

Growing and Harvesting Cultures for Freeze Drying

Only authenticated cultures should be selected for freeze drying. They should be tested again for purity by streaking them out on yeast-mannitol agar (YMA) plates containing congo red (CR) and plates containing bromthymol blue (BTB), as well as by Gram stain. If antisera are available, they should be used as an additional check for strain identity and culture purity. After these tests, the cultures are grown on YMA slants in 50-ml culture tubes at 25–30°C. They should be harvested a few days after their log phase of growth. All work should be done under strict aseptic conditions in a transfer chamber.

Two milliliters of the previously prepared peptone/glucose medium are added to each slant culture. The growth is gently dislodged with an inoculation loop and then transferred into a 10-ml vial. In the case of large batches, the growth from several slants is pooled in a 50-ml culture tube. The suspension is emulsified on a vortex mixer and immediately transferred to the freeze drying ampoules. The cell suspension should contain approximately between 5×10^9 to 1×10^{10} cells per ml. Usually 6–8 ml are sufficient for 30–40 ampoules.

Filling the Ampoules

For this operation, stringent aseptic conditions cannot be over-emphasized. The work should be carried out on a laminar flow chamber that has been cleaned with an antiseptic such as 70% ethanol, and, if possible, irradiated with UV light for 20 min before use. As an additional precaution, we recommend wearing a disposable face mask and sterile surgical gloves.

To avoid a mix-up and/or cross-contamination, only one strain should be handled at a time. Sterile, cotton-plugged Pasteur pipettes with long, fine capillaries and equipped with a 1-ml capacity rubber suction bulb are used to transfer the cell suspensions to the ampoules. Eight drops of suspension, delivered by a Pasteur pipette with a 16-gauge tip will equal a volume of approximately 0.2 ml of material. If each ampoule receives 0.2–1.0 ml, the actual number of cells per ampoule are: $0.2 \times 5 \times 10^9 = 1 \times 10^9$ cells. This is a sufficiently large number for survival.

Loading the ampoules requires a steady hand and practice. Avoid contaminating the upper portion of the ampoule with the cell suspension because this will cause charring during the constriction process. If large batches of ampoules are to be filled, a 1-ml capacity repetitive Cornwall syringe (Baxter Diagnostic, Inc., Scientific Product Division, McGraw Park, IL) is recommended.

After filling, use a sterile glass rod to push a sterile cotton plug into the center of each ampoule. A second sterile cotton plug is used to close the opening. The ampoules are then loaded into a paper towel lined VirTis vacuum jar (available through Baxter Diagnostics, Inc., Scientific Products Division, McGraw Park, IL). The jar holds approximately 50 ampoules. Ideally, freeze drying should be carried out at this stage without delay. We frequently store filled and plugged ampoules contained in a vacuum jar in a freezer overnight without ill effect to the survival of the cultures.

Primary Freeze Drying

We use a LABCONCO no. 12 freeze dryer (Lab Con Co Corp., Kansas City, MO) for the first stage of lyophilization. It is equipped with a large 48-port manifold, a freeze bath, a condenser chamber, and a heavy duty vacuum pump. The machine has two compressors, one for the freeze bath and the other for the condenser. A McLeod manometer is used to monitor the vacuum.

On the night before use, the freeze bath is filled to approximately the 10-cm level with methanol, and its condenser is activated. The bath will reach a temperature of −40°C on the following morning. Vacuum jars containing ampoules may then be placed in the freeze bath. The condenser chamber is closed, and its compressor turned on. The condenser temperature usually drops to −40°C in 20 min. The vacuum pump may then be activated. Fifteen minutes later, the vacuum gauge should indicate a reading below 0.1 torr. The vacuum jars containing the frozen ampoules may then be removed from the freeze bath and attached to the manifold. This should be done quickly to prevent thawing of the ampoules and a subsequent bubbling over of the suspensions. Sufficient time should be allowed for the vacuum to reestablish itself between the attaching of each jar. The paper towel liner in the jar will help to prevent a thawing of the material. As an additional precaution, the jars may be further insulated by wrapping them in paper bags for an initial 30 min, or until the evaporating water is cooling the suspensions in the ampoules effectively. Freeze drying is continued for approximately 6 h. The primary drying is completed when the pressure gauge shows a reading of 1.3×10^{-1} mbar or below.

Constricting the Ampoules

Prior to secondary freeze drying, the ampoules are constricted at approximately 6 cm, as measured from the bottom. The constriction should be done in equal distance from each of the two cotton plugs to avoid charring, which may have a toxic effect on the culture. Constrictions may be done manually over a finely adjusted propane plus oxygen flame. This is a learned skill that requires practice. The ampoule is rotated slightly below the tip of the blue flame so the flame passes over the horizontally held tube but not below it. The rotating is continued until the walls of the heated area have constricted and thickened, and the inner diameter is not more than 2 mm. At this point, the ampoule is removed from the flame and pulled out until the inner diameter measures a little less than 1 mm.

At NifTAL, most ampoules are constricted on an Edward's Ampoule constrictor (Edward's High Vacuum, Manor Royal, Crawley, West Sussex, UK). This machine performs beautifully on a propane plus air flame, provided both the retaining wheels are slightly adjusted from paralleled to toed-in position during the process, and the flame is properly adjusted. Constricting one ampoule requires approximately 1 min.

Secondary Freeze Drying

An Edward's Modulyo freeze dryer is used at NifTAL for the second stage of freeze drying. This unit is equipped with a double manifold, which can hold 96 ampoules; a condenser chamber; and a two-stage vacuum pump. Pressure is measured by a built-in Pirani gage.

The condenser is switched on until a temperature of $-50°C$ has been reached. Then, the vacuum pump is activated and freeze drying is continued for 12–18 h to reduce the moisture level in the ampoules to 1%. At the completion of freeze drying, the reading on the Pirani gage should show a pressure of 2×10^{-2} mbar or less. The ampoules are then sealed with a twin-jet torch (Figure A23.1). This is done by heating both sides of the constriction simultaneously (Figure A23.2), and pulling gently at the bottom of the ampoule with a slight twist until the constricted area has sealed and is disconnected from its upper end, which remains on the freeze dryer. The freeze dryer may then be switched off and air permitted to flow slowly into the chamber. The drain should be opened to remove the condensed water.

The ampoules are checked for the presence of leaks before storage. This is done with an Edward's T2 HF ampoule tester, which is a high-frequency probe. At discharge, a properly sealed ampoule will display a blue flame. Ampoules without vacuum seals will show no color. The spark tester should be used only briefly on each ampoule because each discharge may kill a certain number of bacteria.

Storing the Freeze-Dried Cultures

Ideally, lyophilized cultures of rhizobia should be stored at 4°C and in the dark. Optimal storage conditions are not always available, and storage at room temperature and away from light is an accepted alternative. At NifTAL, cultures are stored within a steel cabinet in an air-conditioned room held at 20°C.

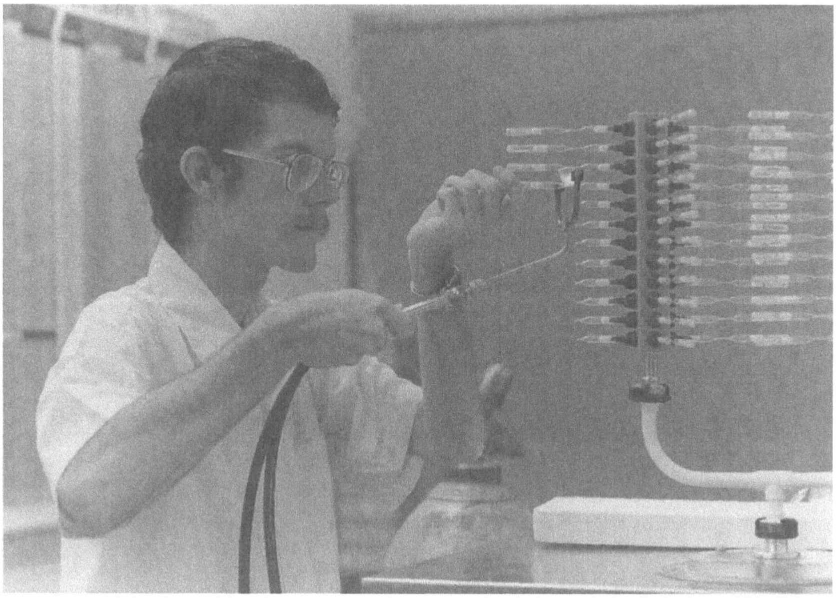

FIGURE A23.1 Sealing ampoules.

Opening Ampoules

Ampoules containing freeze-dried bacteria culture should be opened in an aseptic environment. A mark is filed on the ampoule at about the middle of the cotton wool plug, and a red-hot glass rod is applied to the mark. The ampoule should then crack at the marked area. Care should be taken in opening the ampoule slowly so that the onrushing air will filter through the cotton plug without drawing it into the ampoule. Often, the heated glass rod will not cause the desired crack at the mark. In such a case, two layers of sterile tissue paper are wrapped around the ampoule and minimal pressure is applied to break open the ampoule at the file mark. This method is especially recommended for ampoules that do not contain cotton plugs.

The cotton plug is removed with forceps and discarded because culture may be adhering to it. It should be replaced with a new sterile cotton wool plug. The contents of the ampoule is rehydrated with 0.5 ml of sterile water. Since the number of surviving cells may be low, attempts are made for maximum recovery. A loopful is streaked out on a YMA plate containing CR and on another plate containing BTB. The label, which may contain a large number of cells, is transferred to another YMA plate. The remainder of the culture is then removed with a sterile Pasteur pipette and added to 50 ml of yeast-mannitol broth (YMB) contained in a 125-ml Erlenmeyer flask. Broth and plate cultures are then incubated at their optimal temperatures for growth.

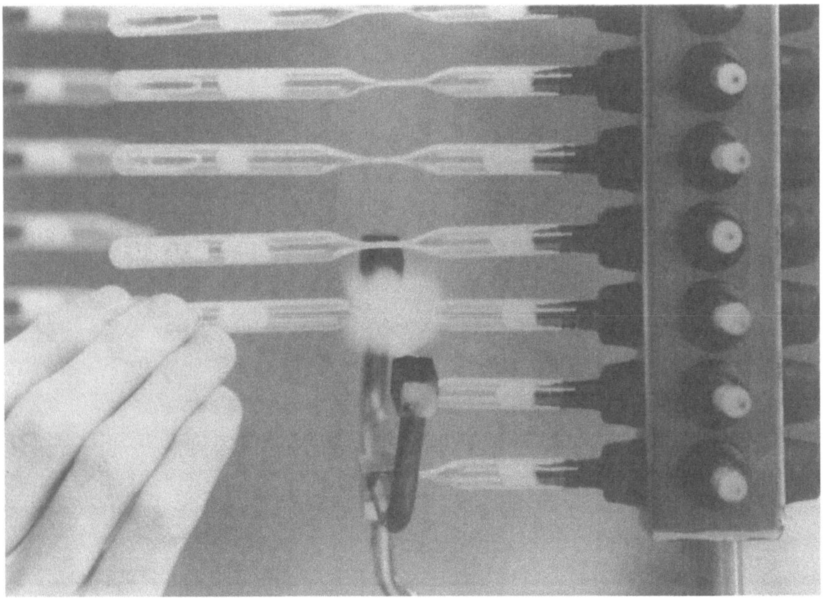

FIGURE A23.2 Sealing ampoules (a close-up).

Appendix 24

Source of Rhizobia

The NifTAL Rhizobia Germplasm Resource is a comprehensive collection of rhizobia for numerous legumes (tropical and temperate) and is maintained at the NifTAL Center (Paia, HI). All strains cited in the various exercises of this book are available on written request addressed to: Curator, Rhizobia Germplasm Resource, NifTAL Center and MIRCEN, University of Hawaii, 1000 Holomua Road, Paia, HI 96779.

The INLIT strains of rhizobia are also available. INLIT is an acronym for NifTAL's International Network of Legume Inoculation Trials in which response to inoculation with rhizobia on 18 species of economically important legumes were tested worldwide. A set of three effective and antigenically distinct strains of rhizobia tested in the INLIT are listed in Table A24.1. Because each strain in the group of three rhizobia recommended for each legume is antigenically distinct, serological methods of strain identification can be used to study competition, persistence, and other ecological aspects.

There are also other laboratories/institutions that maintain collections of rhizobia:

Dr. Carlos Batthyany
Nitrosoil, Florida 622, 4 Piso
Buenos Aires,
ARGENTINA
Rhizobia for Tropical Legumes

Dr. R.J. Roughley
Australian Inoculants Research and Control Service
Horticultural Research Station
P.O. Box 720
Gosford, N.S.W. 2250
AUSTRALIA
AIRCS Strains

Dr. R.A. Date
CSIRO, Div. Tropical Crops and Pastures
Mill Road, St. Lucia
Queensland 4067
AUSTRALIA
Rhizobia for Tropical Legumes

Dr. F. Bergersen
Microbiology Section
CSIRO, Div. of Plant Industry
Canberra, ACT 2600
AUSTRALIA
Rhizobia for Clovers, Medics, and Other Temperate Species

Prof. J.R. Jardim Freire
Rhizobium MIRCEN
IPAGRO
Caixa Postal 776
90000 Porto Alegre Do Sul
BRAZIL
Rhizobia for Tropical Legumes

Dr. D.J. Hume
Crop Science Dept.
University of Guelph
Guelph, Ontario N1G 2W1
CANADA

Rhizobia for Pea, Lupin, Alfalfa, and Soybean

Dr. John Day
Soil Microbiology Dept.
Rothamsted Experimental Station
Harpenden, Herts. AL5 2JQ
UNITED KINGDOM

Rhizobia for Clovers, Alfalfa, Peas, Beans, and Other Temperate Legumes

Plant Diseases Division
D.S.I.R.
Private Bag
Auckland
NEW ZEALAND

Rhizobia for Clovers, Alfalfa, and Lupin

Dr. Peter van Berkum
USDA CCNFL
Bldg. 001, Rm. 309, BARC-W
Beltsville, MD 20705
USA

Rhizobia for Soybean and Temperate Legumes

Dr. O.P. Rupela
Senior Microbiologist
Legumes Program
ICRISAT
Pantancheru, A.P. 502 324
INDIA

Rhizobia for Chickpea, Pigeon Pea, and Peanut

TABLE A24.1 Legumes and Recommended Strains of Rhizobia

Legumes	Rhizobia[1]	TAL No.	Other Designation(s)
Arachis hypogaea	B	1000	none
		169	Nit 176A22 (Nitragin)
		1371	T-1, Nit 8A11 (Nitragin)
Cajanus cajan	B	1127	IHP 38
		1132	IHP 195
		569	MAR 472
Centrosema pubescens	B	651	UMKL 44
		655	UMKL 09
		1146	CIAT 590
Cicer arietinum	R	620	IHP 3889, CB1189
		480	UASB 67
		1148	Nit 27A3 (Nitragin)
Desmodium intortum	B	569	MAR 472
		1147	CIAT 299
		667	CIAT 13, MAR 471
Glycine max	B	102	USDA 110
		377	USDA 138
		379	CB 1809, USDA 136b

Table A24.1 (continued)

Legumes	Rhizobia[1]	TAL No.	Other Designation(s)
Lens culinaris	R	634	Nit 92A3 (Nitragin)
		638	I-2
		640	I-11
Leucaena leucocephala	R	82	none
		1145	CIAT 1967
		582	CB 81
Medicago sativa	R	380	SU 47
		1372	POA 116
		1373	POA 135
Phaseolus lunatus	B	22	none
		169	Nit 176A22 (Nitragin)
		644	CIAT 257
Phaseolus vulgaris	R	182	none
		1797	CIAT 899
		1383	CIAT 632
Pisum sativum	R	634	Nit 92A3 (Nitragin)
		1236	ALLEN 344
		1402	Nit 128C75 (Nitragin)
Psophocarpus tetragonolobus	B	228	none
		1021	Nit 132B13 (Nitragin)
		1022	Nit 132B14 (Nitragin)
Stylosanthes guianenis	B	309	CB 756
		310	CB 1024
		658	CIAT 71
Vicia faba	R	1397	Nit 175F9 (Nitragin)
		1399	Nit 175F12 (Nitragin)
		1400	Nit 175F16 (Nitragin)
Vigna mungo	B	441	UPLB M6
		420	THA 301
		169	Nit1 176A22 (Nitragin)
Vigna radiata	B	441	UPLB M6
		420	THA 301
		169	Nit 176A22 (Nitragin)
Vigna unguiculata	B	209	none
		173	Nit 176A30 (Nitragin)
		658	CIAT 71

[1]B = *Bradyrhizobium*; R = *Rhizobium*. Each group consists of three antigenically distinct strains of rhizobia.

Appendix 25

Absorption of Antisera

Cross-reacting antisera can be made more strain specific by absorbing common antibodies, leaving those specific ones against which the antiserum was prepared. In principle, this is an agglutination procedure in which a heavy suspension of the absorbing is incubated with the antiserum. The mixture is centrifuged after a reaction time of several hours. The supernatant is then tested for specificity for absence of positive reactions with the absorbing strain. It is also tested for ability to react with the strain against which the antiserum was developed. The absorption procedure is repeated until no reaction with the absorbing strain can be observed.

For example, consider a situation in which antiserum A cross-reacted with the antigen of strain B. To obtain strain-specific antiserum A by absorption, proceed as follows: Wash and heat treat strain B as described in Chapter 8 and make a heavy antigen suspension of approximately 5×10^9 cells ml^{-1}. Pipette 2 ml of this suspension into a test tube containing 2 ml of the undiluted antiserum of strain A. Incubate this mixture in a water bath at 37°C for 2 h, then allow the reaction to continue at 4°C overnight. Centrifuge at $5000 \times g$ and test the supernatant against its homologous antigen (strain A) and against the heterologous antigen (strain B) by agglutination after each absorption step. If a cross-reaction is not observed, then the Antiserum A is absorbed and has been made strain (antigen) A specific. The strain-specific antiserum A may now be tested by the other serological method for which the antiserum is intended [e.g., immunodiffusion, fluorescent antibody (FA) technique, or enzyme-linked immunosorbent assay (ELISA)]. Usually, a minimum of two absorption steps is required for antisera intended for the agglutination and the FA techniques. If the antiserum is to be used for a more sensitive technique such as ELISA and Immunoblot, more absorption steps are required to render the antiserum free of interfering nonspecific antibodies.

Supplemental Reading List

ARTICLES

Bohlool, B.B., and E.L. Schmidt. 1974. Lectins: A possible basis for specificity in the *Rhizobium* legume root nodule symbiosis. Science (Washington, DC) 185:269–271.

Brockwell, J. 1981. A strategy for legume nodulation research in developing regions of the old world. Plant Soil 58:367–382.

Brockwell, J., and R.J. Roughley. 1967. An examination of the numbers of nodule bacteria associated with legume seed following commercial multiple inoculation. J. Aust. Inst. Agric. Sci. 33:204–207.

Broughton, W.J. 1978. Control of specificity in Legume-*Rhizobium* associations. J. Appl. Bacteriol. 45:165–194.

DeLey, J., and A. Rassel. 1965. DNA base composition flagellation and taxonomy of the genus *Rhizobium*. J. Gen. Microbiol. 41:85–91.

Gorbet, D.W., and J.C. Burton. 1979. A non-nodulating peanut. Crop Sci. 19:727–728.

Ismande, J. 1981. Exchange of metabolites and energy between legume and *Rhizobium*. pp. 179–188. *In* K.L. Giles and A.G. Atherly (eds.) International Review of Cytology, Suppl. 13, Biology of the Rhizobiaceae. Academic Press, New York.

Johnston, A.W.B., and J.E. Beringer. 1975. Identification of the *Rhizobium* strains in pea root nodules using genetic markers. J. Gen. Microbiol. 87:343–350.

Kandorosi, A., and A.W.B. Johnston. 1981. The Genetics of *Rhizobium*. pp. 191–219. *In* K.L. Giles and A.G. Atherly (eds.) International Review of Cytology, Suppl. 13, Biology of the Rhizobiaceae. Academic Press, New York.

Keyser, H.H., D.N. Munns, and J.S. Hohenberg. 1979. Acid tolerance of rhizobia in culture and in symbiosis with cowpea. Soil Sci. Soc. Am. J. 43:719–722.

Kliewar, M., R. Lowe, P.A. Mayeux, and H.J. Evans. 1964. A biological assay for cobalt using *Rhizobium meliloti*. Plant Soil 21:153–162.

Kurz, W.G.W., and T.A. La Rue. 1975. Nitrogenase activity in rhizobia in absence of plant host. Nature (London) 256:407–408.

Lennox, L.B., and M. Alexander. 1981. Fungicide enhancement of nitrogen fixation and colonization of *Phaseolus vulgaris* by *Rhizobium phaseoli*. Appl. Environ. Microbiol. 41:404–411.

Mahler, R.L., and A.G. Wollum II. 1980. Influence of water potential on the survival of rhizobia in a Goldsboro Loamy Soil. Soil Sci. Soc. Am. J. 4:988–992.

McComb, J.A., J. Elliott, and M.J. Dilworth. 1975. Acetylene reduction by *Rhizobium* in pure culture. Nature (London) 256:409–410.

Moffett, M.L., and R.R. Colwell. 1968. Adansonian analysis of the Rhizobiaceae. J. Gen. Microbiol. 51:245–266.

Munevar, F., and A.G. Wollum II. 1981. Growth of *Rhizobium japonicum* strains at temperatures above 27°C. Appl. Environ. Microbiol. 42:272–276.

Mytton, L.R. 1975. Plant genotype × *Rhizobium* strain interactions in white clover. Ann. Appl. Biol. 80:103–107.

Norris, D.O. 1958. *Rhizobium* needs magnesium not calcium. Nature (London) 182:734–735.

Pagan, J.D., J.J. Child, W.R. Scowcroft, and A.H. Gibson. 1975. Nitrogen fixation by *Rhizobium* cultured on a defined medium. Nature (London) 256:406–407.

Peterson, H.L., and T.E. Loynachan. 1981. The significance and application of *Rhizobium* in agriculture. pp. 311–331. *In* K.L. Giles and A.G. Atherly (eds.) International Review of Cytology, Suppl. 13, Biology of the Rhizobiaceae. Academic Press, New York.

Phillips, D.A. 1980. Efficiency of symbiotic nitrogen fixation in legumes. Annu. Rev. Plant Physiol. 31:29–49.

Roberts, G.P., and W.J. Brill. 1981. Genetics and regulation of nitrogen fixation. Annu. Rev. Microbiol. 35:207–235.

Vincent, J.M. 1962. Influence of calcium and magnesium on the growth of *Rhizobium*. J. Gen. Microbiol. 28:653–663.

Weaver, R.W., and L.R. Frederick. 1974. Effect of inoculum rate on competitive nodulation of *Glycine max* L. Merril. I. Greenhouse studies. Agron. J. 66:229–232.

BOOKS

Advances in Agricultural Microbiology. 1982. Edited by N.S. Subba Rao. Oxford and IBH Publishing Co. New Delhi.

A Guide to Better Pastures for the Tropics and Sub-tropics. 1974. By L.R. Humphreys. Wright Stephenson and Co., Flemington, Australia.

A Treatise on Dinitrogen Fixation, Section III: Biology. 1977. Edited by R.W.F. Hardy and W.S. Silver. John Wiley & Sons, New York.

A Treatise on Dinitrogen Fixation, Section IV: Agronomy and Ecology. 1977. Edited by R.W.F. Hardy and A.H. Gibson. John Wiley & Sons, New York.

The Biology of Nitrogen Fixation, Vol. 33, Frontiers of Biology. 1974. Edited by A. Quispel. North Holland Publishing Company, Amsterdam.

Biology of the Rhizobiaceae, International Review of Cytology, Suppl. 13. 1981. Edited by K.L. Giles and A.G. Atherly. Academic Press, New York.

Exploiting the Legume-*Rhizobium* Symbiosis in Tropical Agriculture, College of Tropical Agriculture Miscellaneous Publication 145. 1976. Edited by J.M. Vincent, A.S. Whitney, and J. Bose. Department of Agronomy and Soil Science, University of Hawaii, Honolulu.

Genetic Engineering of Symbiotic Nitrogen Fixation and Conservation of Fixed Nitrogen,

Vol. 17, Basic Life Sciences. 1981. Edited by J.M. Lyons, R.C. Valentine, D.A. Phillips, D.W. Rains, and R.C. Huffaker. Plenum Press, New York.

Handbook of Legumes of World Economic Importance. 1981. By J.A. Duke. Plenum Press, New York.

Methodologies for Soil-Improving Legumes. 1991. By M. Sarrantonio. Rodale Institute, Kutztown, PA.

Methods for Evaluating Biological Nitrogen Fixation. 1980. Edited by F.J. Bergersen. John Wiley & Sons, New York

Mineral Nutrition of Legumes in Tropical and Subtropical Soils. 1978. Edited by C.S. Andrews and E.J. Kamprath. Commonwealth Scientific and Industrial Research Organization, Melbourne, Australia.

Nitrogen Fixation, Vol. 1, Ecology. 1981. By W.J. Broughton. Clarendon Press, Oxford.

Nitrogen Fixation, Vol. 2, *Rhizobium*. 1982. By W.J. Broughton. Clarendon Press, Oxford.

Nitrogen Fixation, Vol. 3, Legumes. 1983. By W.J. Broughton. Clarendon Press, Oxford.

Nitrogen Fixation, Vol. 4, Molecular Biology. 1986. By W.J. Broughton. Clarendon Press, Oxford.

Nitrogen Fixation in Legumes. 1982. Edited by J.M. Vincent. Academic Press, Sydney, Australia.

Nitrogen Fixation in Tropical Cropping Systems. 1991. By K.E. Giller and K.J. Wilson. CAB International, Wallingford, UK.

Recent Advances in Biological Nitrogen Fixation. 1979. Edited by N.S. Subba Rao. Oxford and IBH Publishing Co., New Delhi.

Symbiotic Nitrogen Fixation in Plants. 1976. Edited by P.S. Nutman. International Biological Programme, Cambridge University Press, Cambridge, MA.

Tropical Crops: Dicolyledons, Vol. 1. 1968. By J.W. Purseglove, John Wiley & Sons, New York.

Tropical Legumes: Resources for the Future. 1979. National Academy of Sciences, Washington, DC.

World Soybean Research Conference II: Proceedings. 1979. Edited by F.T. Corbin, Westview Press, Boulder, CO.

Users Manual for MPNES Most-Probable-Number Enumeration System, ver. 1.0–1990. by J.E. Bennet, P. Woomer, and R.S. Yost. University of Hawaii Niftal Project, Paia, HI.

Index

MIX
Papier aus verantwortungsvollen Quellen
Paper from responsible sources
FSC® C105338

If you have any concerns about our products,
you can contact us on
ProductSafety@springernature.com

In case Publisher is established outside the EU,
the EU authorized representative is:
Springer Nature Customer Service Center GmbH
Europaplatz 3, 69115 Heidelberg, Germany

Printed by Libri Plureos GmbH
in Hamburg, Germany